Urban Horticulture

Urban Horticulture

Edited by

Tina Marie Waliczek, PhD

Texas State University
Department of Agriculture
San Marcos, Texas, USA

Jayne M. Zajicek, PhD

Texas A&M University
Department of Horticultural Sciences
College Station, Texas, USA

CRC Press
Taylor & Francis Group
Boca Raton London New York

CRC Press is an imprint of the
Taylor & Francis Group, an **informa** business

CRC Press
Taylor & Francis Group
6000 Broken Sound Parkway NW, Suite 300
Boca Raton, FL 33487-2742

Printed on acid-free paper
Version Date: 20151027

International Standard Book Number-13: 978-1-4822-6099-1 (Hardback)

Library of Congress Cataloging-in-Publication Data

Names: Waliczek, Tina Marie, editor. | Zajicek, Jayne M., editor.
Title: Urban horticulture / editors: Tina Marie Waliczek and Jayne M. Zajicek.
Description: Boca Raton : Taylor & Francis, 2015. | Includes bibliographical references and index.
Identifiers: LCCN 2015022644 | ISBN 9781482260991 (alk. paper)
Subjects: LCSH: Urban gardening. | Horticulture.
Classification: LCC SB453 .U73 2015 | DDC 635.09173/2--dc23
LC record available at http://lccn.loc.gov/2015022644

Visit the Taylor & Francis Web site at
http://www.taylorandfrancis.com

and the CRC Press Web site at
http://www.crcpress.com

Contents

Preface

Public green spaces within city environments are threatened by urbanization. Technological advances and the commonality of electronic devices have influenced people to spend less and less time in nature. The purpose of this book is to explore the importance of horticulture in the lives, health, and well-being of urban populations. The book examines the types of public and private community, state, and federal programs in horticulture, their history, management and administration, programming, evaluation, funding, and benefits. Readers will learn how passive and active interactions with plants benefit humans by improving physical and mental health, quality of life, and social well-being, as well as neighborhoods and communities.

The research area of human issues in horticulture, sociohorticulture, people–plant interactions, and/or horticultural therapy has developed to include a body of research that is not presented in one text. However, the subject matter is becoming a course option at universities with horticulture programs as an elective, service-learning, or web-based class for undergraduates or graduate students. This book includes past and current research on school, community, public, and prison gardens among others, as well as questions for discussions, example class activities, example case studies, and online access to examples of Power Point presentations, which can provide the basis for a web-based course or a jump-start to lectures for instructors.

The idea for the book grew from the editors' experiences researching the field for over 20 years.

Traditional studies in horticulture have focused on production and landscape horticulture, whereas urban horticulture considers and emphasizes the individuals and communities benefiting from the horticultural landscape. The hope is that this book will raise awareness and appreciation for the value of horticulture for aspects beyond food and aesthetics, and that readers will realize the need and value of nature and plants in our daily lives.

Researching and practicing experts from across the United States were selected to discuss the history, importance, and researched benefits of each topic. The book is written for the general public, as well as those studying general horticulture, future horticultural educators, horticultural therapists, and those involved in public horticulture.

Additional material is available from the CRC website: http://www.crcpress.com/product/isbn/9781482260991.

Editors

Tina Marie Waliczek is currently a professor of horticulture in the Department of Agriculture at Texas State University. She earned her bachelor and master of science degrees at Kansas State University and her PhD at Texas A&M University. She oversees the horticulture degree program at Texas State University and the management of the gardens and greenhouse. Her research interests have focused on the area of urban/plant interactions, including a wide range of studies in the areas of children's gardens, community gardens, studies on the effects of green spaces on people, the influence of gardening on perceptions of quality of life, and interior plants usage on job satisfaction and interior space in buildings. She has also researched teaching techniques in horticulture, such as studying the benefits of integrating service-learning into the horticulture curriculum. Dr. Waliczek is interested in sustainability issues including composting and managing invasive species using large-scale composting systems, as well as the economics of integrating cafeteria composting programs at universities.

Jayne M. Zajicek is currently a professor of urban horticulture in the Department of Horticultural Sciences at Texas A&M University. She earned her bachelor and master of science degrees at the University of Nebraska and her PhD at Kansas State University. Dr. Zajicek teaches in the area of urban horticulture and developed a course titled "sociohorticulture" that fulfills the social science requirement in the core curriculum. Dr. Zajicek also developed the bachelor of arts degree in the department, the first of its kind in a college of agriculture in the United States. Dr. Zajicek's research program titled "The Growing Minds Research Program" develops and conducts sound scientific research to evaluate the effects of gardening and horticulture in populations including school children, at-risk youth, the elderly, mentally and physically disabled individuals, and the incarcerated. Since the beginning of the Growing Minds Research Program, the program has awarded 16 master's of science and 8 PhD degrees.

Contributors

Jennifer Campbell Bradley is a freelance garden writer. She earned her PhD in horticultural sciences and master's degree in agriculture education from Texas A&M University. Dr. Bradley served as an assistant professor in environmental horticulture at the University of Florida where she helped establish public gardens management as a discipline within the department. She was also instrumental in establishing the Children's Gardening Competition in collaboration with EPCOT's Flower and Garden Festival. As a community volunteer, she works with children and youth through scouting, education programs, and the local arts council. Dr. Bradley currently resides in North Carolina, where she is a full-time home educator and mother of two sons.

James Buratti is a PhD candidate in the Department of Geography at Texas State University. He has been involved in organic and local agriculture since 2002 and has been an active participant with the Texas Organic Farmers and Gardeners Association. Mr. Buratti works full time at Texas State University and also co-owns Dobry Den Farm, raising grass-fed lamb for the central Texas market.

Ronald R. Hagelman III is an associate professor and the associate chair in the Department of Geography at Texas State University in San Marcos, Texas. Dr. Hagelman teaches courses at the undergraduate and graduate levels in environmental management, urban environment, and environmental hazards and disasters. He has worked closely with urban food producers in Austin, Texas to design and implement university field trips to Austin's rapidly expanding urban farms and local food markets. His research includes investigations into the geography of Austin's community gardens, motivations of community gardeners, and regulatory landscapes of urban agriculture.

Amy McFarland is an assistant professor teaching introduction to environmental studies and the sustainable agriculture practicum in the Environmental Studies program at Grand Valley State University. She also teaches a first-year interdisciplinary sequence in the Honors College titled "Food for Thought." Dr. McFarland completed a bachelor's degree at Rice University, her master's degree at Texas State University, and a doctoral degree at Texas A&M University. She completed a postdoctoral fellowship with the Smithsonian Institute Gardens in Washington, DC in 2012. In 2014, she received a grant to study agriculture and food in Thailand.

Carolyn W. Robinson began her career with children and gardens as a graduate student at Texas A&M University. She was an extension graduate assistant who helped with the development, writing, teacher trainings, and website for the Junior Master Gardener® program. Her research explored how gardening influenced the life skills of children. Since becoming a faculty member at Auburn University, she has developed a socio-horticulture research program that includes children's studies as well as other areas of people–plant interactions.

Deborah Rutt teaches community-based learning in the Department of University Studies at Portland State University. She is the garden coordinator at Coffee Creek Correctional Facility, Oregon's state prison for women, where she partners with public health and conservation agencies to engage incarcerated women in growing organic food and native plants for habitat restoration. Ms. Rutt received an Audubon/Toyota Together Green Fellowship in 2011, and in 2014 she received a Women Greening Food special recognition award from Women in Conservation.

Sonja M. Skelly is the director of Education & Communications for Cornell Plantations, the arboretum, botanical garden, and natural areas of Cornell University, and an adjunct assistant professor in the Department of Horticulture. She oversees the education, visitor services, and communications programs at Cornell Plantations. Her academic work focuses on aspects of public garden management, educational programming, plants and human well-being, and use of plants to improve schools and communities. She earned BA and MS degrees from Texas A&M University and her PhD from the University of Florida.

Leigh Anne Starling is a registered horticultural therapist and a certified rehabilitation counselor. She is active in the American Horticultural Therapy Association and has served on the Board of Directors. Ms. Starling has worked in the field of horticultural therapy for over 20 years in therapeutic, rehabilitative, and vocational programs serving individuals of all ages and abilities. She offers supervision to horticultural therapy interns and consulting services in horticultural therapy program development, education, and therapeutic programming.

Ann Marie VanDerZanden is a Louis Thompson distinguished undergraduate teacher, director of the Center for Excellence in Learning and Teaching, and professor of horticulture at Iowa State University. In her role as director, Dr. VanDerZanden provides leadership for the Center in meeting its mission to support, promote, and enhance teaching effectiveness and student learning at Iowa State. As a horticulture faculty member she has taught over 25 different courses, and she currently maintains an active teaching and mentoring role in the Department of Horticulture. Her research interests include undergraduate pedagogy, and using new technology to enhance the learning experiences of students and landscape professionals.

1 Introduction

Jayne M. Zajicek

CONTENTS

OBJECTIVES

Upon completion of this chapter, the reader should be able to

- Discuss urbanization in the United States including causes and effects.
- Define the theories explaining the influence of nature on humans.
- Define traditional horticulture.
- Define urban horticulture.
- Discuss the role urban horticulture plays in traditional horticulture.
- Recognize the importance of gardening and personal connections with nature in modern culture.
- Identify specific urban horticulture programs.

KEY TERMS

- Evolution theory
- Exurbs
- Human issues in horticulture
- Learning experience theory
- Overload and arousal theory
- People–plant interactions
- Sociohorticulture
- Traditional horticulture
- Urban horticulture
- Urban sprawl
- Urbanization
- White flight

URBANIZATION IN THE UNITED STATES

The United States, during the early years of development, was primarily rural. According to the 1790 census, 95% of the population of the United States lived in the countryside. Urban areas (places with more than 2500 people) included only 5% of the population that lived mostly in small villages. Only three main cities including Philadelphia, New York, and Boston had more than 15,000 inhabitants. In these cities, the center of the city was the most fashionable place to live. Social and financial classes of the population of people developed. Educated and financially wealthy people lived in the center of the city within walking distances of the places they worked and lived. The middle class people lived on the outskirts of the city and the poorer people lived in the suburbs, away from urban amenities.

During the early 1800s, urban areas grew faster than rural areas and this movement has been termed *urbanization.* By 1890, 35% of Americans lived in urban areas due to the Industrial Revolution which transformed urban life and gave people higher expectations for improving their standard of living. Innovations in transportation, including railroads, streetcars, and trolleys, enabled people to live further from their places of work and other establishments. A complete shift in where people of different socioeconomic classes lived took place. The wealthy established separate neighborhoods on the outskirts of cities or in the countryside. The middle class of white-collar employees in business and industry also became established on the outskirts of cities. The middle-class neighborhoods were developments of similar-looking single-family or multiple-family dwellings. As people of the upper socioeconomic class left the industrialized inner cities, people of the lower socioeconomic class moved in. The inner cities became characterized by low-rent apartment buildings and building tenements that were often poorly maintained and unsanitary.

During the 50 years between 1870 and 1920, the population of cities grew from 10 million to 54 million, and continued to grow into the twentieth century. There was a shift in population origin with 75% of the residents born outside the United States.

Many of these immigrants lived in the inner cities where poverty was prevalent. Conditions in the inner cities became worse during this time due to the fact that local governments could not keep pace with this extreme growth in extending clean water, garbage collection, and sewage systems. In addition to poor services, high rents and low wages contributed to the misery in the poorer areas.

During the late nineteenth and early twentieth centuries, several movements, including the Progressive movement, the New Deal, and the War on Poverty, succeeded in improving conditions in the inner cities by reducing corruption and establishing housing codes and public health measures. However, cost-cutting political movements during several periods in the early 1900s cut funding to these programs worsening the conditions for the lower class.

Urban life was improved during the latter part of the nineteenth century due to upper mobility, home ownership, educational opportunities, and cheap goods. With this improvement, cities in the United States developed beautification programs including the construction of parks and playgrounds. City populations continued to grow during the first half of the twentieth century; however, during the second half of the twentieth century, factories moved to countries where labor was cheaper, and with this move, jobs in the cities disappeared and cities began to shrink.

Millions of middle-class residents moved to the suburbs during the 1940s and 1950s. They left the crowding, high taxes, and poverty of the inner city for cleaner, newer, better, and more secure homes of the suburbs. With this population shift, stores, services, and entertainment also followed the movement to the suburbs. Poverty, mainly due to unemployment, remained in the inner cities even with the unprecedented economic growth in suburbs.

New suburbs, and *exurbs* (rural areas bordering cities) continued to spring up during the latter part of the twentieth century and currently in the twenty-first century. This movement to newer suburbs and exurbs has been referred to as *urban sprawl*. The term urban sprawl is generally used with negative connotations due to the fact that not only the inner cities, but also the suburbs now have to deal with many of the inner city problems. The 2010 census data on urban areas shows that Americans continue to prefer their lower density lifestyles, with both new suburbs and exurbs growing more rapidly than the core inner cities. However, urbanization comes with consequences. Urbanization has many adverse effects on the structure of society as large concentrations of people compete for limited resources. Rapid housing construction leads to overcrowding and slums, which experience major problems such as poverty, poor sanitation, unemployment, and high crime rates.

Plants soften urban areas by providing a positive physical surrounding in which it is more comfortable to live and work by purifying the air, moderating temperatures, reducing glare and noise, removing pollutants from the air, screening unattractive sights, and increasing relative humidity (Nighswonger 1975). Not only do plants provide environmental benefits, research has shown that people prefer nature scenes over urban scenes, and among urban scenes, those with vegetation are preferred to those without (Kaplan et al. 1972; Herzog et al. 1982; Herzog 1989). There are several theories explaining why people prefer nature over buildings and man-made features of the urban setting.

THEORIES EXPLAINING THE INFLUENCE OF NATURE ON HUMANS

Ulrich and Parsons (1992) discuss several theories that help to explain why being around plants can be beneficial. One of these theories is the *overload and arousal theory*. This theory explains that in the complex modern world there is so much noise, movement, and visual stimulation that it can be overwhelming to an individual's senses leading to damaging levels of psychological and physiological excitement. An environment that is dominated by plants is less complex, helping to reduce arousal, in turn, reducing feelings of stress. An example of this type of environment would be an urban condominium with a large park like setting in the back of the condominium and trees and shrubs surrounding the building.

The *learning experience theory*, another theory from Ulrich and Parsons (1992), explains that people's responses to plants are a result of their early learning experiences or the culture in which they were raised. According to this theory, individuals who grow up in the farmlands of Nebraska will have a more positive attitude toward flat lands with sparse, natural vegetation, and cultivated crops, such as sorghum and wheat, when compared to someone from the mountains of Colorado. Ulrich goes on to explain that modern, Western culture conditions people to be one with nature and to have negative feelings about the cities. However, this theory does not explain the similarities in response to nature found among people from different geographical and cultural backgrounds or from different historical periods.

The last theory proposed by Ulrich is the *evolution theory*. The evolution theory maintains that since people evolved in environments comprised primarily of plants, they are genetically predisposed to prefer these types of surroundings. This evolutionary response is seen in an unlearned tendency to pay attention and respond positively to certain combinations of plants and other natural elements, such as water and stone. One hypothesis is that individuals evolved in a more savanna-like setting so possess an innate preference for grasslands mixed with trees (Balling and Falk 1982).

Plants are nearly universally considered to create a positive atmosphere in urban environments. It is important for traditional horticulture to expand into the urban area and to not only understand the botanical and physical factors related to plants in the urban environment, but also to document the social and emotional impacts plants have on urban residents.

THE DEFINITION OF TRADITIONAL AND URBAN HORTICULTURE

Prior to 1980, urban horticulture was not a term that traditional horticulturists included in the definition of horticulture. The traditional definition of horticulture included the science which encompasses the production, utilization, and improvement of fruits, vegetables, and ornamental plants.

However, in the early 1980s, urbanization was happening at a rapid rate and concern about the improvement of urban environments and well-being of humans in these environments started to grow. During this time the term "urban horticulture" was used to define further the science of horticulture.

Urban horticulture not only encompasses the traditional definition of horticulture, but additionally also includes the importance of horticulture in populated areas. The term also begins to explore the benefits of plants to urban environments and to people who live in these environments. Traditionally, horticulturists were focused mainly on the specific cultural requirements needed for growing plants in urban environments, including mitigating problems associated with pollution, water quality and quantity, automobiles, and heat effects. Most horticulturists working on these challenges in the urban environment, referred to as urban horticulturists, were not interested in people, but just plants.

However, it was also important during this time, in order to validate the importance of horticulture in these populated areas, to consider the benefits of plants to the audience "people." Urban horticulture can be divided into two separate entities. One is the entity of plants and how they function environmentally in the urban environment. The other entity encompasses the area of "people/plant interactions."

Because of urbanization and the role horticulture potentially plays in this environment, it is necessary to expand the traditional definition of horticulture to include urban horticulture. This definition includes the art and science of producing, utilizing, and improving fruit, vegetables, and ornamental plants resulting in the enrichment and health of individuals and communities.

By this definition, horticulture encompasses PLANTS, including the multitude of products (food, medicine, and O_2) essential for human survival; and PEOPLE, whose active and passive involvement with gardens and green spaces brings benefits to them as individuals and to the communities and cultures they comprise.

It is important to include "art" in the definition of horticulture, because of the broad areas that horticulture includes. The design areas of floral design and landscape design are overlooked in the traditional definition of horticulture, but are both very important not only in the new definition including urban horticulture, but also in traditional horticulture.

HUMAN ISSUES IN HORTICULTURE/SOCIOHORTICULTURE

As urbanization continues throughout the United States, plants continue to have a major role functionally, socially, psychologically, and physically. This is not a new concept. Horticulture has been used in a treatment context since ancient Egyptian court physicians prescribed walks in palace gardens for mentally disturbed royalty. During the 1800s, horticulture benefits were recorded in treating mental illness. Somewhere along the way the term "horticulture therapy" was coined and recognized as a valid addition to traditional treatments. Because horticulture therapy was an important treatment module, two universities included programs in horticulture therapy including Milwaukee Downer College and Kansas State University in the mid- to late 1900s. In addition to educational programs in horticulture therapy, the American Horticulture Therapy Association (AHTA) was established as a nonprofit organization to promote and advance the profession of horticulture therapy.

Prior to the urbanization movement, people living in rural areas were very connected to nature. They realized the importance of nature as they cultivated the earth, providing food for both their families and their animals. Children lived on acreages where they could run and play freely in nature, helping them to value the importance of nature. Once urbanization began, people no longer produced their own food, and children lost their connection to green spaces.

During the 1970s, pioneers in the area of people–plant interactions, including Diane Relf (Horticulture Department, Virginia Polytechnic Institute and State University), Charles Lewis (author of *Green Nature, Human Nature*), and Rachel Kaplan and Stephen Kaplan (authors of *The Experience of Nature, A Psychological Perspective*), began to express the importance of plants on people in a social science perspective. Charles Lewis used the term "*people/plant interactions*" since the early 1970s to describe the psycho-social benefits of plants in the urban environment (Relf 1998). As this field developed, the term "*human issues in horticulture*" (HIH) became the term of choice to describe the psycho-social impact of plants on people. Diane Relf explains the reasoning behind this terminology in her article "Moving toward a New Millennium in People–Plant Relations" (1998). Her explanation includes (with modifications)

- It clearly describes the area of psycho-social benefits of plants, neither encompassing a much broader concept (people–plant interactions) nor limited to one specific group of people (horticulture therapy).
- The term human issues in horticulture may be easier to conceptualize than other terms since the horticulture community is familiar with the idea of discussing issues in other related fields of horticulture.
- Other professional areas such as engineering and architecture have divisions that address human issues, making it logical for horticulture professionals to also address these issues.

However, Human Issues in Horticulture is not the only term that has been recognized in the people–plant issues area. Dr. Joe Novak from Texas A&M University coined the word *sociohorticulture* to cover the same areas as human issues in horticulture. This term encompasses two concepts. One "socio" from the word *socius* meaning "associate" (L.) or relating to society. The second concept is "horticulture" from the terms *hortus* meaning "garden" (L.), and *cultura* meaning "cultivation" (L.). This term includes both sciences of sociology and traditional horticulture. The terms sociohorticulture and human issues in horticulture have been used interchangeably. A proposed definition of both human issues in horticulture and sociohorticulture is

> The art, science, and application of people and plant interactions resulting in sustainable enrichment and quality of life for individuals, communities, and ecosystems in human created environments.

This definition is much more specific for the people–plant area of horticulture. It still contains the traditional part of horticultural science and urban horticulture, but adds the dimension of the effects of plants on the psycho-social aspects of people. Issues encompassed in this area are focused on the human value and impact of plants and horticulture rather than on the costs of production and management of plants in the urban environment.

BENEFITS OF PLANTS TO URBAN COMMUNITIES

The disconnection between nature and people in modern times, coupled with the economic pressure to only value what has monetary worth, contributed to the need to document the impact of plants on people. As people become increasingly committed to living in an urbanized, high-tech style, it is important to understand that plants have an influence on quality of life, human well-being, and social development.

Research done by Rachel and Stephen Kaplan, environmental psychologists at the University of Michigan, on landscape preferences (Kaplan and Kaplan 1989) were some of the first studies that provided documentation on the importance of parks on the psycho-social well-being of people. The landscape preference studies were just the beginning of a research movement that studied the impact of plants on mental restoration and stress reduction of people.

Roger Ulrich (professor of Architecture at the Center for Healthcare Building Research at the Chalmers University of Technology in Sweden), a pioneer in the area of the beneficial effects of plants on people, published one of the classic research papers in this field. This research reported the health benefits to hospital patients from having a room with a view of trees rather than a view of a brick wall (Ulrich 1984). Results from his publication included

- The hospital stay of patients with a view of nature spent less time in the hospital (7.96 vs. 8.70 days).
- Patients with a view of nature required fewer doses of strong pain relievers.
- Patients with a view of nature received fewer negative comments from hospital staff on their charts.

Roger Ulrich also published results on human psychological and physiological responses to nature as it relates to stress reduction (Ulrich 1979, 1981, 1986; Ulrich and Simons 1986). One of his studies (Ulrich 1981) measured alpha brain wave amplitudes of people viewing different landscape scenes. Alpha brain wave amplitudes are associated with human psychological arousal and attentive relaxation. In this study, people viewed slides of nature scenes with water and vegetation, slides of nature scenes with only vegetation, and slides of urban scenes without vegetation. When people viewed the slides of either category of nature scenes, they exhibited higher alpha brain wave amplitudes than when viewing urban scenes. The higher the brain wave amplitude score the lower the level of psychological arousal and the higher the level of attentive relaxation, which is regarded as a more positive state for people. Another study of landscape preferences was done by Tennessen and Cimprich (1995) testing college students in their own dormitory rooms. Students were either housed in dorm rooms with window views of nature or dorm rooms with window views of hardscape such as buildings or sidewalks. All students were given a series of tasks that required mental concentrations. Students with views of nature performed better on the mental tasks compared to those with views of hardscapes.

These early studies indicating that individuals preferred views of nature and vegetation compared to views of urban hardscapes have opened the doors to further research in the area of people–plant interactions. The current research in this area focuses on the benefits of plants to the health and quality of life of individuals living in urban environments.

Another area of research that indicates that plants can have a significant effect on stress reduction is in the area of violent crimes and domestic violence (Kuo and Sullivan 1996; Snelgrove et al. 2004). Benefits of parks, gardens, and trees to neighborhoods include greater feelings of safety, increased social contact, and communication among neighbors (Waliczek et al. 1996; Kuo and Sullivan 2001), as well as reduced feelings of mental fatigue (Kuo 2001). In addition, studies reported that trees surrounding apartment buildings had an impact on incidences of domestic violence compared to apartments without vegetation. In one study (Kuo and Sullivan 1996), women living in apartments mentioned above were interviewed about domestic violence and child abuse. Women living in apartments with surrounding trees reported fewer incidences of domestic violence (13%) compared to the women living in nongreen apartments (22%). The women living in apartments with surrounding trees also reported less child abuse (3%) compared to women in the nongreen apartments (14%).

Other research indicated that green spaces have a positive influence on urban neighborhoods with residents reporting neighborhood revitalization and a perceived immunity from crime (Gorham et al. 2009). These benefits again contributed to the health of individuals in these communities as well as to their quality of life.

This text book is a comprehensive overview of urban programs using horticulture to provide a multitude of benefits to a variety of populations. Below is an overview of the programs that are covered in the following chapters.

URBAN HORTICULTURE PROGRAMS

Children's Gardens

Children can be exposed to plants, gardens, and plant-based activities in a variety of ways. Some of those ways include gardening at home or with family or neighbors, school gardens, botanical gardens, plant-based curriculum activities in school and activities in class, or field trips to living history museums or cultural fairs. All of these present academic, social, psychological, environmental, physical, and nutritional benefits to children of all ages. There are many ways to incorporate plants resulting in many benefits. Some countries and states have required every school to have a garden.

School gardens have seen quite an increase in popularity in recent years; however, they are not a new concept. School gardens have seen cycles of popularity in the United States since they began in the late 1800s. These gardens were influenced by instructional gardens from Europe and the Middle East that have much earlier beginnings.

The first step in introducing school gardens to the United States was taken by the Massachusetts Horticultural Society. They sent Henry L. Clapp to study the school gardens of Europe in 1890. He became the first to establish a school garden in the United States at the George Putnam School in Roxbury, Massachusetts in 1891. This garden was a wildflower garden with a vegetable garden added later.

More currently, school gardens have regenerated their popularity with the environmental movement of the 1970s and another burst of interest in the early 1990s. During the first decade of the twenty-first century, there was an eightfold increase in teacher requests for school garden materials in libraries, and it is estimated that about one fourth of public and private schools in the United States now have gardens.

Historically, school gardens were used to teach horticulture, increase work ethics, and provide other psychological benefits. However, current research is finding numerous other benefits of youth gardening. Many teachers believe the garden is an effective place to enhance science education, social skills, academic performance, physical activity, language arts, and healthy eating habits (Graham and Zidenberg-Cherr 2005). Children can learn social responsibility because gardens can improve neighborhood appearances, social interaction, participant cooperation (Relf 1998), and access to healthy foods (Armstrong 2000; Macias 2008). Service projects can be incorporated that capitalize on these benefits and help children see the direct support they are giving others, such as allowing the students to give tours and teach community members how to start a garden or donating the produce grown to a local food bank.

Community Gardens

Community gardens are typically defined as any plot of public or private land collectively managed by a group of people. Community gardens can be implemented in urban, suburban, or rural areas, though they are traditionally operated in cities or in shared community settings where citizens have limited access to gardening space. Gardens can be installed in neighborhoods, at schools or universities, at hospitals, apartment complexes, or other multi-family housing situations.

History indicates that some form of community gardens were in existence in Great Britain as early as the mid- to late 1700s. The idea of community gardens in Europe spread to other countries with most countries having documented evidence of community gardening being practiced to help fulfill needs for food, as well as to accommodate other socio-cultural functions such as reducing crime and immorality, among others.

Documentation of people gathering and creating communal gardening areas in cities has occurred in the United States since the Panic of 1893. Urban gardens were organized by social reformers promoting gardens built in vacant lots as a means of providing assistance for unemployed laborers during the economic trials. Historically, community gardening efforts in the United States were a result of economic instability and lack of food security. More often today, the goal of community gardens is to build community while providing space and opportunity for food production.

The benefits of community gardens are numerous and well documented through research. Urban community gardens have shown potential in being able to supplement commercially available fresh produce. Along with the nutritional benefits of fresh produce, gardening has also been reported to have a positive effect on enhancing physical activeness as well as reducing levels of stress and mental fatigue (Armstrong 2000). Gardeners have reported personal satisfaction from producing their own food in addition to feelings of self-sufficiency (Patel 1991). Another major benefit of community gardens is their ability to bring people together to share knowledge, stories, and ideas (Waliczek et al. 1996). This sense of community has been shown to have a positive effect on reducing crime rates in these neighborhoods.

Botanical/Public/University Gardens

A "public garden" is a garden open to the public. Every public garden has a mission which is the organization's reason for existing. The mission of many public gardens includes the cultivation and display of a wide range of plants—its collection. This accessioned collection (cataloged collection of plants) is maintained for conservation, research, education, teaching, recreation, and display.

The first public gardens were primarily used to grow herbs for medicinal purposes and were found in Italy in the fifteenth century. During the sixteenth and seventeenth centuries, exploration and trade expansion led to the creation of public gardens where new species of plants from around the world were displayed and cultivated. Botanical gardens were often set up to display new plants from European colonies and newly visited lands as trophies, and were also important learning laboratories of medicinal plants for clergy.

John Bartram's nursery, established in 1728 in Philadelphia, Pennsylvania, is considered to be the earliest public garden in North America. His collection of plants native to America, which he supplied to European nurseries, was visited by notable figures such as Benjamin Franklin and Thomas Jefferson. The mid-1800s saw the growth of many parks and public gardens. Establishing these gardens was seen by early leaders as important to the development of America and its cities. Frederick Law Olmsted, the father of landscape architecture, embraced this vision by designing more public parks compared to any other designer. His best known work, Central Park in New York City, was designed with the mission to be a place where city workers with no chance to travel to the country could enjoy nature (Hobhouse 1997).

The benefits public gardens provide have greatly expanded from their earliest days of providing a learning laboratory of medicinal plants for clergy. Today, they are centers of global plant conservation and environmental stewardship, they advance knowledge of the plant world and its many associated disciplines, they promote social justice, enhance the well-being of their communities and the individuals who visit them, provide engines for economic development, and provide places of beauty.

Zoo and Amusement Park Horticulture Programs

A zoo is a place where many kinds of animals are kept so that people can see them. Zoos have existed through the centuries, stemming from people's fascination with animals and nature. People have long held a desire to see animals not common to their local geographical area.

The first zoos located in Europe during the nineteenth century were the result of wealthy individuals gathering and keeping animals they found interesting. Little thought was given to the humane handling of animals because the sport of capturing the animals and the curiosity of the animal itself was more important. Animal confinements were small, simple enclosures securely containing the animals, while providing a view to the public. At this time, horticultural displays and plantings were not much of a consideration.

The first zoological park in the United Sates was the Philadelphia zoo started in 1974. As public perception and environmental concerns began shifting, zoos changed the way they displayed animals. Since the late nineteenth century, zoo directors have acted on the belief that the best way to display wildlife was in a natural setting. Horticultural displays were the beginning of establishing the feel of an animal's actual, or perceived, environment. Zoos around the world are helping to preserve not only endangered animal species but also endangered plant species.

An amusement park is a group of elaborate attractions for large numbers of visitors and may include live entertainment, theatrical presentations, and rides. Amusement parks evolved from European fairs developed for people's recreation. A premier example of an amusement park is Disneyland Park which was started by Walt Disney in 1955.

Landscape plantings have become an integral part of zoos and amusement parks. While visitors enjoy the green settings that the plant displays create, these plants and natural materials may also serve as valuable food sources for animals along with environments promoting natural behaviors and animal well-being.

Carefully planned gardens and displays successfully serve as a seamless transition between park areas. In amusement parks, harsh lines of rides and hardscape rails are softened with the use of plant materials. Plantings effectively conceal park infrastructure and beautify the parks as a whole. Plants have become a valuable and expected part of these parks.

Prison Horticulture Programs

Horticulture programs in prisons can cover a wide diversity of different types of programs. Overall, prison horticulture refers to a variety of plant-based activities and programs that take place in any type of detention facility focused on the rehabilitation and life skill development of incarcerated individuals.

Prison horticulture programs have been an element of detention facilities throughout history. The first prison gardens primarily used inmate labor to produce a cheap source of food for the detention facility. The use of horticulture programs in these facilities has evolved to include inmate rehabilitation through vocational training, a source of healthier food, a way to beautify the facility grounds, and provide physical and mental stimulation for the inmates.

Prison horticulture programs can range from small-scale, informal programs to be a more formal programs involving a collaborative effort between inmates, facility staff, volunteers, academic institutions, and possibly other community partners. Many times these larger scale programs are part of the vocational training program at the detention center. In some cases, these programs offer scholarly activity including in-class instruction for managing crops in a greenhouse or garden. Prison horticulture programs have also been found to benefit the inmates in the area of life skills development and improve communication skills and learning to work in team-type situations.

Horticulture Therapy

Horticulturists and practitioners have defined horticulture therapy in several different ways. A comprehensive definition describes horticultural therapy as a professionally conducted client-centered treatment modality that utilizes horticulture activities to meet specific therapeutic or rehabilitative goals of its participants. The focus is to maximize social, cognitive, physical, and psychological functioning and to enhance general health and wellness (Haller 2006). The horticultural activities are primarily based on the cultivation of plants. The therapist is trained in the use of horticulture as a medium to assist the client in achieving their treatment goals.

In the early history of the United States, people with disabilities were placed in asylums, poorhouses, almshouses, and sanitariums. Farming was a means of self-sufficiency and patients were often involved in the growing and harvesting of crops. As early as 1812, Dr. Benjamin Rush, considered the Father of American Psychiatry, acknowledged improvement in male patients who worked in the garden (AHTA 2007). Dr. Rush is credited with being one of the first to document the therapeutic benefits of gardening in a hospital setting.

A turning point in the profession came in the late 1940s and early 1950s with the return of World War II veterans and an increase in the number of Veteran's hospitals. Occupational therapists, volunteers, and garden club members brought gardening and plants into the hospitals to provide activities for returning veterans.

The 1900s saw the use of gardening shift from a form of work to a form of therapy and the 1970s saw the formal establishment of the profession. By the turn of the century, horticultural therapy, as a profession, was moving toward increased professional standards, achievement of professional recognition, and collaboration with allied professions such as landscape architecture and environmental psychology. Today, horticultural therapy is practiced throughout the world.

The process of horticultural therapy occurs through the relationship that develops between the person and the plant, the specific plant-related activities a person does or completes when taking care of plants, and the interactions that occur when people are engaging with plants in a horticultural environment. Ultimately, the therapeutic process is experienced through direct interaction and nurturing of plants and the therapeutic element is the benefit a participant experiences as a result of this interaction (Relf and Dorn 1995). Participants in horticultural therapy may experience emotional, intellectual, social, and physical benefits. These identified benefits also help to understand various types of horticultural therapy programs and settings.

Urban Forestry and Green Spaces

Urban green space can take on many different forms depending upon the size and density of the urban area. One form of an urban space is an urban park. Urban parks can be large and well known, such as Central Park in New York City. These large parks can serve as tourist attractions and destinations, potentially serving as an income source for the cities. In addition, large cities can also have numerous smaller parks which also serve the same ecological and social functions as large parks.

Urban green space can also be found through urban forests. Trees in urban forests have different functions including street trees, trees in residential yards, trees in parks, and trees around businesses. Green belts are a method used to increase green space in urban environments. Originally, green belt is defined as a policy that protected nonsettlement area on the perimeter of an urban area for the purpose of conservation and prevention of urban sprawl (Amati 2008). Urban sprawl is physical growth in cities and includes the development of suburbs. Green belts fell out of favor with policy makers as urban population increased and the need for expansion grew. Urban sprawl began to simply jump over the protected areas. The desire to protect these spaces for conservation remained high, however, and other types of urban green spaces grew out of this movement.

The location of green space in urban areas in the United States dates back as far as the founding of the country. Green space was considered in the designs of many of the earliest cities in the United States. Both George Washington's and Thomas Jefferson's visions of Washington, DC included trees, spacious grounds, and parks dispersed throughout capital buildings (Choukas-Bradley 2008). In fact, in the years following the turn of the nineteenth century, Jefferson himself designed and supervised the installation of trees along Pennsylvania Avenue between the Capital and the White House. The direct oversight by a person of Jefferson's status exemplifies the importance and impact of green space.

The White House itself is well known for its landscaped grounds with many media events occurring within, for example, the Rose Garden. Numerous presidents throughout US history have been involved with improving the green space surrounding the property including John Adams, James Madison, John Quincy Adams, Martin Van Buren, and many others (Choukas-Bradley 2008). The changing nature of the space around the White House is the icon of America—the grounds were used for a Victory Garden during WWII, planted by Eleanor Roosevelt to aid in the promotion of home food production during the war and has been used in a similar fashion by Michelle Obama to assist in the fight against obesity and type II diabetes (Whitehouse 2009; Obama 2012).

Urban population is in need of reconnecting with nature and urban green spaces provide them this opportunity. Other benefits of green spaces on the people that use them and the communities that they are in include reduction in crime, alleviate stress in individuals, improve physical health of individuals, increase social interactions of individuals, increase property values of surrounding areas, and provide environmental and ecosystem services.

Local Food

Local food or the local food movement is a collaborative effort to build more locally based, self-reliant food economies—one in which sustainable food production, processing, distribution, and consumption is integrated to enhance the economic, environmental, and social health of a particular place. The definition of what qualifies as local food is both relative and subjective. Producers, consumers, retailers, nongovernmental organizations, farmers, markets, state and federal governments may all have different and conflicting definitions.

Prior to the nineteenth and twentieth centuries nearly all food consumed in the United States was sourced locally and limited by seasonality in all regions. Consumers were often in direct contact with farmers, ranchers, or dairymen and had knowledge of seasonal availability of various foods. However, throughout WWI and WWII American agricultural food products became industrialized. Building on manufacturing and distribution techniques established before the wars, innovative processing, packaging, preserving and preparation approaches further expanded the reach of food production and consumption. Not everyone welcomed the industrialization and globalization of food. With it came food designed for convenience and profit, instead of nutrition or health.

During the early twenty-first century, there has been an increasing demand for locally produced food in the United States. Consumer research shows a wide variety of motivations to eat locally produced foods. These include distrust of corporations responsible for globally sourced food products, food safety concerns, rising rates of obesity and diabetes, a desire to avoid genetically modified organisms and crops, concerns for the environment and sustainability, the loss of local farms in communities around the country, and the increasing demand for organic foods.

Volunteers

Volunteers are defined as staff who give time and expertise without receiving or expecting monetary pay. Millions of people in America and around the world participate in volunteer activities. They are motivated to share their time and talents with organizations for a variety of reasons, and the end result is that both the organization and the volunteers benefit from this service. Volunteers play a very important role in most urban horticulture programs and institutions.

Volunteers help organizations extend their reach and accomplish their mission which otherwise would not be possible with paid staff alone. Developing a new volunteer program requires careful planning. Successful volunteer programs have a clear set of policies and procedures and an effective orientation and training program to prepare volunteers for their service. Once volunteers begin their service, they benefit from supervision, evaluation, and recognition to ensure they are meeting the organization's needs and that they are having a rewarding volunteer experience. Finally, systematic and regular program evaluation informs volunteer managers on the successes and impact of the volunteer program and can be critical in securing continued or expanded funding.

SUMMARY

In the early years of development, the majority of the people in the United States lived in rural areas. However, during the early 1800s, urban areas grew faster than rural areas and this movement has been termed urbanization. As urban cities grew in population, wealthy and middle socioeconomic classes moved to the outskirts of cities, called suburbs or exurbs, taking with them their wealth along with their businesses. As people of the upper socioeconomic classes left the industrialized inner cities, people of the lower socioeconomic class moved in, and poverty, poorly maintained apartment and tenement buildings, contaminated water, poor garbage collection, and crumbling sewage systems became prevalent. Inner cities continue to struggle today as millions of middle-class residents choose to live in suburbs and exurbs leaving the high taxes and poverty of the inner city for cleaner, newer, better, and more secure homes of the suburbs.

Urbanization can have many adverse effects, not only on the environment, but on the health of the communities and individuals living within these urban communities. Plants and green spaces have many beneficial effects on urban communities including creating a positive community atmosphere. Plants provide a physical condition that makes people proud to be a part of the community in addition to providing opportunities for sharing values, interests, and commitments with other community members. Plants are also extremely important in helping to make the community environment more pleasing by mediating environmental factors such as temperature, noise, and pollution.

With the importance of roles that plants play in the urban environment, came the need to expand the traditional definition of horticulture to include aspects of "urban horticulture." Urban horticulture encompasses not only the science of plant production, utilization, and improvement of plants, but additionally includes the importance of horticulture in populated areas. This new definition of horticulture includes the benefits of plants not only to the urban environment, but also to the people who live in this environment. As urbanization continues throughout the United States, plants continue to have a major role functionally, socially, psychologically, and physically.

During the 1970s, pioneers in the area of people–plant interactions, including Diane Relf, Rachel and Stephan Kaplan, Charles Lewis, and Joe Novak, began to express the importance of plants to people in a social science perspective. During this time, terms were coined to describe the psycho-social impact of plants on people including people–plant interactions, human issues in horticulture, and sociohorticulture.

Past and present research is helping to validate the importance of plants to people in the urban environment. Roger Ulrich, a research pioneer in the area of the beneficial effects of plants on people, published one of the classic research papers in this field indicating individuals preferred views of nature and vegetation compared to views of urban hardscapes. Ulrich's research, in addition to other early studies, has opened the doors to further research in the area of people–plant interactions. The current research in this area focuses on urban programs and the benefits of plants to the health and quality of life of individuals living in urban environments and participating in various urban horticulture programs.

Urban programs, using horticulture, can provide urban community members an opportunity to connect with nature that may not otherwise be available to them. These programs provide a multitude of benefits to the individual including physical, social, and psychological. These benefits are just as important to urban communities as are the functional and physical benefits of plants in these communities.

Horticulturists are recognizing the importance of validating the importance of plants to the psycho-social aspects of people in urban environments. The definition of traditional horticulture has expanded to include the urban environment and the effects of plants on people. As urbanization continues to grow throughout the world, the importance of plants to people in these urban communities will continue to be an important area for horticulturists to address.

REVIEW QUESTIONS

1. List several adverse effects urbanization has on urban communities. Discuss how plants and green spaces can help in reducing these effects and contribute to a more healthy urban community.
2. Discuss the three theories explaining the influence of nature on humans presented in this chapter. Choose the one you think explains why people are influenced by plants the best and give reasons why you support this theory.
3. Discuss the similarities and differences between traditional horticulture, urban horticulture, and human issues in horticulture/sociohorticulture.
4. Choose the term that you think best describes the area of urban horticulture that deals with the relationship of plants and people from a social science perspective. Give reasons in support of this term.
5. Give two benefits of plants to people presented in this chapter. Based on your own knowledge and experience, suggest three additional benefits plants may have to people in urban areas.
6. Name two urban horticulture programs presented in this chapter. Give an explanation why these are considered urban horticulture programs, the definition and history of these programs and the benefits these programs provide.
7. From you own experience and knowledge, list and describe one urban horticulture program that is not discussed in this chapter. Give the definition of this program, and the benefits you perceive this program provides to urban residents.

ENRICHMENT ACTIVITIES

1. Identify areas of urban sprawl in the nearest metropolitan areas to you. Identify impacts on the community from this sprawl.
2. Research to find urban horticulture programs in your own community. Try to find at least one example of each of the types of programs mentioned in the chapter.
3. Describe your own experience with urban horticulture. Do you remember a special green space from your personal history? Explain where it is at and what benefits it provided to you and the community.
4. Describe some properties or programs in the community that you think could be developed further into urban horticulture programs. What kind of programs are these and where do they exist?

REFERENCES

Amati, M. 2008. *Urban Green Belts in the Twenty-First Century*. Burlington, VT: Ashgate Publishing Limited.

American Horticultural Therapy Association. 2007. *AHTA Definitions and Positions*. King of Prussia, PA: American Horticultural Therapy Association.

Armstrong, D. 2000. A survey of community gardens in upstate New York: Implications for health promotion and community development. *Health and Place*, 6, 319–327.

Balling, J.D. and J.H. Falk. 1982. Development of visual preference for natural environments. *Environment and Behavior*, 14(1), 5–28.

Choukas-Bradley, M. 2008. *City of Trees: The Complete Field Guide to the Trees of Washington, DC*. 3rd edition. Charlottesville, VA: University of Virginia Press.

Gorham, M., T.M. Waliczek, A. Snelgrove, and J.M. Zajicek. 2009. The effect of community gardens on incidence of crime in Houston. *Hort Technology*, 19(2), 291–296.

Graham, H. and S. Zidenberg-Cherr. 2005. California teachers perceive school gardens as an effective nutritional tool to promote healthful eating habits. *Journal of the American Dietetic Association*, 105, 1797–1800.

Haller, R.L. 2006. The Framework. In R.L. Haller and C.L. Kramer (eds.), *Horticultural Therapy Methods: Making Connections in Health Care, Human Service, and Community Programs*. Binghamton, NY: The Haworth Press, Inc., pp. 1–22.

Herzog, T.R. 1989. A cognitive analysis of preference for urban settings. *Proceedings of the Longwood Program Seminars*, Vol. 12, Longwood Gardens, PA, pp. 40–45.

Herzog, T.R., S. Kaplan, and R. Kaplan. 1982. The prediction of preference for unfamiliar urban places. *Population and Environment*, 5(1), 627–645.

Hobhouse, P. 1997. *Gardening through the Ages: An Illustrated History of Plants and their Influence on Garden Styles—From Ancient Egypt to the Present Day*. New York, NY: Barnes and Noble.

Kaplan, R. and S. Kaplan. 1989. *The Experience of Nature: A Psychological Perspective*. Cambridge, New York: Cambridge University Press.

Kaplan, S., R. Kaplan, and J.S. Wendt. 1972. Rated preference and complexity for natural and urban visual material. *Perception and Psychophysics*, 12(4), 354–356.

Kuo, F.E. 2001. Coping with poverty: Impacts of environment and attention in the inner city. *Environment and Behavior*, 33(1), 5–34.

Kuo, F.E. and W.C. Sullivan. 1996. *Do Trees Strengthen Urban Communities, Reduce Domestic Violence?* (Technology Bulletin No. 4, Forestry Report R8-FR 55). Athens, GA: USDA Forest Service Southern Region.

Kuo, F.E. and W.C. Sullivan. 2001. Environment and crime in the inner city: Does vegetation reduce crime? *Environment and Behavior*, 33(3), 343–367.

Marcias, T. 2008. Working toward a just, equitable, and local food system: The social impact of community based-agriculture. *Social and Science Quarterly*, 89(5), 1087–1101.

Nighswonger, J.J. 1975. *Plants, Man and Environment*. Cooperative Extension Service Publication C-448. Manhattan, KS: Kansas State University.

Obama, M. 2012. *American Grown: The Story of the White House Kitchen Garden and Gardens Across America*. New York: Crown Publishers.

Patil, I.C. 1991. Gardening's Socioeconomic impacts. *Journal of Extension*, 29(4), 7–8.

Relf, D. and S. Dorn. 1995. Horticulture: Meeting the needs of special populations. *HortTechnology*, 5(2), 94–103.

Relf, P.D. 1998. People–plant relationship. In S.P. Simson and M.C. Straus (eds.), *Horticulture as Therapy: Principles and Practice*. Binghamton, NY: The Haworth Press, Inc., pp. 21–42.

Snelgrove, A.G., J.H. Michael, T.M. Waliczek, and J.M. Zajicek. 2004. Urban greening and criminal behavior: A Geographic Information System perspective. *HortTechnology*, 14(1), 48–51.

Tennessen, C.M. and B. Cimprich. 1995. Views to nature: Effects on attention. *Journal of Environmental Psychology*, 15, 77–85.

Ulrich, R.S. 1979. Visual landscapes and psychological well-being. *Landscape Research*, 4(1), 17–23.

Ulrich, R.S. 1981. Natural versus urban scenes: Some psychophysiological effects. *Environment and Behavior*, 13(5), 523–556.

Ulrich, R.S. 1984. View through a window may influence recovery from surgery. *Science*, 224, 420–421.

Ulrich, R.S. and R. Parsons. 1992. Influences of passive experiences with plants on individual well-being and health. In D. Relf (ed.), *The Role of Horticulture in Human Well-Being and Social Development: A National Symposium*. Portland, OR: Timber Press.

Ulrich, R.S. and R.F. Simons. 1986. Recovery from stress during exposure in everyday outdoor environments. In J. Wineman, R. Barnes, and C. Zimring (eds.), *The Costs of Not Knowing: Proceedings of the Seventeenth Annual Conference of the Environmental Design Research Association*. Washington, DC: Environmental Design Research Association, pp. 115–122.

Waliczek, T.M., R.H. Mattson, and J.M. Zajicek. 1996. Psychological benefits of community gardening. *Journal of Environmental Horticulture*, 14(4), 204–209.

Whitehouse. 2009. *Inside the White House: The Kitchen Garden*. http://www.whitehouse.gov/video/Inside-the-White-House-The-Garden (retrieved September 7, 2014).

FURTHER READING

Gabriel, B. eHow Contributor. 2014. *Causes and Effects of Urbanization*. http://www.ehow.com/info_8413652_causes-effects-urbanization.html (retrieved December 14, 2014).

Kuo, F.E. and W.C. Sullivan. 2001. Aggression and violence in the inner city: Effects of environment via mental fatigue. *Environment and Behavior*, 33(4), 543–571.

Lewis, C.A. 1996. *Green Nature/Human Nature: the Meaning of Plants in Our Lives*. Chicago, IL: University of Illinois Press.

theUSAonline.com. *People Urbanization of America*. http://www.theusaonline.com/people/urbanization.htm, http://www.definitions.net/definition/local%20food (retrieved December 14, 2014).

Waliczek, T.M., J.M. Zajicek, and R.D. Lineberger. 2005. The influence of gardening activities on consumer perceptions of life satisfaction. *HortScience*, 40(5), 1360–1365.

2 Children and Nature

Carolyn W. Robinson

CONTENTS

OBJECTIVES

Upon completion of this chapter, the reader should be able to

- Describe the history of school gardens.
- Understand the need for children to be involved with gardens.
- Explain the benefits of gardens.
- Describe the types of youth gardens.
- Identify resources to get started.

KEY TERMS

- Afterschool programs
- Children's gardens
- Cooperative extension service
- Experiential learning
- Hands-on learning
- Home gardens
- Horticulture therapy
- Intergenerational programs
- Nature deficit disorder
- Public gardens
- School gardens
- Stakeholders

Practitioners have long promoted safe, outdoor playtime opportunities for children, toddlers, and infants. Communication, physical, emotional, social, and personal skills are all areas that can be enhanced with outdoor play. This chapter will focus on how gardens and nature can impact children in these ways and more. There are many ways that children can be exposed to plants and gardens and nature. The use of plant activities inside the classroom is a start, but taking those lessons outdoors is critical.

People generally remember their first exposure to seeds or plants and planting. A familiar activity is placing lima bean seeds along the side of a glass jar and keeping them in place with wet paper towels. The roots and shoots form and push out of the seed. Students learn about the first little leaves that began to grow and why the endosperm is important for the seed to grow into a plant. While years pass, it is very easy for people to remember the hands-on activities they did in primary school instead of the notes they took or textbooks they read. Active, hands-on education is the way to have the greatest long-term influence on the education of children. Research has also shown that hands-on activities with children can have a lasting influence on their attitudes and behaviors (Kolb and Lewis 1986).

Children can be exposed to plants, gardens, and plant-based activities in a variety of ways. Some of those ways include gardening at home or with family or neighbors, school gardens, botanical gardens, plant units in school and activities in class, or field trips to living history museums or cultural fairs. All of these present benefits to children of all ages (Figure 2.1). The benefits are academic, social, psychological, environmental, physical, and nutritional. There are so many ways to incorporate plants and so many benefits; it is very easy to see why whole countries and states have required or suggested every school to have a garden.

HISTORY OF SCHOOL GARDENS AND GARDENS FOR EDUCATION

School gardens have seen quite an increase in popularity in recent years; however, they are not a new concept. School gardens have seen cycles of popularity in the United States since they began in the late 1800s. These gardens were influenced

FIGURE 2.1 A young boy stops to smell the flowers. Even at one year old, he expects flowers to have fragrance.

by instructional gardens from Europe and the Middle East that have much earlier beginnings.

Some believe that education in gardens can be traced back to the days of King Solomon (around 1015 BC). King Cyrus of Persia also created gardens where the sons of noblemen were to be instructed in horticulture. The use of gardens associated with schools dates back to 1525 AD when Italian arboreta were linked to Italian universities in Pisa and Padua. Educators of the sixteenth and seventeenth centuries including Comenius and Rousseau felt that gardens should be connected with schools and could be very educational for students (Miller 1904).

In 1869, the Austrian government passed laws making gardens mandatory in rural schools. Agricultural societies in Switzerland provided financial support to schools to help implement school gardens. In 1882, France set up laws to outline horticulture as a part of the school curricula and later stated that no school would be financially supported by the state unless there was a garden attached. In Belgium in the early 1900s, horticulture was mandatory and no school was without a garden. Germany at this time had some of the most impressive school garden programs; however, they were not mandated by the state. Austria–Hungary had 18,000 school gardens in the early 1900s, while the school garden movement in the United States was barely 10 years old.

The first step in introducing school gardens to the United States was taken by the Massachusetts Horticultural Society. They sent Henry L. Clapp to study the school gardens of Europe in 1890. He became the first to establish a school garden in the United States at the George Putnam School in Roxbury, Massachusetts, in 1891. This garden was a wildflower garden with a vegetable garden added later.

FIGURE 2.2 This is a photograph of children's victory garden in New York, NY taken June, 1944. (Courtesy of the Library of Congress.)

Following the reports of European school gardens by Clapp, the American school garden movement began to grow. The United States Department of Agriculture made estimates of 75,000 school gardens in 1906 with a rise to 80,000 by 1910. One of the larger more successful garden programs of early movements was the National Cash Register Boy's Garden in Dayton, Ohio. This garden was started by J.H. Patterson, the president of the company, to educate and instill work ethics in local boys. Patterson was raised on a farm and felt that his farm experience was one of the reasons for his success (Basset 1979). School gardens had an increase in popularity again around World War I and II. The "War Gardens" of World War I and the "Victory Gardens" of World War II were ways that children showed their patriotism and supported the war efforts (Figure 2.2) (Hayden-Smith 2014).

In the early 1900s, Montessori (1964) developed a philosophy and methods of education based on the natural interaction of children and their environment. She used gardens to work with underprivileged children in Italy and found that these youth had more responsibility, patience, and better interpersonal skills. She also felt that the garden was an important place to teach moral lessons to children. The focus became the plants and harvests rather than differences of the individuals in the garden. Montessori felt that gardening guides children to investigate nature, which brings about awareness and an appreciation for the environment. Montessori-based education came to the United States in 1911 and now serves children in over 4000 private Montessori schools and over 200 public schools with Montessori-styled programs (AMTA 2014).

More currently, school gardens have regenerated their popularity with the environmental movement of the 1970s and another burst of interest in the early 1990s. During the first decade of the twenty-first century, there has been an eightfold increase in teacher requests for school garden materials in libraries, and it is estimated that about one-fourth of public and private schools in the United States now have gardens (Carter 2010). The importance of nature and gardening has even begun to influence legislation in the United States with California's "A Garden in Every School" program. This program encourages every school in the state of California to incorporate gardening into their curriculum and provides resources and supporting information online.

NEED FOR YOUTH GARDENS

Today's parents are very concerned with their children's physical and mental health, personality development, and academic performance. Since the 1970s, parents have sought more and more to protect their children. This and the fact that more women were entering the work force, more structured activities and care were created for children (Mintz 2009). According to the Bureau of Labor Statistics (2014), 60% of families have two working parents. A large percentage (69.9%) of mothers with children under 18 are in the work force. This means that afterschool programs or sitters are a necessity for many. In households with working mothers, 32.7% had their children in organized care (CCAA 2013), but 26% of families with unsupervised children would likely use organized afterschool care if it were available (CCAA 2011). Even when the parents are at home, there are fewer children exploring the neighborhood compared to a few decades ago (Louv 2008). Parents are confronted with the dangers of being a child more today—kidnappings, sexual predators, drug, and alcohol pressures are of serious concern (Valentine and McKendrick 1997).

Historically, children in neighborhoods played across the front yards and parents gathered on front porches to talk with each other and watch the children playing. Today, people do not necessarily know their neighbors as they once did (Dunkelman 2014). Children's afternoons are filled with organized afterschool programs, such as sports practice, music lessons, tutoring sessions, and dance practice. Weekdays and weekends are booked. In all of this worry and hurry, nature is getting left behind. Slowing down to take a walk and watch the spiders spin a web or notice the color changing in the leaves is too slow and not productive enough for most. Activities and time need to be reassessed. Children and parents alike need this time to connect to nature. Society needs to slow down (Grogan 2011). There is so much concern over the environment, the negative impacts of human behavior, and the ways in which people can help. There are recycling programs and environmental clubs in schools

FIGURE 2.3 Two young boys interact when they find turtles in a pond.

and more families trying to be "green" by recycling, eating local, and using less chemicals; but, if children are not connected to the "why" (nature) these things are being done, there will be a next generation that does not feel the need to save that last bit of green space in the town to just be a green space, a park, or a wandering space (World Forum-NACC 2008).

This lack of time in nature has been termed "nature deficit disorder" by Louv (2008). He proposes that nature deficit has decreased children's respect and understanding of the natural, outside environment. He also suggests that disorders like attention deficit disorder or depression may develop based on the lack of being outside. There is a healing and calming effect to being outside. It is not just about learning science or any other subject; it is about connecting and relaxing—something everyone needs to have psychological and spiritual health. Children in contact with nature improve their mental, emotional, and social heath based on the creative and intellectual development, as well as social relationships that are encouraged in outdoor activities (Figure 2.3) (Heerwagen and Orians 2002).

There are problems for children associated with less time spent outdoors including less physical activity and poor diet choices leading to obesity, asthma, higher stress, and lower self-esteem among potential others. The Centers for Disease Control and Prevention (2014) asserts that childhood obesity has doubled for adolescents and tripled for children 6–11 years in the past 30 years—a byproduct of our more sedentary lifestyles and dependence on electronics versus getting active outdoors. Spending more time indoors exposes children to indoor allergens versus outdoor allergens, which contribute significantly to the development of asthma (Halken 2003). Visitors to public gardens have indicated that the visit had a perceived stress reduction (Kohlleppel et al. 2002) from one walk in nature. Imagine what daily interaction could provide.

According to the American Academy of Ophthalmology (2011), a lack of bright light exposure for children contributes to nearsightedness because the exposure to sunlight creates chemical signals that prevent elongation of the eye during the growth process. All these problems could be ameliorated with active playtime outdoors in green spaces.

While children are not getting outside and playing and exploring as much as they did historically, education reform and policies have limited time that students have to spend in scientific inquiry also. Time in the elementary school classroom devoted to science has decreased to an average of 2.3 hours per week (IES 2007). This is the lowest level since 1988. With the increasing demand for students in science, technology, engineering, and math, it is critical to raise children that understand and appreciate science. This idea is supported by the National Science Board (NSB), which is part of the National Science Foundation. The NSB has stated that in order for our country to remain a world leader in science and technology, there are steps that our country must take including an early start to science education among others (NSB 2009). It might not be their academic field of choice, but to understand how to grow a plant or how and why to read the nutrition labels on a package of food can help that child and their family in the future as well as the environment in which they live (FDA 2014; NGA 2015).

Parents' attitudes about nature impact how much time children spend outdoors. The more positive the attitude, the more likely children were to spend time outdoors (Hammond et al. 2011). Parents were asked about the health of their children and researchers found that the more time children spend indoors watching television and playing video games, the more likely they were to experience health problems (Hammond et al. 2011).

WAYS TO GARDEN WITH YOUTH

There are many ways and places to involve youth with plants and gardens including school gardens, afterschool programs, children's gardens, home gardens, and community gardens. All types of gardens for children are of great benefit and have some common goals, as well as different outcomes in mind. For example, a school garden's main goal is typically to educate on school subjects, children's gardens within public gardens want the children to have fun, learn, and want to return, and community and home gardens are usually for vegetable and fruit production for the family. They may have slightly different objectives, but they all have the opportunity to give children a place to learn and play and connect with others.

School gardens are an ideal way to involve academics in the garden setting. These gardens reach many children at once, and there are no restrictions on socioeconomic status, race, or gender. While a school garden is great to have, even doing plant activities in the classroom can be something to attract a child to plants and nature (Figure 2.4). A research study was conducted with third graders participating in a plant curriculum. The curriculum did not require a school garden, but some schools did have and make use of their school garden. The study reported no differences in knowledge of gardening, science, and the environment between the groups that did or did not have a garden; however, there were significant differences in student'

FIGURE 2.4 School children look for insect damage on plants under a light table in the classroom.

attitude toward gardening, science, and the environment (Dirks and Orvis 2005). An additional benefit was an increased interest in gardening with the teachers at several of the schools that did not have gardens. These schools chose to add outdoor gardens to their campuses. These results suggest that indoor plant activities are beneficial, but adding an outdoor garden component can make even more impact.

There are also ways to incorporate accessible gardens and activities with children with special needs. Raised beds, ergonomic tools, and adapting techniques to the specific needs of the child are things to consider. The National Gardening Association has tips and resources to help plan and work in gardens with special needs children included on their website, kidsgardening.org, under "special needs."

There are programs and curricula that can be used in school settings and often they are coordinated to the state standards of learning. There are many resources on the Internet—programs, activities, curriculum sheets, and websites that act as a clearinghouse for garden- and school-related resources. Some of these resources are listed and discussed in the "Youth Garden Resources" section as well as in Table 2.1.

Afterschool programs are used by 10.2 million children according to the Afterschool Alliance (2015a). Afterschool programs, in general, have helped students improve work habits and self-efficiency and have reduced school absences for those participating (Auger et al. 2013). Students have also shown significant improvements in classroom behavior, completion of homework, and participation in class (Kochanek et al. 2011). Afterschool programs are run by schools, daycare centers, community centers, museums, libraries, churches, and community organizations such as YMCA, Boys & Girls Clubs of America, and the 4-H Council. The Afterschool Alliance (2015b) suggests types of activities that would be appropriate for different age groups. Getting outside, interacting with nature, connections to real-world experience, and physical activity are suggested across the ages of 5–18 years. Working in a garden setting can provide an option for these and many others of the suggested activities.

Children's gardens. While school gardens are associated with schools, the term "children's garden" typically refers to the garden spaces dedicated to children within a public garden or arboretum. These gardens are created with children in mind from a size, theme, and aesthetic point of view. The first children's garden in a botanical garden was established in Brooklyn Botanic Garden in 1914. It is the oldest continuously run program in the country (BBG 2014). Many children's gardens exist today with more gardens added each year to bring children and their parents into the garden (Table 2.2).

The themes of different areas within the garden entice children to play and use their imaginations. These spaces often include water features and play structures that promote the children to be physically active and to use all their senses when experiencing the garden. Public gardens and arboreta with children's gardens usually have programming for children as well, including opportunities for school field trips, passive learning stations, summer camps, and weekend learning sessions. Public gardens and arboreta that do not have a dedicated children's garden will often have these programs as well. Many public gardens have an education director or specialist that develops and directs the youth programming. This person can be a great resource for school gardens as well.

Home gardens are another place where children may garden that will allow them to connect to nature and learn about plants, animals, and insects. Gardening with family and friends not only connects them to nature, it can create bonds and memories that will stay with them for life (Francis 1995). Gardens for children are a complex mix of idea, place, and action. Older adults were asked to describe their childhood garden memories. They were able to describe with great detail the garden space and elements within it, and the things they did and places they created (Francis 1995). They even drew complex maps of the garden from memory indicating their favorite elements in the garden—vegetation and natural elements, followed by structures, then specific areas within the garden.

In 2013, 37 million households had home gardens. Fifteen million households with children were participating in home food gardening, which is a 25% increase

TABLE 2.1
Children's Garden Resources Including Activities, Benefits, Grant Writing, and Gardening Tips from the Internet and Books

Internet Resources

Resource Name	Type of Information	Web Address
Bringing Back Outdoor Play	Play initiatives, information on inclusion play, references for nature play	http://www.bringingbackoutdoorplay.com
Childhood in the Garden—A Place to Encounter Natural and Social Diversity	Article detailing benefits of gardening with young children	https://www.naeyc.org/files/yc/file/200801/BTJNatureNimmo.pdf
Children and Nature Network	Latest news and research on children and nature; resources and tools for getting children connected	http://www.childrenandnature.org
Children's Gardens: A Field Guide for Teachers, Parents, and Volunteers	Online book; gardening resource	http://celosangeles.ucdavis.edu/files/96723.pdf
Cooperative Extension Service	Interactive map to find the local Cooperative Extension Service office	csrees.usda.gov/Extension
Cornell Garden-Based Learning: Resources for Gardeners and Educators	Highlights from journal articles supporting garden-based learning	http://blogs.cornell.edu/garden/grow-your-program/research-that-supports-our-work/highlights-from-journal-articles/
Developing Youth Leadership through Community Gardening: Opportunities and Outcomes	Thesis on youth leadership development in community gardens	http://kb.osu.edu/dspace/bitstream/handle/1811/54635/Michelle_Beres_Thesis.pdf?sequence = 1
Edible Schoolyard Project	Resources and curriculum for edible education	edibleschoolyard.org
Farm to School Network	Information, advocacy, networking hub for local foods and agriculture education	farmtoschool.org
Food Corps	Volunteers who connect children to real food	foodcorps.org
Garden ABCs	Garden books, blog and resources for how to, grants, activities, and success stories	gardenabcs.com
GreenHeart Education	Resources for environmental sustainability	www.greenhearted.org
Green Ribbon Schools	Online community of ideas	greenribbonschools.org

(Continued)

TABLE 2.1 (*Continued*)
Children's Garden Resources Including Activities, Benefits, Grant Writing, and Gardening Tips from the Internet and Books

Internet Resources

Resource Name	Type of Information	Web Address
Growe Foundation	Educating children about eating and the environment	growefoundation.org
Growing School Gardens	Online community of resources/webinars	www.edweb.net/schoolgardens
How Our Gardens Grow: Cultivating Nutrition and Learning through Idaho School Gardens	Introduction to school gardens, benefits, connecting gardens to other subjects, school garden food safety, getting started, tasks, types of gardens, school spotlights	http://www.sde.idaho.gov/site/cnp/schoolgarden/docs/Garden%20Booklet.pdf
Kitchen Gardeners International	Education, garden planner, activities, recipes, grants	www.kgi.org
Master Gardeners	Find local Master Gardener groups	extension.org/mastergardener
National Ag in the Classroom	Online resources and curriculum for teachers and students	agclassroom.org
SchoolGrants	Grant writing tips and opportunities	k12grants.org
South Dakota Kids Garden Resource Kit	Compilation of online resources; creating a garden, age specific activities, subject specific activities	http://www.sdstate.edu/hns/outreach/nutrition/upload/kids-garden-resources.pdf
Univ of Arizona Community and School Garden Project	Website for teaching, gardening, and grant resources	sgdschoolgardens.arizona.edu
Western Growers Foundation Collective School Garden Network	Online resources for school gardens; grant program for California and Arizona (lists national grants also)	csgn.org
West Virginia School Gardens	Curriculum resources, activities, gardening resources	http://wvschoolgardens.org
Whole Kids Foundation	Nonprofit foundation; funding for schools	www.wholekidsfoundation.org

(*Continued*)

TABLE 2.1 (*Continued*)
Children's Garden Resources Including Activities, Benefits, Grant Writing, and Gardening Tips from the Internet and Books

Books	
Book Title	**Author/s**
A Child's Garden: 60 Ideas to Make Any Garden Come Alive for Children	Molly Dannenmaier
Garden Projects for the Classroom and Special Learning Programs	Hank Bruce and Tomi Jill Folk
Gardening in School All Year Round	Clare Revera
Gardening with Young Children	Sara Starbuck, Marla Olthof, and Karen Midden
Growing Plants, Functional Skills, and Communication Skills in School Gardens	Tammy Blake, Dawn Leach, and Shannon Fenix
How to Grow a School Garden: A Complete Guide for Parents and Teachers	Arden Bucklin-Sporer and Rachel Pringle
Learning Gardens and Sustainability Education: Bringing Life to Schools and Schools to Life	Dilafruz Williams and Jonathan Brown
Outdoor Classrooms: A Handbook for School Gardens	Carolyn Nuttal and Janet Millington
Ready, Set, Grow! A Kid's Guide to Gardening	Rebecca Spohn

from 2008 (NGA 2014). Gardening allows children to see vegetables grown, harvested, and prepared for consumption. A child is more likely to try something that they have had a part in growing and willingness to taste vegetables is important in establishing a child's food preferences (Morris et al. 2001).

Community gardens are another place where children garden with family and friends. In 2013, there were 3 million community gardens as opposed to 1 million in 2008 (NGA 2014). Community gardens vary greatly on their inclusion of youth from children coming to garden with their family members, to children planning and maintaining areas within the garden (Eames-Sheavly et al. 2007). Community gardens are connected with many societal benefits such as physical fitness, access to healthy foods, community cohesion, leadership skills development, and social interaction (Armstrong 2000; Glover 2003; Shinew et al. 2004).

TABLE 2.2
Public Children's Gardens in the United States

Garden	City, State
Huntsville Botanical Garden—Children's Garden	Huntsville, AL
Old Town Alabama—Children's Garden	Montgomery, AL
Georgeson Botanical Garden—Babula Children's Garden	Fairbanks, AK
Spring Creek Farm—Children's Garden	Palmer, AK
Boyce Thompson Arboretum State Park—Children's Garden	Superior, AZ
Tohono Chul Park—Children's Garden	Tucson, AZ
Tucson Botanical Gardens—Children's Discovery Garden	Tucson, AZ
Botanical Garden of the Ozarks—Children's Garden	Fayetteville, AR
Garvan Woodlands Garden—Evans Children's Adventure Garden	Hot Springs National Park, AR
Children's Ethnobotany Garden	San Diego, CA
Children's Storybook Garden and Museum	Hanford, CA
Conejo Valley Botanic Garden—The Kids' Adventure Garden	Thousand Oaks, CA
The Huntington—Children's Garden	San Marino, CA
San Diego Botanic Garden—Hamilton Children's Garden	Encinitas, CA
San Francisco Botanical Garden at Strybing Arboretum	San Francisco, CA
San Luis Obispo Botanical Garden—Children's Garden	San Luis Obispo, CA
South Coast Botanic Garden—3 Bears' House: Children's Garden	Palos Verdes Peninsula, CA
Turtle Bay's McConnell Arboretum & Botanical Gardens—Children's Garden	Redding, CA
Yerba Buena Children's Garden	San Francisco, CA
Betty Ford Alpine Gardens—Children's Garden	Vail, CO
Cheyenne Botanic Gardens—Paul Smith Children's Village	Cheyenne, CO
Denver Botanic Gardens—Mordecai Children's Garden	Denver, CO
The Gardens on Spring Creek—Children's Garden	Fort Collins, CO
Western Colorado Botanical Garden—Children's Garden	Grand Junction, CO
Mystic Seaport Children's Museum—Children's Zoo Garden	Mystic, CT
Winterthur Museum, Garden, & Library—Enchanted Woods	Winterthur, DE
US Botanic Garden Conservatory—Children's Garden	Washington, DC
US National Arboretum—Washington Youth Garden	Washington, DC

(Continued)

TABLE 2.2 (*Continued*)
Public Children's Gardens in the United States

Garden	City, State
Discovery Gardens—Children's Garden	Tavares, FL
Marie Selby Botanical Garden—Ann Goldstein's Children's Rainforest Garden	Sarasota, FL
Naples Botanical Garden—Children's Garden	Naples, FL
Sarasota Children's Garden	Sarasota, FL
Atlanta Botanical Garden	Atlanta, GA
Garden of the Coastal Plain at Georgia State University—Children's Learning Garden	Statesboro, GA
Spark M. Matsunaga International Children's Garden for Peace	Hanapepe, HI
Na 'Aina Kai Botanical Gardens & Sculpture Park—Children's Garden	Kilauea, HI
Oahu Urban Garden Center—Children's Garden	Pearl City, HI
Idaho Botanical Garden—Children's Adventure Garden	Boise, ID
Chicago Botanic Garden—Grunsfeld Children's Growing Garden	Chicago, IL
The Children's Garden at Illinois State University	Normal, IL
The Children's Garden Project of Elwood	Elwood, IL
Garfield Park Conservatory—Elizabeth Morse Genius Children's Garden	Chicago, IL
Klehm Arboretum & Botanic Garden—Nancy Olson Children's Garden	Rockford, IL
Luthy Botanical Garden—Children's Garden	Peoria, IL
Morton Arboretum Children's Garden	Lisle, IL
Coxhall Gardens + Children's Garden	Carmel, IN
Taltree Arboretum & Gardens—Adventure Garden	Valparaiso, IN
Iowa Arboretum Children's Garden	Madrid, IA
Reiman Garden at Iowa State University—Patty Jischke Children's Garden	Ames, IA
Botanica Wichita—Downing Children's Garden	Wichita, KS
Deanna Rose Children's Farmstead	Overland Park, KS
Kansas Children's Discovery Center—Outdoor Adventure	Topeka, KS
Overland Park Arboretum & Botanical Gardens—Children's Discovery Garden	Overland Park, KS
Wonderscope's Wonder Why Children's Garden	Shawnee, KS
Boone County Arboretum—Children's Garden	Union, KY
Children's Botanical Garden of South Central Kentucky	Somerset, KY
Kentucky Children's Garden	Lexington, KY

(*Continued*)

TABLE 2.2 (*Continued*)
Public Children's Gardens in the United States

Garden	City, State
Western Kentucky Botanical Garden—Moonlite Children's Garden	Owensboro, KY
LSU AgCenter Botanic Gardens—Children's Garden	Baton Rouge, LA
Longue Vue House & Gardens—Discovery Garden	New Orleans, LA
Coastal Maine Botanical Garden—Bibby & Harold Alfond Children's Garden	Boothbay, ME
Merryspring Nature Center—Children's Garden	Camden, ME
Penobscot Landing and Children's Garden	Brewer, ME
Topsham Public Library Children's Garden	Topsham, ME
Adkins Arboretum—The Funshine Garden	Ridgley, MD
Brookside Gardens Children's Garden	Wheaton, MD
Miller Branch Library—The Enchanted Garden	Ellicott City, MD
Berkshire Botanical Garden—Children's Garden	Stockbridge, MA
Gardens at Elm Bank—Weezie's Children's Garden	Wellesley, MA
Heritage Museums & Gardens—Hidden Hollow	Sandwich, MA
Long Hill and Sedgwick Gardens—Children's Garden	Beverly, MA
The Display Garden on Suncrest Children's Garden	Lapeer, MI
Fernwood Botanical Garden & Nature Preserve—Nature Adventure Garden	Niles, MI
Dow Gardens—Children's Garden	Midland, MI
Eastpointe Children's Garden	Eastpointe, MI
Frederik Meijer Sculpture Garden—Lena Meijer Children's Garden	Grand Rapids, MI
Grand Traverse Children's Garden	Traverse City, MI
Leila Arboretum & Children's Garden	Battle Creek, MI
Matthaei Botanical Gardens—Gaffield Children's Garden	Ann Arbor, MI
Michigan 4-H Children's Garden	East Lansing, MI
Slayton Arboretum—Children's Garden	Hillsdale, MI
Windmill Island Gardens—Children's Garden	Holland, MI
Minnesota Landscape Arboretum—Green Play Yard	Chaska, MN
WCROC Horticulture Display Gardens—Children's Garden	Morris, MN
Mississippi Children's Museum—The Literacy Garden	Jackson, MS
Missouri Botanical Garden—Doris I. Schnuck Children's Garden	St. Louis, MO
Tizer Botanic Gardens & Arboretum—Children's Garden	Jefferson City, MT
Lauritzen Gardens—Children's Garden	Omaha, NE
Strawbery Banke Museum—The Victorian Children's Garden	Portsmouth, NH
Camden Children's Garden	Camden, NJ

(*Continued*)

TABLE 2.2 (*Continued*)
Public Children's Gardens in the United States

Garden	City, State
Wagner Farm Arboretum	Warren, NJ
ABQ BioPark Botanic Garden—Children's Fantasy Garden	Albuquerque, NM
Brooklyn Botanic Garden—Children's Garden	Brooklyn, NY
Buffalo & Erie County Botanical Gardens—Wegmans Family Garden & Outdoor Children's Garden	Buffalo, NY
Ithaca Children's Garden	Ithaca, NY
New York Botanic Garden—Everett Children's Adventure Garden	Bronx, NY
The Strong National Museum of Play—Discovery Garden	Rochester, NY
Gateway Gardens—Michel Family Children's Garden	Greensboro, NC
North Carolina Botanical Garden—Children's Wonder Garden	Chapel Hill, NC
The Children's Museum at Yunker Farm—Children's Garden	Fargo, ND
The Allen County Children's Garden	Lima, OH
Cleveland Botanical Garden—Hershey Children's Garden	Cleveland, OH
The Schoepfle Children's Garden	Birmingham, OH
Simpson Garden Park—Children's Discovery Garden	Bowling Green, OH
Wegerzyn Gardens MetroPark—Children's Discovery Garden	Dayton, OH
The Botanic Garden at Oklahoma State University—Children's Garden	Stillwater, OK
Myriad Botanical Gardens—Children's Garden	Oklahoma City, OK
Luscher Farm—Children's Garden	West Linn, OR
Oregon Garden—Children's Garden	Silverton, OR
The Arboretum at Penn State—Childhood Gate Children's Garden	University Park, PA
Longwood Garden—Children's Corner & Indoor Children's Garden	Kennett Square, PA
Phipps Conservatory & Botanical Gardens—Discovery Garden	Pittsburgh, PA
Tyler Arboretum—Stopford Family Meadow Maze area	Media, PA
Welkinweir—Children's Garden	Pottstown, PA
Providence Children's Museum—Children's Garden	Providence, RI
Brookgreen Gardens—A Garden Room for Children	Murrells Inlet, SC
Carolina Children's Garden	Columbia, SC
The Children's Garden at Linky Stone Park	Greenville, SC

(*Continued*)

TABLE 2.2 (*Continued*)
Public Children's Gardens in the United States

Garden	City, State
Chidlren's Museum of South Dakota—Children's Garden	Brookings, SD
Children's Museum of Oak Ridge—Kids Go Green! Environmental Center & Gardens	Oak Ridge, TN
Ijams Nature Center—Jo's Grove	Knoxville, TN
Memphis Botanic Garden—Hyde & Seek Prehistoric Plant Trail	Memphis, TN
Children's Museum of Houston—EcoStation	Houston, TX
Dallas Arboretum & Botanical Garden—Rory Meyers Children's Adventure Garden	Dallas, TX
Fort Worth Botanic Garden—Children's Garden	Fort Worth, TX
SFA Gardens—Jim and Beth Kingham Children's Garden	Nacogdoches, TX
Shangri La Botanical Gardens & Nature Center—Here We Grow Children's Garden	Orange, TX
Texas Discovery Gardens	Dallas, TX
UT Austin Lady Bird Johnson Wildflower Center Children's Garden	Austin, TX
Zilker Botanical Garden—Children's Garden	Austin, TX
Red Butte Garden—Children's Garden	Salt Lake City, UT
Thanksgiving Point—Children's Discovery Garden	Lehi, UT
Amercian Horticultural Society River Farm—Children's Garden	Alexandria, VA
Community Arboretum—Children's Garden	Roanoke, VA
Holly Point Nature Park—Children's Garden	Deltaville, VA
Lewis Ginter Botanical Garden—Children's Garden	Henrico, VA
Meadowlark Botanical Gardens—Children's Garden	Vienna, VA
Norfolk Botanical Garden—World of Wonders: A Children's Adventure Garden	Norfolk, VA
Virginia Living Museum—Children's Garden & Playground	Newport News, VA
Bellevue Demonstration Garden—Children's Garden	Bellevue, WA
Benton/Franklin Demonstration Garden—Children's Garden	Kennewick, WA
Children's Garden at Good Shepherd Center	Seattle, WA
Children's Hands On Museum—Outdoor Discovery Center	Olympia, WA
Magnuson Children's Garden	Seattle, WA
Neototems Children's Garden	Seattle, WA
Green Bay Botanical Garden—Gertrude B. Nielsen Children's Garden	Green Bay, WI
Cheyenne Botanic Garden—Paul Smith Children's Village	Cheyenne, WY

Community gardens create an opportunity for informal science education, multigenerational and multicultural interactions, and participation in community action (ALGA 2015). One such program, the Garden Mosaics program, was created to be a resource for people wanting to promote the understanding of science within the community setting. The mission statement of the program is "Connecting youth and elders to investigate the mosaic of plants, people, and cultures in gardens, to learn about science, and to act together to enhance their community." The Garden Mosaics website has many resources for the public that are based on their mission.

BENEFITS OF GARDENING WITH YOUTH

Historically, school gardens have been used to teach horticulture, increase work ethics, and provide other psychological benefits. However, current research is finding that there are numerous other benefits of youth gardening. Many teachers believe that the garden is an effective place to enhance science education, social skills, academic performance, physical activity, language arts, and healthy eating habits (Graham and Zidenberg-Cherr 2005). Children can learn social responsibility—gardens can improve neighborhood appearances, social interaction, participant cooperation (Relf 1998), and access to healthy foods (Armstrong 2000; Macias 2008). Service projects can be incorporated that capitalize on these benefits and help children see the direct support they are giving others such as allowing the students to give tours and teach community members how to start a garden or donating the produce grown to a local food bank.

Research on the effects of incorporating plants and gardens into youth curriculum has been investigated most within the last 20 years. The majority of that research has focused on science achievement and attitudes, environmental awareness, nutrition knowledge, attitudes and behavior, and life skills (Blair 2009). These benefits can be categorized into six main groups: academic, psychological, social, environmental, physical, and nutrition.

ACADEMIC

Gardens are often used to teach science and environmental education, but they are not limited to use in those areas. Garden activities can be incorporated into many disciplines including mathematics, history, geography, language arts, and art. Basic math, algebra, and geometry can be used to determine length and area of the garden, how many seeds will be used, or how much fertilizer to use for given ratios. Discovering how crops from around the world have moved into new areas and what plants are native to different places can help children give more meaning to history, social studies, and geography. Writing poems, stories, and journals about their experiences in the garden can strengthen their language arts skills. There are thousands of children's books that have gardens and plants as the topic for reading including *Tops and Bottoms* by Janet Stevens, *The Runaway Garden* by Jeffery Schatzer, and *What's in the Garden?* by Marianne Berkes. Plants and flowers are also beautiful and are wonderful subjects for drawings and paintings for children to do themselves and admire the works of renowned artists like Georgia O'Keefe, Claude Monet, or Vincent Van Gogh. Science projects in areas such as earth science, plant science,

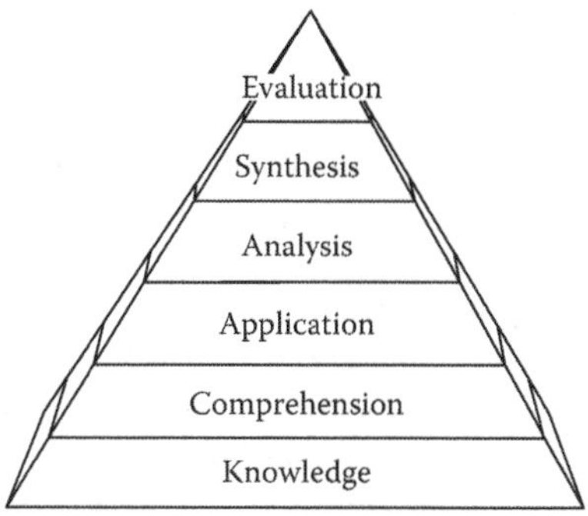

FIGURE 2.5 A diagram of Bloom's taxonomy of higher learning.

entomology, physiology, and physics are too numerous to count as possibilities in the garden. By taking an interdisciplinary approach in the garden, students have shown better writing skills, stronger grasps of abstract concepts, less absenteeism, and lower dropout rates from school.

One of the things that make gardens so successful in educating children on a variety of topics is that children are doing things. They may do some planning at their desk, but they are getting up and being active. They are having fun and not realizing that they are being taught something at the same time. Children comment regularly in gardens that they never had so much fun learning. Gardens incorporate hands-on learning and active participation that make learning fun. Not only is hands-on learning fun, it maximizes what the youth learn and retain through experience. Evaluating, creating, and analyzing are at the top of Bloom's (1984) *Taxonomy of Educational Objectives* (Figure 2.5), and they all play an important role in the garden. Bloom states that there are lower order thinking skills including being able to recall information (knowledge), explain the information (understanding), and use the information (applying). Studying and selecting a seed would start with knowledge—what are the parts of a seed, what type of plant will it produce? It would then move to understanding—why are the parts of the seed important and what do they do? Then students could apply the information by planting a seed at different depths to watch the seed coat break and radicle and shoot emerge. Examples of higher order thinking are to distinguish between different parts (analyzing), support a decision (evaluating), and create a new product or viewpoint (creating) (Bloom 1984). Looking at seed planting depth, students can analyze the germination rates to see what depth worked best. Based on the size and type of seed, students are able to support the decision of planting depth. This creates a new way that students look at seeds and germination—looking at other seeds and estimating how deep they might need to be placed to germinate. Gardening can become a catalyst to learning because children love to explore the unknown. Experiential learning, such as gardening, is a teaching methodology that can add a new breadth to youth education.

Teachers believe that higher science scores and improved academic achievement are both benefits of school gardening (Graham and Zidenberg-Cherr 2005). Children who participated in a garden activity once a week over a 14-week period had higher science achievement test scores from pre- to posttesting (Smith and Motsenbocker 2005). Another group of third through fifth graders added hands-on plant activities to their curriculum and saw their science achievement scores increase by 5.6 points above the control group that did not participate in the activities (Klemmer et al. 2005). Gardens also increased parental involvement in the school, which has been linked to higher student achievement (Xitao and Chen 2001; Jeynes 2007).

The Education Development Center and the Boston Schoolyard Funders Collaborative (2000) developed a study to look at the benefits of schoolyards to education. School grounds learning programs included any education that took place outdoors in the schoolyard. They surveyed educators across the country and found that gardens were the most commonly reported feature of the schoolyard. They also found that over two-thirds of the respondents felt that "their school grounds learning programs improved academic learning." Of the educators who responded, over 90% of them also felt that their school grounds learning programs supported cognitive development.

The State Education and Environmental Roundtable (SEER) evaluated the impact of schoolyards and outdoor learning on education (Lieberman and Hoody 1998). They chose several schools around the United States and found that 92% of the students who had the opportunity to learn in their school surroundings and environments outperformed their peers who were taught in conventional classroom settings. The benefits included better performance on standardized tests, increased excitement for learning, less discipline and management problems, and greater pride of accomplishments.

Psychological

The psychological benefits of gardens and nature have also been widely accepted for thousands of years. Ancient Egyptian physicians prescribed walks in the garden for mental ailments (Davis 1998). *Horticulture therapy*, the use of plants and gardens by a trained therapist to achieve specific documented treatment goals, began in the late eighteenth century with Dr. Benjamin Rush, one of the first psychiatrists. He noted improvements among patients who worked in the fields and gardens of the hospital. However, the psychological benefits of gardens, gardening, and nature are for everyone. With ever-increasing technology and inside time, everyone needs a place to connect to the earth and calm spirits (Louv 2011).

School gardens have shown to be effective in the area of psychological development. Properly planned gardens and activities reduce stress, encourage a sense of well-being and tranquility (Gross and Lane 2007), and increase self-esteem (Robinson and Zajicek 2005). Prolonged childhood stress without protective relationships with caring adults can be detrimental to the mental and physical development of the child (Middlebrooks and Audage 2008). Gardening as a hobby or a special activity including being in natural environments reduces stress in children (Chawla et al. 2014) and adults.

Positive relationships are seen between a child's attitude and behavior and the management of the school grounds. Gardening can impart pride and joy not only to the gardeners, but also to those who experience the garden. School pride is increased (Mitchell et al. 2010) and vandalism is decreased (Skamp and Bergmann 2001) with well-designed and maintained school grounds and gardens.

Plant activities also offer a way to learn delayed gratification, independence, and motivation (Alexander et al. 1995; Ouden and Wee 2011). Allowing children to experience success in a garden setting gives them the feeling of being valued and a sense of belonging. Children are able to influence their environment and watch something grow and develop. This success can contribute to their favorable self-esteem (Montessori 1964; Sarver 1985; Robinson and Zajicek 2005).

Social

In school gardens, children can work individually as well as in teams to accomplish the necessary tasks and have a positive outcome. Research with children participating in a garden program found that students who participated in a garden program saw increases in working with groups, self-understanding, and self-esteem (Robinson and Zajicek 2005). Gardening experiences teach children about citizenship and how to cooperate with others. Studies have shown having students in green environments increases good behavior and patience, while reducing bullying (Dyment and Bell 2006). Gardens are places where students can work together, make decisions, manage problems, and gain a sense of responsibility. Not only do children have the benefits of working with each other, gardens increased parental involvement in schools allowing children to belong to a larger community of influence. Research found that a garden program improved peer relations, sociality, and peer status while the control group scores did not change (Kim et al. 2014).

Gardens are places to incorporate intergenerational relationships. In gardens, older adults can offer a world of wisdom and support for children who might not have that type of influence in their life otherwise. This relationship can be a positive one for both the child and the older adult. Older adults improve socialization, which can help them live longer and boost their immune system. It also stimulates learning for both the youth and older adult. Youth in intergenerational programs are more likely to achieve success in school, have higher self-esteem, have better social development, and have a decrease in negative behaviors such as drugs, alcohol, and absenteeism from school (GU 2007).

Gardens can be a place where students learn about different cultures and countries. Teachers can incorporate foods and techniques from different nations into their gardens. This helps youth gain an understanding and appreciation for people of different cultures. Children can talk and learn about their heritage and culture with one another, and multicultural gardens can be an effective place for sharing and learning different languages (Cutter-Mackenzie 2009). Allowing students to represent their culture in the garden development creates a sense of pride and belonging for the children that are not indigenous to the area and creates cultural sensitivity in the children who are native to the area (Cutter-Mackenzie 2009). School gardens offer a chance for children to give back to their communities, as well, by sharing the gardens with others and giving food to local food banks for those in need.

Environmental

Often children raised in urban environments lack the connection to nature that develops respect and responsibility for the earth and its inhabitants. Working in gardens can teach children admiration and appreciation for living things. Children who garden develop a more personal relationship with the land based on their experiences with nurturing plants (Mayer-Smith et al. 2007). They have a heightened awareness of the impact of humans on the environment and a greater sense of responsibility to the environment. School grounds that have been well landscaped and maintained tend to foster positive environmental attitudes (Fleener 2013), and garden activities have also led to positive environmental attitudes in children (Skelly and Zajicek 1998). Skelly and Zajicek (1998) found that children engaging in more outdoor experiences had more positive environmental attitudes. Teaching children how our actions affect the environment in both positive and negative ways is good, but if children are not connected to nature, the positive behavior could be short lived. Gardening helps children connect. It can provide the link between environmental knowledge and behaviors. In a survey of educators performed by the Education Development Center (2000), over 80% of the educators felt their school grounds learning programs increased environmental stewardship.

Often children think that vegetables and other foods come from a can or from the grocery store. They do not realize the connection to the earth. Gardening also connects children to food sources.

Physical

The physical benefits of gardening have existed since gardening began. The US Department of Health and Human Services (2008) has acknowledged gardening as an effective option for exercising. It involves a wide range of aerobic and muscle strengthening activities through digging, lifting, pulling, and stretching. Recent studies have determined that gardening does meet the American College of Sports Medicine and the American Heart Association's criteria for the recommended physical activity intensity level (Kweon et al. 2004; Park et al. 2008). Green school grounds can play a noteworthy role in promoting physical activity (Dyment and Bell 2007). Park et al. (2013) studied children as they completed 10 gardening tasks. Oxygen uptake, energy expenditure, and heart rate were all monitored. Results showed that gardening tasks were a moderate- to high-intensity physical activity for the children. Gardening can promote a physically active lifestyle in addition to nutritional and psychological benefits.

The increase in physical activity from gardening or being in natural settings can reduce blood pressure, reduce stress and anxiety, and improve mood and self-esteem (Pretty et al. 2005). Moderate exercise can reduce body fat and overall weight, lower cholesterol, increase aerobic capacity, improve flexibility, increase perceived energy levels, prevent bone loss, and improve sleep along with other mental wellness benefits (AHTA 2015). The physical benefits then help reduce risk of cardiovascular disease, stroke, hypertension, type 2 diabetes, osteoporosis, obesity, colon cancer, and breast cancer (Kesaniemi et al. 2001). Green spaces play a vital role in human health and well-being (Maller et al. 2006).

Nutrition

In addition to physical activity, children who garden also tend to have more positive attitudes toward eating, as well as actually eating fruits and vegetables when

FIGURE 2.6 A young child has harvested two grape tomatoes and prepares to eat them for his snack.

compared to children who do not garden (Figure 2.6). The benefits to health from eating fruits and vegetables have long been recognized, with recommendations of daily fruit and vegetable consumption. The US Department of Agriculture, through the Choose My Plate program, suggests amounts that range from one to three cups depending on age, gender, and level of activity (USDA 2014). Increased consumption can result in reduced risk of cancer, cardiovascular disease, stroke, Alzheimer disease, arthritis, diabetes, and cataracts (Boeing et al. 2012). Fruits and vegetables are loaded with nutrients and antioxidants that contribute to a healthier body.

Studies with second-grade students found that including a garden along with nutrition education produced greater knowledge and preferences of fruits and vegetables when compared to students who only received the nutrition education (Parmer et al. 2009). Another study with fourth- to sixth-grade students who participated in garden activities twice a week over 12 weeks found the students' increased consumption of fruits and vegetables and asking for more fruits and vegetables at home (Heim et al. 2009). Students in second through fifth grade participated in a summer garden program and reported healthier snack selection and knowledge of the benefits of fruits and vegetables (Koch et al. 2006).

Horticultural Therapy

Being in nature and gardening can be therapeutic for anyone; however, *horticultural therapy* is a defined area in which a trained therapist works with clients on activities to try to obtain identified goals. These goals could be cognitive, psychological, social, or physical depending on the needs of the client. Horticultural therapy can be used with both adults and children. The American Horticultural

Therapy Association (2007) lists many groups that use horticultural therapy including rehabilitation programs, mental health programs, substance abuse programs, hospitals, palliative care programs, cancer centers, correctional facilities, shelters for homeless and victims of abuse, as well as schools, senior centers, and botanic gardens.

Attention-deficit/hyperactivity disorder (ADHD) is the most commonly diagnosed neurobehavioral disorder of childhood. ADHD is marked by inattention and impulsivity that can be reduced with exposure to natural views and settings (Kaplan 1995; Kuo and Taylor 2004). Horticultural therapy with ADHD has the opportunity to aid in psychological, cognitive, educational, emotional, and prevocational work skills for students (Etherington 2012). It also aids children with autism spectrum disorder by teaching them social skills, physical skills, and self-esteem. For those youth struggling with anxiety, anger, or depression, gardening can be a welcomed distraction. Gardening can be mentally captivating and reduce thoughts that would cause fear or anxiety. Physical exercise related to gardening releases serotonin to calm the mind, and can reduce muscle tension as well (Etherington 2012). A study using horticultural therapy with students with intellectual disabilities found an increase in sociality of the children (Kim et al. 2012).

The benefits of gardening can also aid in the rehabilitation programs for juvenile delinquents. A *juvenile delinquent* is a youth aged 10–17 years that has committed a crime in which the state has declared that a minor lacks responsibility and therefore cannot be sentenced as an adult (Burton 2007). While juvenile crime rates are at their lowest since 1988, the youth who do commit crimes need programs that will help them successfully reenter society as productive young people. Horticultural therapy programs in youth probation facilities and correctional programs have shown positive results. Young people in these situations tend to come into the criminal justice system with various problems such as substance abuse, academic failure, family issues, physical or sexual abuse among others. They often need both mental and physical health support to rehabilitate. Horticultural therapy has been used as a complementary therapy to help in both these areas. Programs have shown to develop pride, thoughts of the future and careers, and social bonding for participants (Flagler 1995). Horticultural therapy also improves social integration (Kweon et al. 1998), provides healthier patterns of social functioning (Kuo et al. 1998), and helps improve group cohesiveness (Bunn 1986).

Research at the New Jersey Department of Corrections and Division of Juvenile Services (Flagler 1995) found that the youth felt they learned job skills, made contacts with people who could help them, and gained ideas about furthering their education in the future. Many (87%) believed that the horticulture program could improve their quality of life once released, and the majority (85%) wanted to further themselves by attending college after the program. The following quotes from participants are not just positive, they are indicative of a person who has made a connection with people and with plants and their future: "If we ever run out of food, I can plant some." "Plants make life possible and make people happy." "Before, I used to kill plants and step on them" and "I am now beginning to understand the relationship between people and plants." Having a plan or thinking about the future in a positive way can have both mental and physical benefits.

STARTING AND MANAGING A YOUTH GARDEN

There are two things that every garden needs to succeed—people and plans—it will not work for long without both. People and plans are needed to develop and implement the process (NGA 2015). Gardens may start in different ways—a single teacher with a raised bed or two, a dedicated parent that sees a need and comes in to help, or a garden club that wants to bring the joy of plants to a younger generation. All these are great starts, but there must be buy-in with those involved to succeed. The people and plans discussed here are based on the creation of a school garden; however, they can be easily adjusted for any youth gardening venue.

People

It takes a group of dedicated people to create a garden for a school or community. Committees can be formed to oversee garden creation and day-to-day activities. Groups of people can help install and maintain the garden. It is important to have the right person in the right position.

Planning committee. Forming a planning committee of *stakeholders* (anyone with interest in or that might be affected by the project) can help eliminate or minimize problems and challenges. Stakeholders in a school garden, for example, could include school administration, teachers, maintenance personnel, child nutrition personnel, parents, students, local horticulturists, cooperative extension service personnel, and local business owners.

A leader should be chosen for the committee. This person can be the liaison between the members, schedule meetings, and delegate tasks for efficiency. This could be a member from any area, but it should be someone that has the time to commit to the group. A parent could be a great resource for the committee in this role if they have the skills and time available. Parents should be involved on the committee whether in the leader role or not. Parental involvement is good for the school in general and helps children feel the importance of education (Giles 1998).

If the garden will be on school property, one of the first people that should be approached is the school principal. School administrators know the rules and constraints that will need to be followed and addressed to implement and maintain a garden. They may also have access to resources that will facilitate the garden creation. They should have input in the location, design, creation, and use of the garden. Their consent must be received before creation of the committee.

Teachers are a pivotal part of this committee. All teachers can use a garden, but there should be a few that have bought into the idea of having the garden that would be willing and able to serve on the planning committee. These teachers would serve as liaisons to the other teachers within the school. They may also serve as instructional leaders for those teachers who are less experienced with gardening.

Do not forget the students. Including a few key students in the planning committee also helps give them ownership in the garden and develops leadership skills. They are the best source of information for what children like and want in the garden. Children often look at plans and problems differently than adults and could provide new perspectives. Only a few children would be needed for the committee, but classrooms full of children can help plan. The children on the committee would learn to be the voices of the larger group.

If a maintenance person for the school will be impacted by the garden, they should be included in the meetings also. They can give input and voice concerns over garden plans and maintenance. Maintenance for the garden is one of the biggest challenges, so planning ahead can make everyone's experience better.

A horticulturist from the community would be an excellent resource for this committee as well. This member will have plant knowledge and skills that will be instrumental in planning, creating, and maintaining the gardens. *Master Gardeners*, garden club members, local garden center employees or landscapers, or knowledgeable home gardeners are all potential candidates. The Master Gardener program is run through the state Cooperative Extension Service. Members are trained with 40–80 hours of horticulture education and also must serve volunteer hours to become certified and to maintain certification (Extension 2015). The horticulturist could also be employed by the *cooperative extension service*. The cooperative extension service is a US Department of Agriculture program that aims to improve people's quality of life and economic well-being. Each state and territory has an office at its land-grant university as well as a network of local or regional offices.

Day-to-Day Operations

A separate team can manage day-to-day activities in the garden. This team should also have a program leader to coordinate the garden creation and activities. It should include some members from the planning committee to assure that the goals and objectives of the garden are put into practice. Key areas for this committee include teacher leaders, garden trainers, media coordinator, fundraising/grant writer, maintenance leader, and volunteer coordinator (NGA 2015).

The committees are important to achieving goals but others need to be included in all steps of the garden program process. Students in the classroom can develop budgets, draw out plans, and select plants all while working on science, math, art, and life skills. By including students and other teachers in the planning stages, ownership and excitement are both created which can help lead to a successful program.

Plans

Planning the garden project should be done before a shovel touches the ground. The formation of the planning committee starts the program development by showing a need and garnering support for the garden project. The planning committee can gather support from other groups as well to determine who will be involved in different stages of the garden development. Plans should include teacher/volunteer training, funding and resources, site selection, garden design, the physical garden creation, and garden maintenance. Too many times gardens fall into disarray because they were rushed into without plans.

The planning committee should also develop a policy guide for the garden. At a school, the policies should include school and garden safety. Schools often have very specific visitor/volunteer policies that should be communicated to all off-campus volunteers. Other examples of safety issues to address include the use of fertilizers and pesticides and tool safety. Students will be handling the plants and soil. Any added substances whether chemical or organic should be discussed and applied properly if chosen. Students will be using garden tools like clippers, shovels, and rakes. Everyone should be trained on the proper ways to use and store the tools to avoid accidents.

Teacher/Volunteer training. Teachers and volunteers sometimes feel unprepared or overwhelmed by school garden projects. The local extension service, Master Gardeners, or a garden club could possibly provide teacher training. This would give teachers a bit more confidence in the planting and care of the plants. Teachers could also train the volunteers in how to work with children. This could help the volunteers who do not regularly work with children, feel comfortable, and possibly be more effective with the children. Coordinating committees for planning and day-to-day committees can also help with teachers' time. Someone other than a teacher could be the lead for maintenance, which would give teachers more time to use the garden for educational activities. School gardens should not be perceived as a burden or they will not be used or supported.

Funding and resources. What needs to be purchased and what can be donated to the program? Many times local businesses can donate supplies including seeds, plants, soil, landscape timbers, rocks, weed cloth, and mulch. If not obtained through donations, they may need to be purchased. Determine if the school will cover utility costs including water. Another thing to consider is summer maintenance costs. If volunteers cannot be found, it might be possible to hire someone on a part-time basis. See Table 2.3 for resources to help with funding opportunities.

Funding sources. Federal, state, and local groups should be contacted for help in funding including large corporations, garden clubs, fertilizer companies foundations, native garden/plant groups, and local hospitals. Because gardens benefit so many areas (health, academics, environmental education, etc.), there is a wider array of organizations that will fund garden projects. Table 2.3 lists current sources for funding opportunities with youth gardening.

Site selection. An ideal garden site will have access to 6–8 hours of full sun and be located near the classrooms that will use it the most. Access to water is also very important. If the soil is not conducive to growing plants, raised beds with imported topsoil can be built. Determine if the ideal location will be large enough for the groups that desire to use it. Have a design and start small to help ensure success.

Garden design. Include students in the process of garden design. If a local designer will donate their services, he or she could provide very valuable insight to the plans to ensure a highly functional and aesthetically pleasing design. The designer can work with the students to develop fun-themed areas and give the students ownership in the garden. Accessibility for those students with disabilities needs to be considered so that all students can enjoy the garden. Ramps, raised beds, or special tools are potential ways to accommodate these students. In addition to accessibility, other things to consider are types of plants (edible only versus edibles and ornamentals), food allergies (avoid planting plants known to be highly allergenic), and shade areas to offer relief from the sun and heat (trees or built structures). In any garden, choosing xeriscape plants (drought-tolerant and low maintenance plants) for the landscape will give the best chance for long-term success.

TABLE 2.3
A Sample of Funding Opportunities for Youth Gardening

Funding Source	Website for More Information	Information Provided
Ag in the Classroom State specific mini-grants	agclassroom.org/affiliates/state_websites.htm (click on your state to find your state's application)	Financial support
Annie's Grants for Gardens	www.annies.com/giving-back/school-gardens	Financial awards
American Heart Association Teaching Gardens and Bonnie Plants	bonnieplants.com/community/aha-teaching-gardens/	School garden manual, lesson plans, soil, raised beds, plants, guidance
Captain Planet Foundation	www.captainplanetfoundation.org/apply-for-grants/	
Corning Incorporated Foundation	http://www.corning.com/inside_corning/our_commitment/community_foundation.aspx	Financial awards
Fiskars' Project Orange Thumb Community Garden Grants	http://www2.fiskars.com/Community/Project-Orange-Thumb	Tools, financial awards, and resources
Fruit Tree Planting Foundation	www.ftpf.org/apply.htm	Fruit trees and shrubs, equipment, orchard design and oversight, horticulture workshops, aftercare training, and manuals
Lowe's Toolbox for Education	www.toolboxforeducation.com/	Financial awards
National Education Association	www.nea.org/grants/grantsawardsandmore.html	Financial awards
National Gardening Association	Grants.kidsgardening.org	Financial awards, equipment, tools
Whole Kids® Foundation	www.wholekidsfoundation.org/schools/programs/school-garden-grant-program	Financial awards

Garden installation. If any heavy machinery will be needed to start the garden, this is another service that could possibly be donated. There may be parents with skills or businesses that could provide help in this area. Weekend planting dates that include parents, other family, and friends as well as the students will also provide ownership into the garden. Planting days for the students during the week should also be planned to involve as many students as possible. If teachers are unsure about the construction or planting, contact local landscape businesses or the local cooperative extension service for resources and help.

Garden maintenance. Maintenance in the garden varies greatly based on the types of gardens or environments that are created. Care of the garden needs to be planned year round, but especially during those times of the year when limited or no personnel are at the school. Civic groups, garden clubs, or Master Gardeners might be able to help during times when children and teachers are not on campus. Families might be able to "adopt-a-week" during the summer to care for the plants. If raised beds are used, plants could be removed after harvest and beds covered for weed control while not in use. If trees and shrubs are planted, they should be watered well until established.

Challenges. Some challenges for school gardens are having year round maintenance, fundraising, insect and disease problems, untrained teachers, and lack of teacher time. The planning committee should devote some initial time into brainstorming potential obstacles. By being proactive and having plans, many of the listed challenges could be lessened or avoided altogether.

YOUTH GARDEN RESOURCES

There are many ways and places to garden with children, and there are multitudes of different gardening activities and curricula available to teachers, public gardens, as well as parents. Garden curriculums are available to provide specific activities that teachers can use. This will save the teachers' time and effort, making the garden easier to use and, therefore, more likely to be used. A few popular options are highlighted here; however, many more are available. Each of these websites has links to other programs and resources and should be used as a starting point for further investigation of other gardening resources.

Life Lab©. Life Lab© (lifelab.org) is a resource for educators and families that began in 1978 as a science-gardening-nutrition program at Green Acres School in Live Oak, California. In 1979, Life Lab© became a nonprofit organization and The Growing Classroom curriculum was initiated and first published in 1982. To date, The Growing Classroom is in its fourth edition and what started as a local program has since grown into an award winning nonprofit organization reaching thousands of students. Their website includes many links to activities, literature, workshops, field trip information, camps, and a store for purchase.

The Junior Master Gardener Program® (JMG). The Junior Master Gardener Program® (jmgkids.org) was created by Texas Cooperative Extension and the Texas A&M University System. JMG is considered a 4-hour program. This program offers youth the chance to gain certifications in specific areas (Golden Ray Series^{sm} Certification) or overall certification (Junior Master Gardener Certification). The Level One curriculum is geared toward grades three through five and is divided into eight chapters that cover all aspects of gardening as well as a life skills and career exploration chapter. The news There are also three thematic units for Level One including Health and Nutrition from the Garden, Wildlife Gardener, and Literature in the Garden. The newest curriculum project is Learn, Grow, Eat & Go! There are currently two units for Level Two (geared toward Middle or Junior High School students) which include one on plant growth and development and another on soils and water. The JMG program's website has links for teachers and youth. It also has a link to purchase the curriculum as well as recognition pieces for the participants. There is also a blog that has regular updates with helpful tips for gardening and other activities.

Gardening with Kids©. Gardening with Kids© (Kidsgardening.org) is an online educational resource created in 1982 by the National Gardening Association (NGA) for anyone wishing to garden with youth, obtain activity ideas or purchase books, kits, or other products. There are many products available and a great deal of information provided. There are also links on grant opportunities and fundraising ideas to help cover the costs of creating youth gardens. NGA has supported over 10,000 school and youth garden programs around the world.

American Horticultural Society. The American Horticultural Society has a history of supporting gardening with youth and they showcase this on the Youth Gardening page of their website (ash.org/gardening-programs/youth-gardening). They host the National Children and Youth Garden Symposium each year at various locations around the country. This conference brings together teachers, garden designers, botanical garden staff, community leaders, and anyone who is involved with getting youth connected to the natural world. This group also partnered with the JMG program to create the "Growing Good Kids—Excellence in Children's Literature" award program. This program recognizes children's literature that is effective in promoting gardening, nature, and the environment. All winners are shown on the site with a synopsis of the award winning books.

The Youth Gardening page also features a "Find a Garden" link helping connect family and school groups with botanical gardens. The main source of current youth gardening information is through their Youth Gardening Gazette, which is a daily online newspaper. The newspaper has several headings including leisure, science, environment, arts and entertainment, education, and health. This is also a very helpful website with great information submitted from around the world.

While these are just a few sources, the aforementioned programs and websites are a great start into the world of youth gardening. The programs have links to many other sources of information, funding opportunities, and success stories from around the world. Table 2.1 includes a list of additional resources including books and websites.

CASE STUDY 1 Youth Horticultural Therapy Programs

The Chicago Botanic Garden has a well-known horticultural therapy program that provides services to people of all ages and many types of needs. Clients can come to the garden and experience the Buehler Enabling Garden, which has been designed for all abilities and ages. Therapists from the Garden program also regularly go out into the Chicago area to provide programs and assistance to groups in need.

Two closely related groups that the garden helps are children with sensory processing disorders (SPD) and those with autism spectrum disorder (ASD). Sensory processing disorder is a condition in which the sensory signals such as touch, sound, or movement do not organize into the appropriate responses. Children may exhibit motor difficulties, behavioral problems, anxiety, or depression among other symptoms if not treated (SPD Foundation 2015). While many children with ASD have sensory processing difficulties, ASD is also characterized by difficulties in social interaction, verbal and nonverbal communication, and repetitive behaviors (Autism Speaks 2015).

The Garden therapists have conducted both on- and off-site programs with children with SPD and ASD. Barbara Kreski, the Director for Horticultural Therapy Services at the Garden, created a program called "Plants & Me." This program is an

8-week program using concepts from the "How Does Your Engine Run?" alertness program written by Williams and Shellenberger (2015). The Alert Program was developed to help leaders support self-regulation and understand sensory integration. The Plants & Me program was piloted off-site in a classroom of children (aged 9–11 years) with different emotional and behavioral disorders including ASD.

The children in the program were both interested and intrigued with the plants. Kreski made a point of showing the children how different plants need different types of care. By showing a child how one plant might need a lot of water and another plant might need very little water, they were able to better understand why one child in the classroom might need different care or treatment than another. Both Kreski and the school administrators and teachers were very pleased with the outcome of the program. The classroom teacher felt that the program gave her solid tools to address issues in the future. Only the lack of funding has slowed the program and the benefits it can offer to other children.

The Horticulture Therapy Services department has also consulted with the Events department of the garden to develop concerts that are comfortable and enjoyable for children with SPD and ASD. The concerts are advertised as SPD-friendly so different behaviors are expected and tolerated by those in attendance. Having the concerts outdoors minimizes some of the crowd noise, and instructions on how to move around and through the venue are provided to reduce distractions. The performers for the concert are from the Old Town School of Folk Music. This group has been chosen carefully as they currently offer artistic classes for children and adults of "all ages, interests, abilities, and backgrounds" including those with sensitivities and special needs (Old Town School 2015).

The Garden also conducts horticultural therapy sessions in the Buehler Enabling Garden for five local organizations that serve youth with ASD. These programs are usually 1-day programs that focus on sensory tolerance and social skills. These programs could be as little as 1 hour, so it is very important that they have a successful experience in a short amount of time. Children in these programs typically propagate a plant or transplant a seedling, which allows them to take something back to the classroom with them. This gives the teachers the opportunity to extend the experience beyond the garden visit.

Programs are tailored to each child as best as possible. Some children, for example, have a difficult time with sensory processing of touch. To facilitate a successful experience for this child, the soil is warmed and gloves and tools are provided to meet the child where they are in regards to their comfort level. Also, directions are provided step by step and in sequential order. Photographs and examples are shown to minimize frustration. Some children with sensory processing can be uncomfortable just being outdoors. For these children, preparing them with what they might experience before going outside is very important. Being preemptive and telling them about the bees that they will see around the plants, for example, can help them exhibit more self-control when they see a real bee.

The Garden is currently developing information for their website on what to expect when visiting the garden for children and adults with SPD and other special needs. This webpage will provide information specifically related to possible issues with touch, sound, and movement to prepare guests and their caregivers. If the visitors can be prepared for what they will experience, hopefully the experience itself will be more positive. It might include items such as the fact that there are only hand dryers in the restrooms and no paper towels. The hand dryers are very loud and can be a problem for those with hearing sensitivity. These guests can be prepared ahead of time about the dryers or caregivers will know to bring an alternative for drying hands. These are just a few of the ways that the Garden and other horticultural therapy programs are meeting the needs of children with various abilities and sensitivities.

CASE STUDY 2 Children's Garden at Huntsville Botanical Garden, Huntsville, AL

Huntsville Botanical Garden's (HBG) Children's Garden is a two-acre wonderland for children of all ages that opened to the public in 2006. There are eight themed areas including (1) dinosaur garden, (2) space garden (Huntsville is also home to the US Space and Rocket Center), (3) rainbow garden, (4) storybook garden, (5) maze/tree-house garden, (6) international garden, (7) bamboo garden, and (8) Half-Acre Wood. Within those areas are fossils, a real Space Station node, the Pollywog Bog wading pool, bamboo products, and a Root View exhibit among many other displays. Each of these areas draws children into interactive play in the educational exhibits. In the summer, there are water activities in each of the areas for cooling and play. There is also a large open grass area in the center of the different gardens for free play and games.

A Nature Center adjacent to the Children's Garden provides an excellent opportunity to teach children about ecosystems, insects, amphibians, birds, and reptiles. There are ponds, streams, and a waterfall in the Nature Center as well as lush vegetation and flowering plants. This center houses the nation's largest open-air butterfly house.

The children's garden was one of the five original areas in the master plan for HBG. When the time came to create the children's garden, ideas were solicited from the community, staff, and volunteers of the garden. These ideas were given to Landscape Architect, Carol Lambdin to incorporate into the design. HBG raised $500,000 for the children's garden to be installed. Designs were completed in October of 2005, construction began in December 2005, and the grand opening was in June 1, 2006.

A safe and engaging atmosphere for the garden was a goal from the beginning. One of the safety features of the children's garden is fencing around the entire space. There is one way in and out of the garden that gives parents a sense of security and allows the children to explore in a safe setting. Some of the original ideas have been redesigned over the years to make them more durable, safe, or engaging based on feedback and observations of the children and facility managers. Lambdin's advice on designing a children's space was to build it very sturdy and make sure it is interactive.

Huntsville Botanical Garden (HBG) brings hundreds of children from local area schools each year. Field trips, classes, and tours are provided for children as young as 2 years. The classes cover a variety of topics including learning about plants, the environment, insects, and animals. Each of the school field trip programs is linked to the Alabama Science Standards and helps teachers plan and meet school requirements. Some of the program descriptions online include previsit and post-visit activities to engage the students in nature activities before and after their visit.

Soozi Pline, Director of Education, manages a team of employees who are charged with developing and implementing programs, camps, and classes. These classes are geared toward children, families, and adults to connect and educate them about the garden. When asked what she is most passionate about in the children's garden, Pline responded, "Allowing children to enjoy being outside without the rhythms and structures set in place by an adult schedule has intrinsic value not easily measured, but it's what I love the most about our Children's Garden!"

CASE STUDY 3 School Garden: Balboa Middle School, Ventura, CA

A few years ago, Balboa Middle School teacher, Steve Roth noticed an unused area on the school grounds that was overgrown and in need of care. He began a garden club that met at lunch once a week to begin the transformation of the space into a garden as it had been 50 years ago. The vice-principal of the school noticed the club's efforts, and Roth was encouraged to turn the project into an elective credit for the students.

For the past 4 years, a one-semester elective course called Environmental Horticulture has been offered and taken by mainly seventh and eighth grade students. The course has grown in popularity over the years. In the beginning, a few students were placed in the class that did not choose the class. Roth noted that even those students that were not excited about being there quickly changed their minds and enjoyed their time in the garden. He has had high school students come back to visit the garden that they helped care for during the early years. Enrollment in his course has grown to 36 students this semester and 8 of those are repeat enrollment. It is both a credit to the garden and the course that the students want to come back and learn in the garden. Roth said that the students' love for being outdoors and being able to experience something new each day is part of the lure of the course.

The garden space is about a half acre with 28 $5' \times 5'$ boxes and 6 $8' \times 4'$ boxes for mainly vegetables and herbs. There is also a space that showcases drought-tolerant plants and another for a small fruit orchard. The beds have been established with drip irrigation and mist irrigation. Roth separates the students into crews of about six students with a team leader that work together to achieve different tasks in the garden each week. His weekly schedule typically has them learning in the classroom on Monday, cooking with something they have grown on Wednesday, and working out in the garden on Tuesday, Thursday, and Friday. He has developed a fall curriculum as well as a spring curriculum to meet the needs of the changing seasons.

The fruits and vegetables that are grown in the garden are used in several different ways. The produce is cooked and consumed in the class, taken to the school kitchen, or taken home by the students. Roth has also started a school farmer's market to sell produce to parents and the community. The market is held after school on Fridays once a month or as the produce is available. Money that is made at the market goes back into garden for supplies. Produce has also been donated to FOOD Share, the Ventura County Food Bank.

While the garden has full administrative support, the financial aspect falls on Roth's shoulders. He has spent a good deal of time outside the classroom finding donations and resources to make the garden possible. The Parent–Teacher-Organization (PTO) has helped with fundraising for the garden. For one particular event, they sponsored a pumpkin patch for the garden with proceeds returning to the garden fund.

Roth has a degree in aquatic biology and teaches five periods of life science each day. Creating a garden has been and continues to be a learning experience for him as well. The statewide science teacher conference and the California Foundation for Agriculture in the Classroom have both proved to be excellent resources for information and curricula. Roth also relies on local experts to be guest speakers in the classroom on topics such as composting and beneficial insects.

The challenges of the Balboa MS garden are mainly the challenges of most gardens—insects, birds, and small mammals; however, there has been a disconnect between the garden and the district's facilities department. The garden lost water for a period of time because facilities was unaware there were seedlings that could die in a short amount of time. This is a good point for all school gardens to consider when planning—how are the grounds maintained and how can a garden representative be added to a contact list when resources need to be limited.

The school garden is used daily by the Environmental Horticulture class, but it has also seen use by the drama club, English teachers for poetry writing, and by the art teacher. Roth hopes to add tables to the space to encourage further use of the garden space by other teachers. Steve Roth is a great example for other teachers—you do not have to have a plant background or a horticulture degree to start a school garden. One of the very things that make this a great opportunity for his students, make it a great opportunity for Roth as well—they can both learn together. Another way that he is a great example is that one teacher can make a huge impact on not only his students, but also with others and his environment. A teacher does not need to have everything set up in a garden when a student walks in the door. Teach them to build the garden from the ground up—let them design, build, plant, maintain, and harvest the garden. These are lessons they will take with them forever.

SUMMARY

Gardening and being in nature have been recognized as very important for human development for hundreds, if not thousands, of years. From ancient days, garden/nature therapy has provided mental peace, physical movement, and a connection to the environment. The benefits of gardening spread to the academic world in Europe in the 1800s followed by the United States in the early 1900s. School gardens were very prevalent, and while they have seen periods of decline, they are currently increasing in popularity as a way to connect children with nature and the outdoors.

Many children have limited exposure to nature whether from location or busy afterschool schedules or parents' fear. Free-time outdoors is needed for mental and physical health. School gardens and structured garden programs add to mental and physical wellbeing with benefits associated with academics, social development, environmental attitudes, and nutrition attitudes and behaviors. These benefits have been spoken for years, but now current research is providing the verification to support the anecdotal evidence. The lessons children are learning from experiential design have the potential to make the most impact, as they are typically the most memorable. Children involved with plant care and maintenance benefit from this experience, whether the experience is in a school garden, children's garden, or inside the classroom.

Horticultural therapy is also presenting great developments with youth. Children with intellectual disabilities, anger, fear, anxiety, autism, and ADHD have shown improvement after being outside. Working with plants allows them to be on a level field with other children. It lets them take their minds off of their problems and relax mentally and physically. Horticultural therapy has also helped juvenile offenders in these ways while teaching them skills in which they can find confidence and possibly a future career.

Planning a garden program takes a lot of people and planning to be successful. There is not one plan that will work for every location. The best plan is to start small and work toward adding more space or areas. This will help give success and momentum to the garden project. Plans should include identifying funding and resources, site selection, garden design, garden implementation, and garden maintenance. There are many curricula and website resources for teachers and program planners to help ensure success. Gardens are work, and they need care and monitoring. They need planning and maintenance, and replanting and watering. However, the benefits children realize from a garden program make all the work extremely meaningful.

REVIEW QUESTIONS

1. Describe the early school garden movement in Europe.
2. How did school gardens spread to the United States? Describe one of the early garden success stories.
3. What are three reasons supporting the need for youth garden programs?
4. What cultural and societal pressures have impacted the amount of time children spend in nature?
5. What makes a garden a good place to teach academic lessons to children?
6. Describe how gardens and nature can have a positive impact on psychological and social well-being.
7. How do garden programs/nature affect the environmental attitudes of children?
8. What are four different ways children can be exposed to plants/nature (formally or informally)?
9. What two variables are critical for school garden programs to get established and be maintained?

10. What are three challenges school gardens face and how can these be overcome?
11. Who are the stakeholders of a school garden?

ENRICHMENT ACTIVITIES

1. Contact a local school and ask for permission to teach a short plant lesson to a group of students. Be sure to have content, an activity, and supplies prepared before visiting the classroom.
2. Find a children's garden at a botanical garden. Visit the garden and make note of the different activities for the children. How can these activities affect them socially, mentally, and physically?
3. Interview a school teacher currently participating in a school garden program. Find out how and why he or she got involved with gardening. What benefits do they see with the children in their classes? What obstacles do they face?
4. Prepare a brochure, poster, or presentation to show the benefits of nature and gardening to educators and parents.
5. Prepare a list of local/regional resources for educators to get their own garden program started.

REFERENCES

Afterschool Alliance. 2015a. Afterschool programs keep kids safe, engage kids in learning and help working families. http://www.afterschoolalliance.org/documents/National_fact_sheet_10.07.14.pdf (accessed January 29, 2015).

Afterschool Alliance. 2015b. What to look for in an afterschool program. http://www.afterschoolalliance.org/myCommunityLook.cfm (accessed January 29, 2015).

Alexander, J., M. North, and D. Hendren. 1995. Master gardener classroom garden project: An evaluation of the benefits to children. *Children's Environ*, 12(2), 256–263.

American Academy of Ophthalmology. 2011. More time outdoors may reduce kids' risk for nearsightedness. http://www.aao.org/newsroom/release/20111024.cfm (accessed November 22, 2014).

American Community Gardening Association (ACGA). 2015. Garden Mosaics. https://communitygarden.org/programs/garden-mosaics/ (accessed August 7, 2015).

American Horticultural Therapy Association. 2007. AHTA Position Paper. http://ahta.org/sites/default/files/Final_HT_Position_Paper_updated_409.pdf (accessed January 26, 2015).

American Horticultural Therapy Association. 2015. Ahta.org (accessed March 13, 2015).

American Montessori Teachers' Association (AMTA). 2014. Introduction to Montessori Education. http://www.montessori-namta.org/About-Montessori (accessed December 16, 2014).

Armstrong, D. 2000. A survey of community gardens in upstate New York: Implications for health promotion and community development. *Health & Place*, 6, 319–327.

Auger, A., K. Pierce, and D. Vandell. 2013. Participation in out-of-school settings and student academic behavioral outcomes. Unpublished paper at the annual meeting of the American Educational Research Association, San Francisco, CA.

Autism Speaks. 2015. Autismspeaks.org (accessed March 13, 2015).

Basset, T.J. 1979. Vacant lot cultivation: Community gardening in America, 1893–1978. MS Thesis, University of California, Berkeley.

Blair, D. 2009. The child in the garden: An evaluative review of the benefits of school gardening. *The Journal of Environmental Education*, 40(2),15–38.

Bloom, B. 1984. *Taxonomy of Educational Objectives*. White Plains, New York: Longman.

Boeing, H., A. Bechthold, A. Bub, S. Ellinger, D. Haller, A. Kroke, E. Leschik-Bonnet, et al. 2012. Critical review: Vegetables and fruit in the prevention of chronic diseases. *The European Journal of Nutrition*, 51, 637–663.

Brooklyn Botanic Garden (BBG). 2014. Children's Garden History. http://www.bbg.org/discover/gardens/childrensgarden#/tabs-3 (accessed December 17, 2014).

Bunn, D. 1986. Group cohesiveness is enhanced as children engage in plant stimulated discovery activities. *Journal of Therapeutic Horticulture*, 1, 37–43.

Bureau of Labor Statistics. 2015. Employment Characteristics of Families Summary. www.bls.gov/news.release/famee.nr0.htm (accessed August 7, 2015).

Burton, W. 2007. Burton's Legal Thesaurus, 4E. http://legal-dictionary.thefreedictionary.com/juvenile + delinquent (accessed January 30, 2015).

Carter, C. 2010. Transcript of School Gardens with Constance Carter–Journeys and Crossings Pages. https://www.loc.gov/rrlprogram/journey/schoolgardens-transcript.html (accessed August 7, 2015).

Centers for Disease Control and Prevention. 2014. Childhood Obesity Facts. http://www.cdc.gov/healthyyouth/obesity/facts.htm (accessed January 26, 2015).

Chawla, L., K. Keena, I. Pevec, and E. Stanley. 2014. Green schoolyards as havens from stress and resources for resilience in childhood and adolescence. *Health and Place*, 28, 1–13.

Child Care Aware of America. 2011. After-School Programs. http://www.naccrra.org/sites/default/files/default_site_pages/2011/after-school_programs_march2011.pdf (accessed December 17, 2014).

Child Care Aware of America. 2013. Who's minding the kids? Child care arrangements: Spring 2011. 17 December 2014. http://www.naccrra.org/sites/default/files/default_site_pages/2013/census_bureau_fact_sheet_april_2013.pdf (accessed December 17, 2014).

Cutter-Mackenzie, A. 2009. Multicultural school gardens: Creating engaging garden spaces in learning about language, culture, and environment. *Canadian Journal of Environmental Education*, 14, 122–135.

Davis, S. 1998. Development of the profession of horticultural therapy. In S.P. Simpson and M.C. Straus (eds.), *Horticulture as Therapy: Principles and Practices*. New York, NY: The Hawthorn Press, Inc., pp. 3–20.

Dirks, A.E. and K. Orvis. 2005. An evaluation of the junior master gardener program in third grade classrooms. *HortTe chnologl*, 15(3), 443–447.

Dunkelman, M.J. 2014. *The Vanishing Neighbor: The Transformation of American Community*. New York, NY: W. W. Norton & Company.

Dyment, J. and A. Bell. 2006. Our garden is colour blind, inclusive and warm: Reflections on green school grounds and social inclusion. *International Journal of Inclusive Education*, 12(2), 169–183.

Dyment, J. and A. Bell. 2007. Active by design: Promoting physical activity through school ground greening. *Children's Geographies*, 5(4), 463–477.

Eames-Sheavly, M., K. Lekies, L. MacDonald, and K. Wong. 2007. Greener voices: An exploration of adult perceptions of participation of children and youth in gardening planning, design, and implementation. *HortTechnology*, 17(2), 247–253.

Education Development Center and Boston Schoolyard Funders Collaborative. 2000. Schoolyard learning: The impact of school grounds. http://www.schoolyards.org/pdf/Schoolyard%20Learning-The%20impact%20of%20School%20Grounds.pdf (accessed December 19, 2014).

Etherington, N. 2012. *Gardening for Children with Autism Spectrum Disorders and Special Educational Needs*. London: Jessica Kingsley Publishers.

Extension. 2015. Beeoming an Extension Master Gardener. http://www.extension.org/pages/27285/becoming-an-extension-mastergardener#.VcVP OLd4s44 (accessed March 2, 2015).

FDA. 2014. Nutrition facts label: Read the label youth outreach campaign. http://www.fda.gov/Food/IngredientsPackagingLabeling/LabelingNutrition/ucm281746.htm (accessed January 21, 2015).

Flagler, J. 1995. The role of horticulture in training correctional youth. *HortTechnology*, 5(2), 185–187.

Fleener, A. 2013. Cultivating life: A study of a school landscape project. PhD dissertation, Auburn University.

Francis, M. 1995. Memory and meaning of gardens. *Children's Environments*, 12(2), 183–191.

Generations United (GU). 2007. Fact sheet: The benefits of intergenerational programs. Washington, DC: Generations United.

Giles, H. 1998. *Parent Engagement as a School Reform Strategy* (ED419031). New York, NY: ERIC Clearinghouse on Urban Education.

Glover, T. 2003. Community garden movement. In K. Christiansen and D. Levinson (eds.), *Encyclopedia of Community*. Thousand Oaks, CA: Sage, pp. 264–266.

Graham, H. and S. Zidenberg-Cherr. 2005. California teachers perceive school gardens as an effective nutritional tool to promote healthful eating habits. *Journal of the American Dietetic Association*, 105, 1797–1800.

Grogan, P. 2011. Carpe diem and slow down. *Bulletin of the Ecological Society of America*, 92, 281–284.

Gross, H. and N. Lane. 2007. Landscapes of the lifespan: Exploring accounts of own gardens and gardening. *Journal of Environmental Psychology*, 27(3), 225–241.

Halken, S. 2003. Early sensitization and development of allergic airway disease-risk factors and predictors. *Paediatric Respiratory Reviews*, 4, 128–134.

Hammond, D.E., A.L. McFarland, J.M. Zajicek, and T.M. Waliczek. 2011. Growing minds: The relationship between parental attitudes toward their child's outdoor recreation and their child's health. *HortTechnology*, 21(2), 217–224.

Hayden-Smith, R. 2014. Sowing the seeds of victory: American gardening programs of World War I. Jefferson, NC: McFarland & Co, Inc.

Heerwagen, J.H. and G.H. Orians. 2002. The ecological world of children. In P.H.J. Kahn and S.R. Kellert (eds.), *Children and Nature: Psychological, Sociocultural, and Evolutionary Investigations*. Cambridge, MA: MIT Press, pp. 29–64.

Heim, S., J. Stang, and M. Ireland. 2009. A garden pilot project enhances fruit and vegetable consumption among children. *Journal of the American Dietetic Association*, 109(7), 1220–1226.

Institute of Education Sciences (IES). 2007. Changes in instructional hours in four subjects by public school teachers of grades 1 through 4. http://nces.ed.gov/pubs2007/2007305.pdf (accessed November 24, 2014).

Jeynes, W.H. 2007. The relationship between parental involvement and urban secondary school student academic achievement: A meta-analysis. *Urban Education*, 42(1), 82–110.

Kaplan, S. 1995. The restorative benefits of nature: Toward an integrative framework. *The Journal of Environmental Psychology*, 15, 169–182.

Kesaniemi, Y., E. Danforth, Jr., D. Jensen, P. Kopelman, P. Lefebvre, and B. Reeder. 2001. Dose-response issues concerning physical activity and health: An evidence-based symposium. *Medicine and Science in Sports & Exercise*, 33(6), S531–S538.

Kim, B., S. Park, J. Song, and K. Son. 2012. Horticultural therapy program for the improvement of attention and sociality in children with intellectual disabilities. *HortTechnology*, 22(3), 320–324.

Kim, S., S. Park, and K. Son. 2014. Improving peer relations of elementary school students through a school gardening program. *HortTechnology*, 24(2), 181–187.

Klemmer, C. D., T. M. Waliczek, and J. M. Zajicek 2005. Growing minds: The effect of a school gardening program on the science achievement of elementary students. *HortTechnology*, 15(3), 448–452.

Koch, S., T.M. Waliczek, and J.M. Zajicek. 2006. The effect of a summer garden program on the nutritional knowledge, attitudes, and behaviors of children. *HortTechnology*, 16(4), 620–625.

Kochanek, J.R., S. Wraight, Y. Wan, L. Nylen, and S. Rodriguez. 2011. Parent involvement and extended learning activities in school improvement plans in the Midwest Region. (Issues & Answers Report, REL 2011–No. 115). Washington, DC: U.S. Department of Education, Institute of Education Sciences, National Center for Education Evaluation and Regional Assistance, Regional Educational Laboratory Midwest. http://ies.ed.gov/ncee/edlabs (accessed January 29, 2015).

Kohlleppel, T., J.C. Bradley, and S. Jacob. 2002. A walk through the garden: Can a visit to a botanic garden reduce stress? *HortTechnology*, 12(3):489–492.

Kolb, D.A. and L.H. Lewis. 1986. Facilitating experiential learning: Observations and reflections. *New Directions for Continuing Education*, 30, 99–107.

Kuo, F., M. Bacaicoa, and W. Sullivan. 1998. Transforming inner-city landscapes: Trees, sense of safety and preference. *Environment and Behavior*, 30(1), 28–59.

Kuo, F. and A. Taylor. 2004. A potential natural treatment for Attention-Deficit/Hyperactivity Disorder: Evidence from a national study. *The American Journal of Public Health*, 94(9), 1580–1586.

Kweon, B., W. Sullivan, and A. Wiley. 1998. Green common spaces and the social integration of inner-city older adults. *Environment and Behavior*, 30(6), 832–858.

Kweon, H., E. Matsuo, J. Choi, T. Ogaki, and K. Shibuya. 2004. Exercise intensity of horticulture as physical activity. *Acta Horticulturae*, 639, 277–280.

Lieberman, G. and L. Hoody. 1998. *Closing the Achievement Gap: Using the Environment as an Integrating Context for Learning*. Poway, CA: Science Wizards.

Louv, R. 2008. *Last Child in the Woods: Saving our Children from Nature-Deficit Disorder*. Chapel Hill, NC: Algonquin Books.

Louv, R. 2011. *The Nature Principle: Reconnecting with Life in a Virtual Age*. Chapel Hill, NC: Algonquin Books.

Macias, T. 2008. Working toward a just, equitable, and local food system: The social impact of community-based agriculture. *Social Science Quarterly*, 89(5):1087–1101.

Maller, C., M. Townsend, A. Pryor, P. Brown, and L. St Leger. 2006. Healthy nature healthy people: 'Contact with nature' as an upstream health promotion intervention for populations. *Health Promotion International*, 21(1), 45–54.

Mayer-Smith, J., O. Bartash, and L. Peterat. 2007. Teaming children and elders to grow food and environmental consciousness. *Applied Environmental Education and Communication*, 6(1), 77–85.

Middlebrooks, J.S. and N.C. Audage. 2008. The effects of childhood stress on health across the lifespan. Atlanta, GA: Centers for Disease Control and Prevention, National Center for Injury Prevention and Control.

Miller, L.K. 1904. *Children's Gardens for School and Home: A Manual of Cooperative Gardening*. New York, NY: D. Appleton and Co.

Mintz, S. 2009. The changing state of childhood: American childhood as a social and cultural construct. http://www.usu.edu/anthro/childhoodconference/pages/reading_material.html (accessed January 8, 2015).

Mitchell, M.M., C.P. Bradshaw, and P.J. Leaf. 2010. Student and teacher perceptions of school climate: A multilevel exploration of patterns of discrepancy, *Journal of School Health*, 80(6), 271–279.

Montessori, M. 1964. *The Montessori Method*. New York, NY: Schocken Books.

Morris, J., A. Neustadter, and S. Zidenberg-Cherr. 2001. First-grade gardeners more likely to taste vegetables. *California Agriculture*, 55(1), 443–46.

National Gardening Association (NGA). 2014. Food gardening in the U.S. at the highest levels in more than a decade according to New Report by the National Gardening Association. http://assoc.garden.org/press/press.php?q=show&pr=pr_nga&id=3819 (accessed December 16, 2014).

National Gardening Association (NGA). 2015. Getting to know plants. http://www.kidsgardening.org/node/36 (accessed January 21, 2015).

National Science Board (NSB). 2009. STEM education recommendations to the Obama administration. http://www.nsf.gov/nsb/publications/2009/01_10_stem_rec_obama.pdf (accessed January 21, 2015).

Old Town School of Folk Music. 2015. Oldtownschool.org (accessed March 13, 2015).

Ouden, K.D. and B.S-C. Wee. 2011. Cultivation of social responsibility through school community gardens. In J. Lin and R. Oxford (eds.), *Transformative Eco-Education for Human and Planetary Survival*. Charlotte, NC: Information Age Publishing, Inc., pp. 75–86.

Park, S., C. Shoemaker, and M. Haub. 2008. A preliminary investigation on exercise intensities of gardening tasks in older adults. *Perceptual and Motor Skills*, 107, 974–980.

Park, S., H. Lee, K. Lee, K. Son, and C. Shoemaker. 2013. The metabolic costs of gardening tasks in children. *HortTechnology*, 23, 589–594.

Parmer, S., J. Salisbury-Glennon, D. Shannon, B. Struempler. 2009. School gardens: An experiential learning approach for a nutrition education program to increase fruit and vegetable knowledge, preference, and consumption among second-grade students. *Journal of Nutrition Education and Behavior*, 41(3), 212–217.

Pretfy, J., J. Peacock, M. Sellens, and M. Griffin. 2005. The mental and physical health outcomes of green exercise. *International Journal of Environmental Heslth Research*, 15(5), 319–337.

Relf, P. 1998. People-plant relationship. In S. Simson and M. Strauss (eds.), *Horticulture as Therapy: Principles and Practices*. Binghampton, NY: Hawthorn Press, pp. 157–197.

Robinson, C.W. and J.M. Zajicek. 2005. Growing minds: The effects of a one-year school garden program on six constructs of life skills of elementary school children. *HortTechnology*, 15(3), 453–457.

Sarver, M. 1985. Agritherapy: Plants as learning partners. *Academic Therapy*, 20(4), 389–396.

Sensory Processing Disorder (SPD) Foundation. 2015. Spdfoundation.net (accessed March 12, 2015).

Shinew, K., T. Glover, and D. Parry. 2004. Leisure spaces as potential sites for interracial interaction: Community gardens in urban areas. *Journal of Leisure Research*, 36(3), 336–355.

Skamp, K. and I. Bergmann. 2001. Facilitating learnscape development, maintenance and use: Teachers' perceptions and self-reported practices. *Environment Education Research*, 7(4), 333–358.

Skelly, S.M. and J.M. Zajicek. 1998. The effect of an interdisciplinary garden program on the environmental attitudes of elementary school students. *HortTechnology*, 8(4), 579–583.

Smith, L.L., and C.E. Motsenbocker. 2005. Impact of hands-on science through school gardening in Louisiana public elementary schools. *HortTechnology*, 15(3), 439–443.

U.S. Department of Agriculture. 2014. ChooseMyPlate.gov. http://www.choosemyplate.gov/food-groups/vegetables.html (accessed December 17, 2014).

U.S. Department of Health and Human Services. 2008. 2008 Physical Activity Guidelines for Americans. http://www.health.gov/paguidelines/guidelines/ (accessed December 17, 2014).

Valentine, G. and J. McKendrick. 1997. Children's outdoor play: Exploring parental concerns about children's safety and the changing nature of childhood. *Geoforum*, 28, 219–235.

Williams, M.S. and S. Shellenberger. 2015. The Alertness Program. www.alertprogram.com (accessed March 11, 2015).

World Forum-Nature Action Collaborative for Children. 2008. Re-connecting the world's children to nature—Call to action. http://www.worldforumfoundation.org/wf/nacc/call_to_action.pdf (accessed January 21, 2015).

Xitao, F. and M. Chen. 2001. Parental involvement and students' academic achievement: A meta-analysis. *Educational Psychology Review,* 13(1), 1–22.

3 Gardens and Community

Tina Marie Waliczek

CONTENTS

OBJECTIVES

Upon completion of this chapter, the reader should be able to

- Define and compare different types of community gardens.
- Describe the development of community gardens throughout history.
- Identify key characteristics of successful community garden programs.
- Identify typical obstacles of community garden programs.
- List and describe benefits of community gardens to individuals, the community, and the region.
- Recognize potential funding sources for community gardens.

KEY TERMS

- Allotment garden
- Allotments Act of 1925
- Anti-inflation gardens
- Broken window syndrome
- Collective efficacy
- Community development block grants
- Crowd funding
- Eutrophication
- Food deserts
- Food security
- Greenway
- Guerilla gardening
- Heat islands
- Land trust
- Liberty gardens
- Potato patch gardens

- Relief gardens
- Social capital
- Stakeholder
- Vacant lot gardens
- Victory gardens

TYPES OF COMMUNITY GARDENS

While definitions of community gardens vary, the term "community garden" is typically defined as any plot of public or private land collectively managed by a group of people. Community gardens can be implemented in urban, suburban, or rural areas, though they are traditionally operated in cities or in shared community settings where citizens have limited access to gardening space. Gardens can be installed in neighborhoods, at schools or universities, at hospitals, apartment complexes, or other multifamily housing situations. Community gardens may target a certain population within a community, such as a specific group of residents at a facility, or may be open to anyone who wants to participate within a geographical area.

Community gardens may include a large parcel of land where everyone tends one set of crops and the bounty is then divided among gardeners, or the land may be divided into plots which are then managed by individual gardeners. Often, in situations where the garden is one large parcel, labor is exchanged for a share of the produce. In contrast, in the situation where the community garden is divided into plots, gardeners will often be charged a seasonal or annual fee in order to rent the plot. Many community gardens have other shared resources which may include water, compost, mulch, hoses, wheelbarrows, and/or other tools. Perennial crops such as fruit trees may also be on the property and may be managed collectively with the produce divided among the gardeners.

Glover (2004) dubbed community gardens as "plots of land on which community members can grow flowers or foodstuffs for personal or collective benefit" and limited community gardens to those which had shared resources of space, tools, and water. Some regions have their own definitions which further limit community gardens to those which are only formal organized efforts involving membership, thereby excluding one-time beautification plantings and *guerilla gardens*, where gardeners have not obtained permission to access the site where they are gardening, and/or limiting community gardens to those only focusing on producing foodstuffs. Austin, Texas, city government defined city-supported community gardens as being those spaces "used by a group of four or more participating gardeners either on separate plots or farmed collectively by the group to grow, produce and harvest food crops for personal or group use, consumption or donation by the nonprofit organization or cooperatively for the benefit of its members (City of Austin 2011)."

Some views support wider definitions of community gardens and include Community Supported Agriculture (a network of individuals supporting local agricultural producers often through a subscription system in exchange for a share of seasonal harvests), school gardens, botanical gardens, neighborhood landscape plantings, and even backyard gardens as additional community garden efforts.

To differentiate, Pudup (2008) offered the term "organized gardening project" as an alternative term to describe these efforts.

One unique form of community gardening is guerilla gardening in which gardeners are not legally permitted to access or utilize a property, but adopt the area for short- or long-term cultivation of ornamental or food crops. The site may be private or public property, an abandoned site, or a piece of land which is not seen as properly maintained. Oftentimes, guerilla gardeners access the sites under the cover of night and see their actions as a means to beautify a forgotten or otherwise underutilized area. They will typically use ornamental plants rather than edible species. Many guerilla gardeners use the act of gardening as a form of activism often with the intention of beautifying their community and as a means to improve the quality of life for the greater good. Additionally, some guerilla gardeners use plants in the urban setting as a palette of materials to create art and beauty in unexpected places and unusual ways. Guerilla gardening may include passively dropping seeds (or seed "bombs" made from clay, compost, and wildflower or ornamental seed) in median areas, or more actively planting flower transplants in areas.

The idea of guerilla gardening is thought to have emerged in urban derelict areas of Bowery Houston in New York in 1973 (Reynolds 2008). However, some people cite Johnny "Appleseed" Chapman as the first guerilla gardener in the United States. Guerilla gardening has occurred all over the world with some famous examples including a 15,000 square foot circular garden in the Lower East Side of Manhattan called Adam Purple's Garden of Eden which was tended from the mid-1970s until 1986 when it was bulldozed by the City of New York (Zukin 2010), or People's Park in Berkeley, California, which was land acquired by eminent domain by the University of California, but then claimed by the people as a park when it was left undeveloped. International Sunflower Guerilla Gardening day is celebrated on May 1 as an international event in which guerilla gardeners plant sunflowers in public places perceived to be neglected and in need of beautification.

Historically, the goal of community gardens was to provide opportunities to people to produce food. More often today, the goal of community gardens is to build community while providing space and opportunity for food production.

HISTORY OF COMMUNITY GARDENS

European Community Gardens

Records indicate *allotment gardens*, a term used to describe a section of land rented or leased by a group of people in which they plant a garden, were in existence in Great Britain as early as the mid- to late 1700s. The idea of allotment or community gardens in Europe spread to other countries with most countries having documented evidence of allotment gardening being practiced to help fulfill needs for food, as well as to accommodate other sociocultural functions such as reducing crime and immorality, among others.

Allotment gardens were said to have emerged in rural areas from the "Open Field System" of agriculture common to the eleventh and twelfth centuries in England in which a portion of field was allocated to a villager for food production and where livestock was shared. In 1770, Lord of Manor devoted 25 acres near Tewkesbury to be used by the poor. His efforts were successful in reducing the numbers of poor citizens. Neighboring towns noticed his success and followed suit.

During the 1800s, allotment gardens were provided more often to help alleviate crime instead of being a source of food and income for the poor. The plots were provided for a small rental fee and were said to help promote independence, self-respect, and feelings of moral responsibility. Initial allotments were set up in more rural areas, but as residents moved to more urban areas, so did the need for land to garden in cities.

The *Allotments Act of 1925* was a law ensuring the availability of parcels of land for use for gardening. The act facilitated "the acquisition and maintenance of allotments and to make further provision for the security of tenure of tenants of allotments." While the law still exists, there were periods of fluctuations of interest in the gardens throughout the decades. In 1965, a committee evaluated needs and recommended there be a minimum of 15 full sized plots per 1000 households. Today, there is estimated to be 300,000 plots with a recent surge in demand for plots being due to fears of pesticides on produce and a taste for varieties of crops not available commercially. Additionally, many residents live in flats or apartments which lack land to garden otherwise. Because of the growing need, quarter plots have been issued to new gardeners where they can experiment with the idea of producing, and observe the requirements and upkeep of a garden (Poole 2006).

COMMUNITY GARDENING IN THE UNITED STATES

Potato Patch Gardens

Documentation of people gathering and creating communal gardening areas in cities has occurred in the United States since the Panic of 1893. Urban gardens were organized by social reformers promoting gardens built in vacant lots as a means of providing assistance for unemployed laborers during the economic trials. Historically, community gardening efforts in the United States were a result of economic instability and lack of food security. In the 1890s, the mayor of Detroit, Haze Pingree, advocated the use of vacant land as food gardens for the unemployed. These vacant areas were called Potato Patch Gardens and as many as nearly 1000 Detroit families were cultivating vegetables to supplement their meager diets (Bassett 1979).

Liberty Gardens

During World War I, *Liberty Gardens* were planted in American backyards as well as communally in neighborhoods as a mean of showing support for the troops. The National War Garden Commission advocated any resource saved on the mainland meant more resources would be available to be shipped to the troops overseas. Propaganda from the National War Garden Commission promoted the idea that citizens gardening were "home soldiers" in the war effort (Bassett 1979). Promotions of gardening efforts substantiated significant food production. In 1917, it was reported that 3,500,000 gardens produced $350,000,000 worth of food, while in 1918, 5,000,000 gardens produced $525,000,000 worth of food (Bassett 1979).

Great Depression Relief Gardens

During the Great Depression of the 1930s, subsistence gardens and cooperative farms were initiated from both private and public sources and made available to families as a means to make ends meet and quell economic uncertainty. These gardens were called *Relief Gardens*. While the Potato Patch Gardens of the 1890s served primarily the unemployed, the Relief Gardens went beyond serving just those on welfare to aid a greater portion of the community who were dealing with extreme circumstances of the Great Depression (Bassett 1979).

Victory Gardens

During World War II, *Victory Gardens* were used again as a means to support war efforts with civilians being encouraged to raise food crops. During World War II, it was reported that 20 million gardeners produced enough food in Victory Gardens to supplement the nation's food supply by 40% (Victory Seeds 2015). One of the earliest Victory Gardens was Boston's Fenway Gardens which are still in existence and include 500 gardens on 7.5 acres (Reid 1996).

THE COMMUNITY GARDEN MOVEMENT OF THE 1970S AND 1980S

The *Anti-Inflation Gardens* of the 1970s were community gardens, modeled after the *Potato Patch Gardens* of the late nineteenth century and the Relief Gardens of the Great Depression (Bassett 1979) to help produce food to reduce food costs. Discounted seeds were offered through county offices and shortages of canning supplies to preserve harvests were reported in some regions (Burger 1975).

One notable community guerilla garden built during the 1970s and 1980s was Adam Purple's Garden of Eden in New York City's Lower East Side. Adam Purple watched neighborhood children play on the rubble of a demolished building outside his window and was inspired to take action (Environmental Design and Research Association 1985). In 1973, he designed and built his "Garden of Eden" using brick dust, scavenged pieces of other demolished buildings in the area, and amended the soil with manure he reportedly bicycled in from Central Park (Environmental Design and Research Association 1985). The garden was designed in a circular ying–yang pattern and children used it as a place to discover the earth. The City of New York bulldozed the Garden of Eden in 1986 to allow room for affordable housing to be built (Not Bored n.d).

In the United States, *vacant lot gardens* were established through grassroots efforts in neighborhoods after the 1970s when many cities in the nation were being abandoned due to "white flight." Empty, rubble-filled lots were magnets for trash, drugs, and crime and sometimes dumping grounds for noxious chemicals. Groups such as the Green Guerillas began beautification efforts in the 1970s in New York. One of these vacant lots, on the corner of Bowery and Houston streets, became the first community garden in the city in April 1974, the "Bowery Houston Community Farm and Garden." After the success of the garden attracted positive attention, the Green Guerillas hosted workshops to teach other communities and gardeners about implementing similar programs (New York City Department of Parks and Recreation 2014). In 1978, because the city recognized the benefits of the grassroots efforts of gardeners managing the vacant lots, a garden initiation assistance program, The Green Thumb program, was offered through the city and funded by Community Development Block grants. The program is still in existence and is the largest urban gardening program in the nation. It also provides support to 600 gardens, 20,000 gardeners, and 32 acres of parks in the city by offering 10-year leases for city-owned land for parcels which are actively maintained.

Another historically noteworthy New York community garden is the Clinton Community Garden in the city's west side. In the early 1980s, the garden received national attention when the property where the garden was housed was slated for auction. Gardeners teamed with the Trust for Public Land, Housing Conservation Coordinators, and the Green Guerillas, and started the Square Inch Campaign through the nonprofit New York Restoration Project (NYRP), "selling" a piece of the garden for a $5.00 donation (Clinton Community Garden n.d., History of Garden, para 3). The square inch campaign raised attention and $70,000. The mayor transferred the garden to the Parks and Recreation Department of the city making Clinton Community Garden the first New York City community garden to be deemed permanent parkland (Clinton Community Garden n.d., History of Garden, para 4). The efforts and events allowing for the preservation of the Clinton Community garden provide a hopeful model for other community gardens in dire situations and in need of preservation.

COMMUNITY GARDENS 1990s TO THE PRESENT

One of the most famous gardens established in the 1990s was constructed after the Los Angeles riots in 1992. A 14-acre plot of land was built as the South Central Farm community garden in South Central Los Angeles. The garden was seen as an oasis in one of the most blighted urban areas in the country, providing fresh food for immigrants, poor, and the local food bank. Approximately 350 gardeners from around the world, and especially from Mexico, worked small parcels of land growing food, herbal remedies, and plants known for spiritual importance from their home countries. The community garden was demolished in a controversial land deal between the landowner and city in 2006. The land still remains vacant (Harris 2011). The South Central Farm is another example of the potential threats to community gardens even when they are successful in achieving the established goals of the city.

More recently, the community garden movement has renewed interest with the local and urban food movement. An increased number of people are interested in growing their own food or supporting local producers. For many community gardeners, having access to a garden plot provides a sense of ownership and access to ground that would otherwise not be possible to those who rent or are in transient situations. The benefits of community gardens to communities and individuals are beginning to be noticed by the general public. For instance, in 2009, First Lady Michelle Obama planted the largest White House Garden in an effort to have access to more fruits and vegetables in her daughters' diets. Community garden groups are recognizing the need to measure their impact in order to bring about more opportunities for resources such as increased funding and land. Local and state policy changes influencing zoning regulations and the cottage food industry, those small-scale farmers producing on small parcels of their own land and with their own equipment, are sweeping the nation affecting the urban food movement. In some cities, established community garden groups are forming partnerships with the urban agriculture movement, helping to involve and educate a larger audience on everything from gardening and horticulture to leadership and management of community gardens and farmers markets. In the United States and Canada, the American Community Garden Association (ACGA) reported 18,000 registered community gardens (ACGA 2014).

BENEFITS OF COMMUNITY GARDENING

Community gardens provide benefits beyond the produce reaped. Vacant lots were cultivated in the United States during the Depression of 1893. Land was provided to the poor in many cities across the United States in order for them to produce their own food. There are several documents indicating that the plots also helped combat problems with poor nutrition, crime, domestic violence, alcoholism, and gambling while providing agricultural worker training (Lawson 2005). Poor people who were given a plot of land were occupied and involved in an activity where they gained benefits from their labor and were, therefore, less likely to be instigating problems in other arenas. It was also an avenue where they could learn a skill that could potentially employ them later.

Need for Community Gardens

In cities, vacant land can be seen as either an eyesore or a resource. While in some locations, vacant properties can lead to weeds and trash, in other cities, the vacant property can provide a much needed green respite. On average, 15% of the land in American cities is classified as vacant (Pagano and Bowman 2000). It is reported that Los Angeles has only 1.106 acres of park land per 1000 residents, which is a fraction of the standard recommendation of 10 acres per 1000 residents set by the National Recreation and Parks Association (NRPA) (Environmental Defense 2002). Per capita green space has been associated with quality of life of American cities (The Trust for Public Land 2007).

Food, Security, Harvest, and Income Benefits

Researchers maintain that community gardening is most often an urban occurrence because of the need in city settings for greater food security (Newman 1997). *Food security* refers to the availability and ease of access of safe, nutritious food, particularly by low-income people (Newman 1997; USDA 2006). *Food deserts* are geographic regions where fresh, nutritious, and affordable food is scarce. Some motivations for community gardening include gardeners preferring to produce food

FIGURE 3.1 Volunteers at the Dunbar Community Garden harvesting potatoes for the Food Bank.

locally, and the sustainability and accessibility of organically grown produce (Mast 2012). Community gardens often provide surplus to local food banks helping them to bestow the bounty beyond their own family and friends, but to the overall neighborhood (Figure 3.1).

The United States is fortunate to have some of the finest and most productive soils in the world (United States Bureau of Soils 2014). However, areas of productive land are being lost due to development, compaction, erosion, and high salinity. At the same time, there are food shortages all over the world including Asia, Africa, and Latin America. The human population of the world is rising and farmers are being expected to produce more food in poorer soils and on less land. Some estimates are that 6 billion people are eating food produced on 11% of the earth's surface and just 3% of that land is considered fertile soil.

Additionally, less people produce food. Based on Environmental Protection Agency statistics, of the 313 million people in the United States, less than 1% are responsible for producing the food supply. Of the 1%, nearly half (45%) of farmers in the United States are also farming only part time. Although numbers of farmers have declined, agricultural product demand has increased with needs being met through agricultural mechanization, increased use of pesticides and fertilizers, and improved crop varieties that are less susceptible to problems. These changes have also reduced the need for field labor. One concern for the United States and family farms in the country is the aging farmer. Approximately 60% of farmers in this country are 55 years or older (Bureau of Labor Statistics 2015). Some see the aging farmer as a threat to the food supply and, in turn, the availability of local, organic food, a diverse food supply not directed by commercialization (The Lampert Report 2014).

Urban community gardens have shown massive potential in being able to supplement commercially available fresh produce. Philly's 501 community vegetable gardens in Philadelphia are part of the Philadelphia Green, a program developed to assist community groups to garden in their neighborhoods on vacant lots and along the street. Records show 2812 families produced $1,948,633 worth of fruit and vegetables in 1994 (City Farmer 2001). Another study of community gardens nationwide found that food grown in community gardens was especially important to those unemployed and produced grocery savings of at least $150 per season to 48% of gardeners surveyed. Some estimates show that gardens of at least 64 ft^2 saved gardeners $600 in food costs per year (Vallianatos et al. 2004).

Nutritional and Physical Exercise Benefits

Gardening's effect on nutrition and exercise has been of interest in part because of America's recent struggle with obesity. Currently, approximately 34.9% of adults and 17% of children in the United States are obese (CDC 2014). In a New York study, researchers found that community gardening had a positive effect on enhancing physical activeness while also reducing levels of stress and mental fatigue (Armstrong 2000). Exercise conducted within gardens was shown to have physical benefits on par with walking and weightbearing exercises. For instance, Park et al. (2014) determined the typical garden tasks of digging, raking, mulching, and planting transplants, among others, ranked as being moderate- to high-intensity physical activities when they measured gardeners' heart rates and blood pressure. Gardening was, in some ways, better when compared to more traditional types of exercise for enhancing bone density. Yard work and weight training were strong predictors of bone density, while bicycling, walking, and dancing were moderate predictors and swimming and jogging were weak predictors (Taylor et al. 2002). Additionally, gardens were touted as keeping gardeners active by providing outdoor recreation to neighborhood residents (Voicu and Been 2008).

Additionally, gardens had a positive effect on nutrition and food choices for the whole family. Children who gardened increased fruit and vegetable consumption by three times what they had historically eaten, while adults increased consumption by four times (Carney et al. 2011). Gardening not only increased food choices but also nutrition knowledge levels of children (Langellotto and Gupta 2012) and of adults (Waliczek et al. 2005).

Psychological/Mental Benefits

Community gardeners are deriving needs beyond general harvests of food. Community gardeners rated benefits on a Likert scale that related to each of the levels on Maslow's hierarchy of human needs (Maslow 1943). Gardeners reported benefits related to food, safety, social, esteem, and self-actualization. The garden was particularly important in meeting these needs in low-income minority gardeners who may not have as many opportunities or resources to allow them access to other means to achieve all these benefits. Community gardens appeared to be a resource serving a diverse community (Waliczek et al. 1996). Another study found that in addition to the economic value of food produced in community gardens, gardeners derived personal satisfaction from producing their own food and maintained feelings of self-sufficiency (Patel 1996).

Benefits to Community: Social Values and Community Gardens

Glover stated, "Community gardens are less about gardening than they are about community" (2004, p. 143) and, in that statement, spoke of the enormous impact community gardens can have on building relationships and the positive outcomes from those relationships. Community gardens have historically been a way for people to come together and create a strong presence within their neighborhoods building *social capital.* The 1961 urban planning book, *The Death and Life of Great American Cities,* by Jane Jacobs, introduced the concept of social capital. Social capital "refers to connections among individuals-social networks and the norms of reciprocity and trustworthiness that arise from them" (Putnam 2000, p. 19). A study looking at the benefits of community gardens found that community gardens are valued for their ability to bring people together to share knowledge, stories, and ideas (Waliczek et al. 1996) (Figure 3.2).

Property Values and Community Gardens

Small parks, green spaces, and gardens in neighborhoods provide benefits such as neighborhood stabilization, aesthetic values, economic values, and psychological and social values for residents. However, the economic impact of the community garden to the overall neighborhood property values and the revenue generated from those property values is another important consideration. In areas with community gardens nearby, there are indicators of improved neighborhood stabilization with an increase in owner-occupied dwellings, and an increase in reported resident incomes. This was believed to be from attracting more wealthy homeowners, which also led to rent increases in areas surrounding community gardens (Whitmire Study 2008).

Community gardens in New York City contributed to positive effects on property values especially in the poorest, most disadvantaged neighborhoods, where

FIGURE 3.2 A volunteer reflects while watering plots at Dunbar Community Garden, San Marcos, Texas. (Photo by Leah Gibson.)

they often substituted for inaccessible city parks. Properties within 1000 feet of the garden resulted in increased sales prices with the impact increasing over time. The types of gardens also had an effect, with gardens of a higher quality having the greatest impact. Garden quality was ranked using qualitative data on variables such as cleanliness, landscaping quality, fencing, permanence, existence of social spaces, and overall condition of the garden. The higher property values led to greater tax revenues for the city over time (Voicu and Been 2008). After 54 community gardens in St. Louis were installed in neighborhoods, median rent, median housing costs and the homeownership rate in the immediate vicinity of the gardens increased when compared to the surrounding neighborhood (Tranel and Handlin 2006).

Similar results occur with homes located near parks and green spaces. Homes near parks or green spaces sold for 1.43%–30% more, depending on the homes, neighborhood, and quality of the park (Bolitzer and Netusil 2000; Espey and Owasu-Edusei 2001; Lutzenhiser and Netusil 2001; Pincetl et al. 2003). Building a *greenway*, or strip of land left undeveloped, also increased the value of adjacent properties by 6.9% (Hobden et al. 2004).

Green spaces can influence crime, though the quality of the green space is an important consideration. Poorly planned, implemented, or maintained green spaces can also have a detrimental effect on neighborhoods. Signs of deterioration within a neighborhood can cause fear of victimization and result in less socialization. Community gardens have a positive effect since they often claim vacant, littered lots and clean them up. "Visible physical decay may spark fear of crime, because Americans have come to associate it with higher levels of risk. Like observable social disorders, physical decay is taken by many as a 'sign of crime'" (Skogan 1990, p. 47). Sometimes called "broken window syndrome," Wilson and Kelling (1982) suggested disorder in the form of vacant buildings, vacant littered lots, graffiti, and broken windows leads to more disorder (Skogan 1990). This, in turn, leads to low feelings of safety and a lack of community (Schweitzer et al. 1999). In Italy, researchers found that the fear of crime is more widespread than crime itself; among some of the best predictors of fear of crime are urbanization and degradation of residential areas (Miceli et al. 2004).

Crime Prevention and Community Gardens

While in some urban areas, excess vegetation is sometimes seen as detrimental since it offers a screen for criminal activity, recent research has found that certain levels and types of greenery is beneficial.

Benefits of parks, gardens, and trees to neighborhoods include greater feelings of safety, increased social contact, and communication among neighbors (Waliczek et al. 1996; Kuo and Sullivan 2001b), as well as reduced feelings of mental fatigue (Kuo 2001). Apartment buildings in low-income neighborhoods with trees on their grounds were less prone to domestic violence when compared to domestic violence levels of similar apartment buildings located in the same poor urban neighborhoods that were barren of greenery (Kuo and Sullivan 2001a).

In Austin, Texas, there was an inverse relationship between the incidence of crime committed in the greater metropolitan area and the amount of vegetation in the area in which the crimes occurred. Areas with less than the average mean greenness level in Austin had an increased amount of crime. Results showed "83% of all crimes occurred in areas that had greenness values below 34%" (Snelgrove et al. 2004, p. 6).

Anecdotal evidence has historically shown that community gardens located in urban areas have not only helped clean up the trash heaped on forgotten urban lots, but have helped create pride and reduce vandalism and other forms of criminal activity (Hynes 1996). Interviews conducted with community garden representatives in Houston, Texas, showed that community gardens appeared to have a positive influence on neighborhoods with residents reporting neighborhood revitalization, perceived immunity from crime, and neighbors emulating gardening practices they saw at the community gardens (Gorham et al. 2009). The community garden also reportedly increased social interaction among neighbors which, in turn, created a sense of a safer environment.

Community gardens' influence on crime has been connected with the idea of *collective efficacy*, or the "mutual trust among neighbors, combined with willingness to intervene on behalf of the common good, specifically to supervise children and maintain public order" (Sampson et al. 1998, p. 18). Collective efficacy is "the most powerful influence keeping violent crime low" (Sampson et al. 1998, p. 18). Community gardens are seen as a place where collective efficacy is built among members of the garden.

Social Capital and Community Gardens

The term "social capital" refers to the "connections among individuals-social networks and the norms of reciprocity and trustworthiness that arise from them" (Putnam 2000, p. 19). There are two types of social capital that can be affected by the act of community gardening. These include bonding and bridging social capital (Putnam 2000). Bonding social capital is the type where connections are made and then reinforced within a group. Bridging social capital occurs when groups come together who might not otherwise meet because of their inherent initial differences. Community gardens have traditionally been known to be places where people of both similar and different cultures and backgrounds can come together on common ground.

Firth et al. (2011) conducted a study in Nottingham, UK, where they found that some gardens and gardeners are location based, while others are interest based. In other words, some gardeners are naturally drawn to gardens where commonalities such as faith or ethnic background attract them in addition to the activity of gardening. Gardens have traditionally offered a refuge for those recently immigrating as a gathering place where they can grow similar plants and produce to that which they had in their native countries. Three nonprofit Asian gardens in Dallas, Texas, attract Cambodian and Laotian refugees. Community gardens in California attracted diverse ethnic backgrounds including 60% Caucasian, 24% Asian (Japanese, Filipino, and Malaysian), 12% Hispanic (Mexican or Puerto Rican), 4% African-American, and 2% American-Indian gardeners (Gordon and Dotter 1996).

Environmental Education Benefits

Community gardens have been recognized as a place to teach children and adults about many concepts including science and the environment (Waliczek et al. 1996;

Mast 2012), as well as to create ecosystems for wildlife within urban areas (Mast 2012). Gardeners often state one of the benefits of gardening as the ability to teach and learn intergenerationally. The National Wildlife Federation has recognized the value of small gardens as pockets of nature capable of building ecosystems to support habitat in their Certified Wildlife Habitat program (Danforth et al. 2008). Communities, schools, and individuals can certify a habitat by providing food, cover/shelter, water, and cover for wildlife. Food sources include planting native plants providing seeds, nuts, fruits, and nectar. Water sources are provided through natural streams, ponds, or artificial water gardens and birdbaths. Cover is provided with dense canopies in trees and shrubs, or through nesting boxes. They estimate that over 176,000 habitats are certified in the nation at this time.

Benefits to the Environment

Green space and gardens in urban areas provide several environmental benefits. The economic value of the environmental impact of these spaces is substantial. The US Forest Service (McPherson et al. 2005) estimated the economic value given the environmental benefits of one tree during a 50-year lifetime was $162,000. They estimated the value to soil erosion at $31,250, water at $37,500, air pollution control at $62,000, and oxygen at $31,250.

Green spaces help mitigate *heat islands*, concrete areas that trap, store, and reflect heat and light, keeping cities cooler. One tree can produce the cooling effect of 10 room-sized air conditioners through the evaporation of moisture from its leaves. Additionally, leaves absorb, rather than reflect, solar radiation.

Green spaces reduce rainwater runoff providing more percolation into aquifers and also reducing potential flooding in cities. Impervious concrete surfaces of sidewalks, roads, parking lots, and even roofs impede percolation of water into the soil. Plant canopies delay the descent of rainfall to the ground, allowing for more penetration into the soil and increasing the amount of water that soaks into root systems, while also reducing the impact of raindrops that would cause soil erosion. This reduces the flow of water into stormwater facilities and reportedly allows for smaller, less expensive stormwater management systems. Garland, Texas, encouraged private property owners to plant more trees by assessing fees to impervious cover on each property based on the volume of stormwater the property would generate. Due to the reduced volume of stormwater, the additional tree cover planted saves the city from handling 19 million cubic feet of stormwater which, otherwise, would cost the city approximately $38 million. While urbanization has resulted in at least 30% less natural tree cover in cities nationwide, some estimates show that trees could save cities as much as $400 billion in building stormwater retention facilities.

Green spaces filtering rainwater helps to keep lake, river, and groundwater clean. Landscapes and gardens help hold soil to reduce erosion from both rain and wind. Leaves, stems, trunks, and roots remove small particles from rain and plants, absorb nutrients before they move into streams, lakes, and rivers. Left unchecked, excess chemical nutrients flowing into waterways cause algae blooms and a proliferation of aquatic plants creating more competition for sunlight, space, and air in waterways, a process called *eutrophication*. Community gardens and other green spaces in cities help mitigate potential contaminants from water in largely concreted areas.

Plants and landscapes filter air, reducing air pollution and sometimes allergens, as well as restore oxygen through gas exchange in both the leaves, understory plants, and the microorganisms within the soils. In a study of New York City in 1994, trees removed 1821 metric tons of air pollution. Trees can remove up to 15% ozone, 14% sulfur dioxide, 8% nitrogen dioxide, and 0.05% carbon monoxide, as well as 13% particulate matter depending on the level of canopy in a region.

Other Benefits

Many community gardeners will garden as a means of connecting to place. Some anecdotal observations have spoke of gardeners developing "place attachment" (Hynes 1996). *Place attachment* is an important concept within environmental psychology which refers to the feelings of emotion related to a particular geographic locale or landscape, and triggers positive feelings of comfort and contributes to identity. Place attachment is sometimes associated with the feeling of connection gardeners develop from the pride associated with their work in the gardens, and/or memories or experiences established there. In this way, gardens can sometimes help shape a community's identity. One study conducted with community gardeners in Korea and the United States found they were attracted to the opportunity to bond socially as well as with nature and this had a positive effect on place attachment (Lee 2011) (Figure 3.3).

FIGURE 3.3 Veronica and Zach Halfin lead an educational program at the Dunbar Community Garden, San Marcos, Texas. (Photo by Leah Gibson.)

Tourism Benefits

In recent years, "garden tourism" is emerging as an important component of economic development for some communities. Richard W. Benfield, author of *Garden Tourism*, reports that the number of people traveling specifically to visit parks, gardens, and/or flower-themed festivals is increasing. Destinations are both private and public. In the United States, garden destinations include private gardens such as the statewide Virginia garden tour. Popular public garden destinations include Epcot International Flower and Garden Festival and Buffalo Olmsted Parks. Benfield reports in his book that gardeners are usually drawn to visit other cultural attractions such as historic sites and museums while their visiting gardens having a significant economic impact within the community.

Memories/Heritage Benefits

Intergenerational exchange, teaching, learning, and working toward common goals are often reported by gardeners as being beneficial. When young people work beside adults, community is built through working toward common goals (Sherer 2006). Often gardeners report fond memories of gardening with grandparents and reflect on gardening as a means of connecting to those memories. Many gardeners will garden as a means of connecting to memories/heritage.

INITIATING AND RUNNING A SUCCESSFUL COMMUNITY GARDEN PROJECT

Building Relationships and Stakeholders in the Community

When initiating the idea of starting a community garden, it is easy for one to set out trying to establish the physical site and groundwork as the first step in achieving the goal. However, to lay a good foundation for the garden, one must instead coordinate the *stakeholders* and community of volunteers. The stakeholders are those who have an interest or concern in the garden, take on ownership of the project, and allow for not only success in installing the garden, but also for its sustained presence in the community. The garden will be more sustainable over time if it belongs to the entire community, rather than to just one person or group.

Supporters for the community garden project should include the public or private landowner who is allowing access to the land, neighbors in the surrounding community, the agricultural extension office and Master Gardeners (volunteers educated by the county extension office on gardening and horticulture), city government officials, and the city parks and recreation office. In many communities, there are several other stakeholders who would be enthusiastic supporters of a garden project. Researching potential supporters within the school district, from other farmers and growers within the community, horticulturally oriented businesses such as Farmer's Markets, and garden centers and/or botanic gardens or arboreta is critical to long-term success. Several cities in the United States also have organizations interested and involved in helping citizens with installing and maintaining community and school gardens and/or green spaces in specific cities. Examples include Green Spaces Alliance in San Antonio, Texas and Harvest Pierce County in Tacoma, Washington.

Teaming with educationally oriented groups such as colleges and universities can offer resources such as help and experience with collecting valuable data to document benefits of the gardens, student interns, volunteers, and new gardeners. Local shop owners such as restaurants, nutritionists, and/or health food markets may be complementary supporters who value local, fresh produce. Other helpful stakeholders will include local media representatives and/or businesses who are potential donors or funding partners. Nearby communities and citizens with successful community gardens are also valuable partners since they can offer advice and suggestions for documentation of success. Figure 3.4 provides an example checklist for developing stakeholder relationships and potential prospects for funding and development of a community garden.

Once supporters and stakeholders are identified, a task or advisory group may be formed to help guide the formation and eventual management of the garden. The stakeholders will help develop a mission statement (Table 3.1) and develop ideas on how the garden will help the community and its citizens. This information should be put in writing, tablegraphs or presentations, or provided otherwise for referencing when needed and updated regularly. The information should be made available in public forums or electronically.

COORDINATING THE INITIAL GARDEN

Finding an Appropriate Parcel of Land

Identifying the potential site for a garden requires looking into both short- and long-term goals and obstacles. How will the garden and surrounding community grow this season and over time? A checksheet of considerations for building a garden is included in Table 3.2.

A site needs to be identified which will provide productive garden plots, but also be comfortable, accessible, and safe for the users. In the best situations, the site is accessible by car or truck in order to be able to move supplies easily. However, this is not critical. The site should also be accessible throughout the day for gardeners and garden coordinators. Occasionally urban sites may be heavily trafficked and allow limited accessibility during certain times of the day or week.

Identifying a site that is not deemed for other uses in the near or far future is important for long-term success. Areas adjacent to parkland have been used in many communities. Private vacant land is also sometimes available. Occasionally, long-term leases can be arranged with private landowners to ensure the availability of the property over time. A sample garden lease agreement is shown in Figure 3.5.

Most vegetable production requires sites with at least 8 hours of sunshine, so a sunny, open, site will work best. The site should be well drained and, while a level slope is ideal, a slight slope can help keep sites from having low, soggy areas. Moderate to extremely sloped areas can be terraced. Soil on the site should be tested for nutrients, as well as for potential contamination. Gardens with low areas or poor

Community and School Garden Planning and Development Checklist

Getting Started / Thinking Ahead Stage	yes/No	Action Steps
Do you know who the garden will serve?		
Does your garden project have a contact person?		
Does your garden have approval by the landowner or school?		
For community gardens—do you have a written lease?		
Does your garden have a fiscal agent that can provide insurance?		
For school gardens—do you have a summer maintenance plan?		
Does your garden have a reliable source of water?		
Can your garden operate on a noncash basis?		
Can rototilling or plowing (if needed) be donated?		
Can gardeners provide their own tools and equipment?		
Can stakes, lumber, mulch, and materials be scavenged?		
Can a garden steering committee be formed?		
Will you seek donations of seeds, plants, and supplies?		
Will you apply for small grants available to ad hoc groups?		
Will your garden project recruit community volunteers?		
Will you seek help or advice from Master Gardeners?		
Is your project being documented via photos and records?		
Does your garden project have an identity (e.g., a sign)?		
Grassroots Fundraising / Budgeting Stage	**yes/No**	**Action Steps**
Are cash resources necessary to sustain your garden project?		
Are steering committee members willing to take on roles?		
Can the steering committee reach consensus in decisions?		
Are ideas and input being sought from stakeholders?		

Vermont Community Garden Network

FIGURE 3.4 Example community garden planning and development checklist.

(Continued)

Grassroots Fundraising / Budgeting Stage (continued)	yes/No	Action Steps
Can the committee develop plans and goals for fundraising?		
Is a cash or checking account established for your garden?		
Can the committee develop plans and goals a fundraising?		
Will the committee ensure that cash expenses are covered?		
Are volunteers and sponsors acknowledged and thanked?		
Does the steering committee have a meeting schedule?		
Does the garden have a budget for revenues and expenses?		
Will a cash balance be carried over from one season to the next?		

Institutional Fundraising / Permanence Stage	yes/No	Action Steps
Does your garden project/program have a timeline in place?		
Does your garden project/program have a presence on the web?		
Does your garden program have a brochure or newsletter?		
Is a scrapbook maintained for your garden project/program?		
Will your garden project/program seek publicity in the media?		
Do you have a mailing and email list of contacts and supporters?		
Does your garden network with other community-based gardens?		
Will your garden seek grants available to schools or nonprofits?		
Does your garden project/program give back to the community?		
Does your steering committee attract & welcome new members?		
Can your steering committee effectively transition leadership?		
Does your garden have a maintenance plan for soil fertility?		
Does your committee have a long range plan for sustainability?		

This is a list of questions to get you start thinking about your garden project. Not all the categories will apply to every garden and it's OK not to have all the answers!

Vermont Community Garden Network

FIGURE 3.4 (*Continued*) Example community garden planning and development checklist.

TABLE 3.1
Example: Community Garden Mission Statement

The mission of San Marcos Neighborhood Gardens is to enhance the health, food security, and community bonds of San Marcos residents by providing training, resources, and space to grow food, flowers, and herb using organic gardening methods.

soils can still be used if raised beds are incorporated into the site rather than gardening directly in the existing soil.

Locating Other Resources (Water, Tools, Supplies, Fencing)

Above and belowground utilities should be located by the utility company in order to identify other resources like electricity, but also to identify areas where gardeners will not be able to dig safely. Water will be essential for irrigation of the plots. Water sources can include city tap water, but also ponds, lakes, rivers, or wells. Sites without easy water access can establish rainwater collection tanks or rain barrels. A shelter can be built to provide shade and allow for programming related to gardening or food preparation. However, access to nearby restrooms or a portable or composting toilet is also ideal.

Often, the garden is in need of relatively small amounts of money for items like tools, supplies, fencing, and building supplies. Sometimes, there are specific corporate grants earmarked for community projects. Local building supply companies may be willing to help support their communities. Keeping funding organizations involved and aware of successes and needs can help provide ongoing funding and support. Promoting the garden successes and needs by utilizing local media and sending press releases of newsworthy items such as garden work days, harvests, and garden walk days can help raise funds for tool and supplies. Local gardening and home improvement businesses may also be willing to donate tools and supplies in exchange for small tokens of recognition such as a sign or placard in the garden.

Documenting Successes over Time

Obstacles will occur both in the short and long term and may include a lack of a site/space and/or a lack of volunteers. In order to show progress over time, a proactive approach in efforts by developing annual reports and tables/graphs of garden development and harvests can help community gardens overcome these obstacles. Collect data on gardeners in order to document the benefits to the community. Information such as demographics, amounts of harvest, level of gardening experience, and other benefits perceived by the community gardeners are recommended. Sample survey inventories can be obtained from the American Community Gardening Association.

Community Garden Organization/Coordinating Volunteers

Managing the garden over time requires good leadership skills and problem-solving capabilities (Reid 1996). Communicating rules of the garden with signage and signed

TABLE 3.2
Example: Checklist of Considerations When Building a Community Garden

Construction

- □ Locate Site
  - Survey residents
  - Coordinate with City
- □ Make Plan, Get Soil Test
- □ Get Permission from City
- □ Contract Survey
- □ Build Beds
  - Layout
  - Build bed frames
  - Collect cardboard, signs, and mulch
  - Install 4 layers cardboard in beds
  - Fill beds with mixture compost and topsoil
  - Install signs and mulch in paths
- □ Manage Subcontractors
  - Create plan
  - Get bids
  - Select contractor and sign contract
  - Supervise work
  - Arrange payment
- □ Contract Fence
- □ Contract Waterline and Hose Bibbs
- □ Contract Shed
- □ Paint Shed, Install Shelves and Racks
- □ Contract Sign
- □ Install Locks and Picnic Table
- □ Clean up Worksite

Finances

- □ Identify Possible Funding Sources
- □ Create Budget
- □ Procure Funding
- □ Obtain Liability Insurance
- □ Pay for and Track Operating Costs
- □ Submit Paperwork, Receipts, and any Required Reports to City

Community

- □ Identify Prospective Gardeners
- □ Identify Garden Leaders
- □ Hold Orientation Meeting
- □ Create Website or Facebook Page
- □ Sign Plot Leases (and Obtain Income Verification if Necessary)
- □ Organize and Hold Workdays
- □ Track Volunteer Hours
- □ Continue to Recruit Gardeners
- □ Organize and Hold Groundbreaking Ceremony and Planting Festival
- □ Optional: Publish Newsletter, Hold Potlucks

Agreement between the
Visiting Nurse Association of Chittenden and Grand Isle Counties
and
Grow Team ONE
for
Use of Land as a Neighborhood Garden

This agreement is entered into between Visiting Nurse Association of Chittenden and Grand Isle Counties (on behalf of the Champlain Long-Term Care Coalition), (hereinafter referred to as VNA) and the Grow Team ONE co-coordinators, on behalf of Grow Team ONE for the use of land as a neighborhood garden. The purpose of this garden is to provide an opportunity for interested residents of the Old North End Neighborhood of Burlington to come together and learn about gardening.

I. ***Term:*** This agreement shall be for a trial period of June 15, 2007 to November 30, 2007, after which both parties agree to meet to evaluate this agreement and upon mutual consent renew this agreement for a two-year period.

II. ***Responsibilities of the VNA:*** VNA agrees to:

A. Provide use of a parcel of land owned by the VNA to Grow Team ONE for use as a neighborhood garden at a price of $1.

B. Maintain the portion of the site not used by the garden; and

C. Identify a VNA staff person to be a primary contact for Grow Team ONE in matters related to this contract.

III. ***Responsibilities of Grow Team ONE:*** Grow Team ONE agrees to:

A. Oversee all arrangements for operations of a neighborhood garden including:

1. obtaining funding for or donations of all plants and materials needed for the garden;
2. building raised beds for growing flowers and plants;
3. making improvements to the soil including bringing in top soil and compost;
4. designing a garden; and
5. engaging residents of the neighborhood to plant and maintain the garden.

B. Obtain approval from the VNA prior to making any structural improvements to the property. Once constructed Grow Team ONE will maintain such structures in sound condition and repair with an attractive appearance;

Archibald Neighborhood Garden 2007 14

FIGURE 3.5 Sample lease agreement between a community garden group and a land provider.

(Continued)

C. Monitor and ensure that the property is kept in good order. This includes keeping the property:
    1. clean and free of trash and debris, including keeping the fence in place, and the plants watered and well maintained;
    2. free from any obstacles or hazards that might affect the safety of neighbors and pedestrians; and
    3. free of any noises or odors that might diminish the quality of life of neighbors.

D. Require gardeners participating in the project sign an agreement clearly documenting the use of the site, their responsibilities, and the following language: "I understand that neither Grow Team ONE nor the Visiting Nurse Association of Chittenden and Grand Isle Counties are responsible for my actions. I therefore agree to hold harmless Grow Team ONE and the Visiting Nurse Association of Chittenden and Grand Isle Counties for any liability, damage, loss or claim that occurs in connection with the use of the garden by me or any of my guests.

E. To the best of your ability limit access to the site:
    1. to only, individuals who have signed and agreed to the terms of the waiver and their guests (only when accompanied by the participant). Grow Team ONE will do this employing means such as site design and signage to indicate that this is a neighborhood garden and not a public park. Clear communication about how to sign up and participate in the garden project should further be used to indicate the limited access of this site; and
    2. to use during day light hours.

G. During the period of this agreement, repairing any damage to the site caused by gardeners to at least its current condition;

H. Allow representatives from the VNA on-site at any time during the duration of this agreement to inspect the site to ensure that the terms of this agreement are being met.

I. Identify a point person to represent Grow Team ONE in all matters related to this agreement;

J. Obtain the applicable city permits and permissions prior to installing any signage on the site.

K. Include the VNA name on any signage or materials distributed about the site indicating that "space for the neighborhood garden has been provided by the Visiting Nurse Association of Chittenden and Grand Isle Counties", obtaining prior approval of the VNA Manager of Community Relations of any print materials using the VNA name and any media contacts and materials pertaining to this project.

Archibald Neighborhood Garden 2007 15

FIGURE 3.5 (*Continued*) Sample lease agreement between a community garden group and a land provider.

L. Upon the end of this agreement, return the property to its original condition which includes:

1. removing any raised beds, stones or other structural improvements unless otherwise agreed upon; and
2. restoring the grass surface.

IV. Additional Responsibilities and Terms of the Agreement:

A. Both the VNA and Grow Team ONE agree to abide by all applicable provisions and regulations of the federal, state, and city laws and ordinances.

B. Both the VNA and Grow Team ONE are acting as independent contractors with independent and severable liability.

C. **Insurance:** Grow Team ONE shall indemnify and hold the VNA, its Trustees and employees harmless from any claim by any person, for any loss, injury or damage, resulting from any activity set out in this agreement or any act or omission by Grow Team ONE or any of its employees or agents.

D. **Amendment:** This agreement may be modified or amended if the amendment if made in writing and signed by both parties.

A. **Termination:** This agreement may be terminated immediately if the conditions of the contract are not met and the party in error is given at least 30 days to remediate any infractions.

B. **Negation of Joint Venture:** Nothing contained in this Agreement shall constitute or create a partnership or joint venture between the parties.

H. **Applicable Law:** This agreement shall be governed by the laws of the State of Vermont.

I. **Entire Agreement:** The Agreement is the entire and only agreement between the parties and supersedes all prior understandings and practices between the parties.

Acknowledged, Accepted and Approved

By: ______________________________________ Date: ________________

Visiting Nurse Association of Chittenden
and Grand Isle Counties

Archibald Neighborhood Garden 2007 16

FIGURE 3.5 ***(Continued)*** Sample lease agreement between a community garden group and a land provider.

By: ____ Date: ________________
Co-Coordinator
Grow Team ONE

By: ____ Date: ________________

Co-Coordinator
Grow Team ONE

Grow Team ONE contact information:

FIGURE 3.5 **(*Continued*)** Sample lease agreement between a community garden group and a land provider.

documents help to keep garden users aware of goals, guidelines, and resources available. Example garden guidelines are in Figures 3.6 and 3.7 and a sample participant agreement for gardeners is in Figure 3.8. An example of one garden's overall maintenance schedule of activities throughout the year is demonstrated in Figure 3.9.

Garden organization and governance vary among groups with some community gardens having one individual in charge as a director or similar role, while others will organize under committees (ACGA 2004). Working through problems with a common goal often brings gardeners closer, especially when the problems are associated with a project for which so many become passionate.

Umbrella nonprofit organizations guide community gardens. There are several examples of these types of organizations throughout the United States. Urban Harvest serves Houston's community gardens by hosting farmer's markets, offering classes on starting up community gardens, designing web pages and garden signage (Urban Harvest n.d.). Other umbrella organizations which help similarly in other cities include the Green Guerillas in New York, Philadelphia Green in Philadelphia, PA, Boston Community Garden Council in Boston, Massachusetts, and the San Francisco League of Urban Gardeners in San Francisco, California (Green Guerillas 2002; Boston Natural Areas Network 2014; The Pennsylvania Horticultural Society 2014; Urban Community Gardens n.d.). The Trust for Public Land (2007) and the American Community Gardening Association (ACGA 2004) are national organizations which have been influential in the success of community gardens nationwide.

Finding and Coordinating Volunteers

Researching and establishing local, regional, and national relationships help secure the success of many gardens. Local gardening clubs, city, and county government, local nonprofits and foundations have all played a part in the success of community gardens. One such example is the Peace in the Valley Community Garden which received materials and volunteers from the City of Chicago's Green Corps and the Openlands Project (Small 2002). The Bay Area Rapid Transit allowed the Peralta Community Garden in Berkley to lease property for their garden (Bacigalupi 2002). Certainly, establishing relationships with diverse community groups fosters positive collaborative success.

Cultural activities outside those of actual gardening have been known to help attract a broader base of the community and also build common ground, which Karl Linn, Architect and Psychologist, referred to as a "commons area" for a neighborhood (Bacigalupi 2002). Examples may include hosting potlucks, poetry readings or storytellings, artists or art projects. At the community garden in Peralta, California, various artists donated sculptures, while others offered building expertise for arbors and a tool shed. The cultural implications arising from opening the garden to all users are endless.

Community Garden Protection/Preservation

Because community gardens are not typically situated on land owned by the community garden organization, many can become threatened by the development when a neighborhood is revitalized. Often this is due in part to the community

Sample community garden guidelines

- Visit your garden plot often and get to know your fellow community gardeners. Find ways to pitch in to help coordinators in maintaining your site.

- Share questions, ideas, or concerns with your site coordinator. Community gardening often involves finding creative ways to work together to meet challenges.

- Keep your plot *and* the adjoining pathways well cared for. Use a garden rake on unplanted areas, and hoe weeds between rows while they're small. Allowing your plot to become overgrown with weeds may jeopardize your present and future participation in the community garden.

- Keep the lawn areas and garden paths free of rocks, weeds, and plant debris. Pick up piles of debris and hoses so as not to create a hazard to trip over or hit with the lawnmower.

- Please conserve water by watering plants in the morning or evening, rather than in the heat of the day. Apply water at the base of plants where possible. Keep hoses and connections in good repair. Use mulch (hay, straw, or grass clippings) to reduce evaporation from the soil.

- Use water wisely and conserve where possible. Unattended watering via sprinklers and/or underground watering systems is not allowed.

- Black (or other colored) plastic may only be used by permission of the site coordinator. Landscape fabric is not allowed as it is less effective for weed control and is difficult to remove after use.

- Please clean garden tools and carts and put them back in the tool shed after use.

- Place only organic materials in the compost piles. Help care for the garden site by participating in community work projects and by picking up litter as found.

- Garden organically to preserve soil fertility and avoid damaging the garden ecosystem. Organic gardening involves three main principals: 1) Feed the soil by using compost and cover crops to add organic matter and nutrients. (Chemical fertilizers such as 5-10-5 are not used in organic gardening.) 2) Apply natural mulches (such as hay or straw) to suppress weeds, reduce water loss, and add organic matter to the soil. 3) Use natural, botanical, and biological insect controls to avoid harming beneficial organisms and pollinators. (Don't use synthetic pesticides, as they can have unwelcome side effects to plants, wildlife, and people.)

- Organic gardening produces healthier soils and safer ecosystems. Organic insect controls include Bt (abacterium), rotenone, pyrethrium, Neem, copper soaps, and various homemade remedies. Please use caution and read instructions before applying any pesticide. A good source for organic pest controls is Gardener's Supply Company (www.gardeners.com).

- Gardeners are responsible to clean up their garden plot by the closing date for the garden site.

- These guidelines are intended for the health and safety of all. If you have questions, suggestions, or are experiencing a problem, please talk with your garden site coordinator. Thanks for your help, and have a great season in the garden!

Vermont Community Garden Network

FIGURE 3.6 Sample community garden guidelines.

Mission: The Upper Dummerston Road Community Garden in Brattleboro VT has been established to provide garden space for people who do not have adequate land at home to grow food.

PARTICIPANT GUIDELINES

As a Community Garden participant, I agree to abide by the following rules.

Participants must complete an application form and sign up for one of the tasks on the attached sheet, necessary to the upkeep of the common space, when taking a plot. The application and fee of \$20 are due by April 15th to reserve a 10'x 20' garden plot for the 2008 season.

Participants are required to attend the initial meeting scheduled for Tuesday, March 25th, 7:30 pm, in the 2nd floor meeting room of the Brooks Memorial Library. ___

The garden committee will do everything possible to assign plots for successive growing seasons if our lease is renewed, though this is not guaranteed.

Gardeners may sow only non-genetically modified (GM) seed or plants.

Only organic herbicides, pesticides, and fertilizers may be used.

Plots must be worked by June 1st and cleaned up, with plant structures taken down, by November 1st. Cool-weather crops may remain, as long as they are tended.

No trees, large bushes, or illegal plants may be planted.

Plots should be kept neat, with weeds under control. Plants should be kept within the confines of participants' plots. Garden plots that appear to be abandoned may be reassigned.

Gardeners are expected to maintain the pathways around their plots.

Participants must remove any trash they bring on-site. Trash barrels will not be provided.

Structures other than plant supports will not be allowed.

Participants must supply their own tools. A garden cart will be supplied.

A water trough will be filled from the spigot on the cottage next to the site.
Gardeners may fill their water buckets from the trough to water plants.

On-site parking will be restricted to the designated parking area.

Children are to be supervised at all times.

No alcohol, drugs, smoking, fires, radios, boom boxes, or pets will be allowed on-site.

The bulletin board is for all to use; please post garden-related notices only.

For questions, comments, further information, or if problems arise, please contact Garden Coordinator.

2/08

FIGURE 3.7 Sample community garden guidelines for garden participants.

garden's beautification efforts and overall community success. One example is when New York community gardens were under siege in 1998 with 114 of 700 community gardens slated to be auctioned. Public opposition and the media negotiated a deal with the city allowing for some gardens to be sold to Trust for Public Land while others were sold to Bette Midler's New York Restoration Project (Englander 2001).

Appendix 2: *Participant gardening agreement*

Archibald Neighborhood Garden
Participant gardening agreement

Participation in the Archibald Neighborhood Garden is open to any Old North End resident free of charge. Access to the site is limited to those who have signed this agreement and their guests. No pets, please.

Gardeners will work cooperatively to establish and maintain two group planting beds, an herb garden and plantings for bioremediation of the site. The next few months will involve your time, energy and commitment, but through it all you will be working with a supportive community of gardeners and harvesting the healthy, delicious produce.

Please read and sign this agreement:

I agree to garden respectfully with my neighbors and follow the garden guidelines.

I will work to keep the garden a happy, secure and enjoyable place where all participants can garden and socialize peacefully in a neighborly manner.

I will garden according to organic principles and use no synthetic fertilizers, pesticides or herbicides.

I understand that neither Grow Team ONE nor the Visiting Nurse Association of Chittenden and Grand Isle Counties are responsible for my actions. I therefore agree to hold harmless Grow Team ONE and the Visiting Nurse Association of Chittenden and Grand Isle Counties for any liability, damage, loss or claim that occurs in connection with the use of the site by me or any of my guests.

I have read and understand terms stated above for the participation in the Archibald Neighborhood Garden.

SIGNATURE DATE

NAME (please print) PHONE NUMBER/E-MAIL

ADDRESS

GROW TEAM ONE

Archibald Neighborhood Garden 2007 13

FIGURE 3.8 Sample community garden participant agreement.

MAINTENANCE ACTIVITIES SCHEDULE

ITEM	ACTIVITY	BY WHOM	FREQUENCY					MONTH IN WHICH ACTIVITY IS PERFORMED											
			D	W	M	Y	Q	J	F	M	A	M	J	J	A	S	O	N	D
PEST CONTROL	Garden plots	G	X							X	X	X	X	X	X	X	X		
	Common plantings	IS, OC, C, G			X							X	X			X	X		
STRUCTURES	General inspection	IS, OC, C, G				X					X	X	X	X	X	X	X	X	
	Metalwork repairs	IS, OC					X				X	X	X	X	X	X	X	X	
	Carpentry repairs	IS, OC					X				X	X	X	X	X	X	X	X	
	Paint	IS, OC, G					X					X	X	X	X	X	X		
SITE FURNISHINGS	General inspection	C, G									X	X	X	X	X	X	X	X	
	Carpentry repairs	IS, OC									X	X	X	X	X	X	X	X	
	Metalwork repairs	IS, OC									X	X	X	X	X	X	X	X	
	Paint	IS, OC, G										X	X	X	X	X	X		
MAJOR EQUIPMENT	General inspection	IS, OC, C, G				X		X	X	X	X						X	X	X
	Preventative maintenance	IS, OC, C, G				X		X	X	X	X	X	X	X	X	X	X	X	X
	Repairs	IS, OC, C, G					X	X	X	X	X	X	X	X	X	X	X	X	X
PLAYGROUND AREA & EQUIPMENT	General inspection	IS, OC, C, G		X				X	X	X	X	X	X	X	X	X	X	X	X
	Safety protocols	IS, OC		X				X	X	X	X	X	X	X	X	X	X	X	X
	Repairs	IS, OC						X	X	X	X	X	X	X	X	X	X	X	X
	Surface clean-up	G	X					X	X	X	X	X	X	X	X	X	X	X	X
	Surface renewal	G					X				X	X	X	X	X	X	X	X	
SIGNS	General inspection	C, G				X		X	X	X	X	X	X	X	X	X	X	X	X
	Repairs	C, G					X	X	X	X	X	X	X	X	X	X	X	X	X
OTHER																			

***Note*: This is a listing of standard maintenance activities. Additional activities that gardeners and caretakers are accustomed to doing or are necessary because of specific garden situations should not be discontinued; add them to this schedule.**

Distributed by Boston Natural Areas Network
UPDATED 2005

Q = As Required
O = Owner
IS = In-House Specialist

OC = Outside contractor
C = Coordinator
G = Gardeners

FIGURE 3.9 Sample annual maintenance list of activities for a community garden.

In some US cities, *land trusts* were established to address the need for protection for community gardens. Land trusts are secured by private or nonprofit groups to obtain and/or conserve land often for the benefit of another group such as a community garden. For example, in Philadelphia, the National Gardening Association purchases land where existing gardens are located to help ensure long-term preservation of community gardens. Additionally, land trusts often provide liability insurance for the community gardeners maintaining properties adding another source of protection for the groups (City Farmer 2001, para 6 NeighborSpace 2004).

EXAMPLES OF SUCCESSFUL COMMUNITY GARDENS

Examples described below illustrate successful old and new gardens, as well as those situated on different types of parcels. The garden managers report potential challenges at various stages of garden establishment.

CASE STUDIES 1 and 2, San Marcos, Texas

San Marcos, Texas, is located midway between Austin and San Antonio, Texas, and has been recently recognized as the fastest growing city in the United States. It is a university town with the pulse of the city being the spring-fed crystal clear waters of the San Marcos River, which meanders through the middle of town.

The river is home to several endangered species and the headwaters are touted as being one of the longest continuously occupied places in North America. The mild weather and earthy community make it a great place to garden year-round. The growing population of the city, combined with increased construction of housing, has created the need for more community gardens to be built. When decisions were made to locate and spawn new gardens, Sustainable San Marcos, a local nonprofit organization, founded in 2008, promoting sustainability and green policies in San Marcos, was charged with locating the gardens. They worked with a GIS specialist to map neighborhoods near interested communities, population, and demographics, as well as the existing gardens to find the best intersections within neighborhoods to target.

Alamo Garden and Dunbar Neighborhood Garden, San Marcos, Texas

The Alamo Garden and Dunbar Neighborhood Garden are two relatively new gardens in San Marcos, Texas. These new gardens offer fresh insight from garden coordinators on implementing community gardens. Both gardens were organized and built by the San Marcos Neighborhood Gardens project, a partnership between the City of San Marcos and Sustainable San Marcos, and are located on leased city property at $10 for 10 years of use. The city required the gardens comply with city community garden rules and guidelines, carry insurance and properly maintain the sites.

Funds for both gardens were acquired from the Federal Community Development Block Grant. Funding through this program requires at least two beds to be designated for production for the Hays County Food Bank. In order to receive funding from the Community Development Block Grant, at least 51% of garden plots must be managed by families in the low to moderate income range.

Raised beds at both locations were built from reclaimed wood which was originally a pedestrian walkway bridge in city parkland. Gardens have access to water, but there is no electricity, nor are there bathrooms on site. These San Marcos garden plots are rented for $40 per year. Gardeners also provide a $30 bed deposit to assure proper maintenance, as well as a $20 tool fee. Requirements for plot rental include participation in monthly garden work days. Applicants must be 16 years old to be issued a plot. Additionally, plots should be planted and managed using organic methods (Figure 3.10).

Both gardens have a garden manager contact who coordinates a garden chore list, keeps track of new gardener applications, makes sure users are following garden guidelines, and organizes work days and maintenance schedules and inventory tools and supplies. While the garden manager is a volunteer appointment, the manager receives a garden plot free of charge. Both gardens have curfews and are open from 6 a.m. until 9 p.m. daily throughout the year.

Gardeners produce flowers, herbs, fruits, and vegetables and are free to sell extra produce in the local Farmer's Market on Tuesdays or Saturdays. After work days, gardeners will often host pot lucks to share harvests. They also have a seed swap once annually.

The Alamo Garden is a one-fourth acre garden on the corner of a neighborhood just one block from the campus of Texas State University on a site where a water tower

FIGURE 3.10 Harvest sits next to raised beds at a local community garden in San Marcos, Texas. (Photo by Leah Gibson.)

once stood. Groundbreaking for the Alamo Garden took place in January, 2013, and took 5 months to build and $19,867 in funding. A fancier rod iron fence (instead of a lower cost chain-link option) was required by the city because the garden is in a high visibility location, and was expensive considering the overall budget. The garden organizer (Robertson, 2014) also mentioned the potential high cost of environmental testing that was required and how this was alleviated through researching local resources available and making good connections in the community. The property was approved for use by Native American tribes and was investigated for any potential contamination from neighboring sites. Good contacts with environmental specialists reduced associated costs from $8000 to a fraction of the cost.

The Dunbar Neighborhood Garden was built in January, 2012, on 2 acres of parkland. Partnerships with the city led to grant and foundation funding opportunities. The garden was ultimately funded with $14,000 provided through four private anonymous foundation grants. The garden was built in 6 months and includes 24 · 10′ × 20′ beds for rental and 2 · 20′ × 20′ beds for food bank production. A food forest of fruit trees was also planted as a result of a partnership with a local permaculture group.

The community garden managers reported some lessons learned including the importance of having a good slate of stakeholders before garden construction (Robertson 2014; Gibson 2014). Additionally, the garden manager volunteer can have a big impact on the overall garden seasonal success so the person should be chosen with some consideration. Also, it is important to become familiar with neighbors who own adjacent property to the intended garden location and try to include them in the planning of the garden. Finally, including produce food preparation tips or cooking ideas, or fostering partnerships with chefs or nutritionists would be helpful to many gardeners and perhaps increase neighborhood participation and help draw in new gardeners.

San Marcos/Saint John's Community Garden, San Marcos, Texas

The original San Marcos Community Garden is located behind Saint John the Evangelist Catholic Church. The site is enclosed in chain-linked fencing and is approximately 1 acre with 40 plots. Plots are typically 800 (40′ × 20′) to 1,000 (40′ × 25′) square feet. Beds are constructed with mounded soil and are, therefore, somewhat harder to maintain and keep free of competing weeds. Compared to a site with raised beds. The current annual plot fee is $20 with an additional one-time $20 registration fee. Fees include access to shared garden tools.

This garden is unique in that it was developed in the 1970s, partly on the land donated by a parishioner and partly on city railway line land. The garden was an early grassroots effort before any other community gardens were developed in the city. Former community garden managers (Adams 2014; Fields, 2014) have reported times of flourish and harvest and times where the garden has been forgotten, overgrown, and an eyesore. The garden is threatened when it is neglected by the need of the nearby Catholic Church for more parking. Both managers suggested the need for energetic and involved coordinators in helping to recruit and keep plots occupied. Kevin Adams also mentioned how an overabundance of harvest from the San Marcos Community Garden led to the development of the Saturday downtown Farmer's Market in town where gardeners could sell their produce. Because the community garden manager is a volunteer position, the funds earned by the manager from produce grown in his/her own plot also helped support and provide incentive for this role.

One of the priorities mentioned by the former garden manager was the need for an active, knowledgeable, strict garden manager who could recognize and police problems before they get out of hand (Fields 2014). It was noted that a manager might overlook a pest problem in one plot during one season, which could then lead to the problem multiplying in the following seasons because of the pests being harbored in the soil. The garden could then develop a reputation for being a place with terrible pest problems where it is difficult to garden successfully. Managers who scout plots and teach gardeners to manage pest problems are indispensable.

Teaching gardeners was mentioned by another garden manager as being important to the community garden (Adams, 2014). He stated that he typically would include planting and maintaining some community shared plots with crops such as lettuce and onions. Crop rotations and cover crops in the garden would be monitored. The garden was considered a success if there was a high turnover rate in plots since this suggested that gardeners were learning skills and gardening at home rather than at the community garden. Through the high turnover, another opportunity was opened to reach a new gardener.

CASE STUDY 3 Eagle Heights Community Garden, Madison, Wisconsin

The Eagle Heights Community Garden in Madison, Wisconsin is one of the oldest and largest community gardens in the United States. The garden is situated in the University of Wisconsin Lakeshore Nature Preserve property and was established in 1962. The garden includes 580 plots at two locations on 10 acres and involves a reported 900–1000 gardeners. While many of the gardeners are affiliated with the university, not all of them are students. Reportedly, 60% are students, 20% are faculty/staff, and 20% are alumni (who often started gardening at the site while still students). The garden is open to residents of the community when plots are available. A garden representative (Dentine 2014) stated that the involvement of the community members was especially important to the maintenance of the garden since these gardeners are those who are usually more stable volunteers and keep the garden running from year to year. The garden typically has a waiting list of gardeners at the beginning of the season, but because many of the gardeners are students, there is approximately a 40% turnover of gardeners each year. No gardener is allowed to have a plot for more than eight consecutive years.

Large 25′ × 30′ plots and small 20′ × 25′ plots are available for $16–$32 per year. Community tools, carts, compost, leaf mulch, some seeds, and a water source are provided. Gardeners must participate in a garden work day or pay additional fees if they choose to not participate. Plot fees finance all supplies, three part-time garden staff, and one registrar who handles the money, assignments of tasks, and communications. Additionally, the garden hires two field staff to maintain plumbing and equipment, and coordinate workdays. The total annual garden budget is approximately $25,000.

The garden's success is said to be due to intensive management (Dentine 2014). "Weed juries" of fellow gardeners mark plots which are seemingly abandoned and insist they be cleaned. If marked plots continue to lack proper maintenance, they must be surrendered.

The garden is a multicultural environment with more than 60 languages spoken collectively, and international crops not available in local grocery stores are grown. The garden allows international students to have access to familiar foods while studying away from their families and traditions. The garden also functions as a teaching and learning environment with experienced gardeners teaching less-seasoned gardeners how to garden.

FUNDING IDEAS

Most gardens are funded through multiple sources. Funding and resources for gardens often come from fees (from plot rentals), donations, grants, fundraisers, and even just scavenging/recycling of tools and materials. When applying for outside funding, it is important to remember that there are regular grant deadlines throughout the year for small gardening grants from local, regional, national, and international sources. However, there may be larger grants available to communities partnering with city or county governments and/or universities for grants for nutrition, health, obesity, food safety, and/or STEM education. Being creative and proactive in partnering opens up access for more lucrative funding opportunities.

City/Parks/Fee-Based Funding

To cover costs for shared supplies, maintenance and management costs, and sometimes leases, many community gardens charge a fee for plot rentals seasonally.

Depending on the gardening program and region, some gardens are accessible year-round while others may only be available to gardeners during a portion of the year. This can affect decisions on the amount of fees charged and revenues earned for the garden. Discounts are often given for low-income gardeners who can document need. Sometimes there is a reduced fee for gardeners who sign up for more than one season or for more than one plot. Fees may also be reduced for less popular plots based on their accessibility to amenities like water, picnic tables, parking or play area, or based on their smaller size. Typical plot fees range from \$6 (4′ × 8′ plot at West Hartford Garden, West Hartford, CT) to \$100 (400′ × 400′ plot at Hawthorn Garden, Boulder, CO) and depend on demand for plots in the region. Often, community gardens will require a refundable deposit which will help ensure an incentive for gardeners to follow rules of the garden, maintain and return borrowed tools, and keep their plots tidy.

Private/Foundation Funding

Several private or foundation donors are available for community gardens. Advantage to applying for private funding is that these agencies will often have more flexible deadlines. Additionally, there may be less competition for funds. A frequent disadvantage with funding through private entities is the specific list of criteria which will need to be met when applying. This may include that the money may be earmarked for a particular region of the country or that specific demographic groups be served. A few examples of private donors include the Darden Foundation, which owns and operates the Red Lobster restaurants among others. The Darden Foundation offers grants for community gardens with the goal of benefiting low-income families through the production of locally grown produce. Similarly, Seeds of Change is an organic seed and plant company offering grants of \$10,000 and \$20,000 to community and school gardens, respectively (Figure 3.11).

Nonprofit Group Funding

Nonprofit groups will sometimes offer grant opportunities to help fund gardening projects, or have ideas for potential funding solutions. For example, the largest community garden association in the country, the American Community Gardening Association (ACGA) has many resources available on their website, as well as ideas for funding garden projects. Another nonprofit organization, Katie's Krops, was started by 14-year-old Katie Stagliano. Grants from this organization are targeted specifically to children hoping to create their own community garden projects to help combat hunger. The National Gardening Association offers awards to school or community gardens (National Gardening Association 2015).

Corporate Grant Funding

Local and regional funding from corporations are often available. Some chain-store operations will often provide small grants for tools and supplies for community garden initiatives. Examples include Walmart, Lowe's, Home Depot, and Target. Local

FIGURE 3.11 Donors to the Sustainable Living Institute of Maui Community Garden are recognized prominently on signage outside the garden gate at this community garden on the campus of University of Hawaii—Maui College. (Photo by Tina M. Waliczek.)

supporters may also offer sponsorship programs in exchange for fresh produce or advertising at the garden.

Several companies offer private grant programs. For example, Mantis Corporation, in combination with the National Gardening Association, annually offers grants for 25 Mantis™ tillers (valued at $350), while Fiskars supports community gardens through their Project Orange Thumb (http://www2.fiskars.com/Community/Project-Orange-Thumb) program to provide "garden makeovers." Grants provide cash, tools, and materials.

Seeds are available from Syngenta Grow More Vegetables Seed Grant Program (https://co1.qualtrics.com/SE/?SID = SV_0wvQ9eQLwGjQH2t). GRO1000 Grassroots Grants offer $1500 to local communities growing edible or flower gardens or developing green spaces and are sponsored by Scotts Miracle-Gro™.

Governmental Funding

Larger grants focusing on nutrition or STEM education can be found through governmental grants such as the United States Department of Agriculture (USDA). Environmental benefits such as watershed protection derived from the garden may be documented and allow for research through the Environmental Protection Agency (EPA). The EPA also regularly funds grants related to environmental education which may offer opportunities for community gardens to partner with schools, universities, or the city. Governmental grant opportunities can be researched at www.grants.gov.

Federal Community Development Block Grants and Community Food Projects Competitive Grants Program (USDA 2014) are also known sources of governmental grant funds for community garden programs. *Community Development Block* grants provide communities with funds to support community development needs such as providing housing, expanding economic opportunities, or creating "suitable living environments" for low- to moderate-income citizens.

Crowd Funding

Defined as a means to raise funds for a project from a large number of people, *crowd funding* gathers supporters via the Internet. The website and environmental nonprofit organization, ioby.org, successfully piloted a program in the New York City that fully funded almost 2000 parks, gardens, composting, and chicken coop projects in the city with most donors giving about $35. The purpose of the project, according to ioby, is to help connect citizens to environmental efforts and projects within their own neighborhoods. Donors typically lived within 2 miles of the sites they were supporting and were already regularly visiting or volunteering with the projects. Crowd funding offers small community garden projects the opportunity to reach a broad audience of additional interested individuals by utilizing the power of social media.

SUMMARY

Successful community gardens all across the country have provided evidence of the value of the gardens to both the individual and the community at large. For a relatively small initial investment, many people derive various quality-of-life benefits from being involved in a community garden, such as food security, access to nutritious food, social needs, self-esteem needs, and safe environment needs. Vacant urban lots are transformed from dumping grounds for trash and criminal behavior into a common area for socialization among neighbors, a science lab, and supermarket for children and families.

Definitions of community gardens vary, but the term "community garden" is typically defined as any plot of public or private land collectively managed by a group of people. Community gardens are implemented in urban, suburban, or rural areas, and are most often operated where citizens have limited access to gardening space.

Community gardens are not a new concept and have been used historically to achieve some of the same benefits which research has recently documented. Community gardens were planted during the Depression and during various war efforts as a means to grow food, save money and resources, and act as a symbol of pride and unity. Community gardens are providing the same benefits today while also allowing urbanites to connect with nature. Community gardens today provide food security, nutritional welfare, green spaces and habitat, environmental and educational aid, tourism, and psychological and mental benefits. Community gardens have been tried to crime prevention, place attachment, and association with memories and heritage of gardeners.

A site needs to be identified which will make for productive garden plots, but also be comfortable, accessible, and safe for the users. Identify a site that is not deemed for other uses in the near or far future. Areas adjacent to parkland have been used in many communities. Private vacant land is also sometimes available. When locating a garden, sunny, open, level, well-drained site with good soil and access to water is ideal. However, modifications can be made to sites if necessary. Equal if not more attention should be devoted toward developing the organization as to developing the land.

Managing the garden over time is one of the biggest challenges to community gardens. A strong garden manager can make a difference in keeping gardeners active. Their enthusiasm is a key component in recruiting new gardeners over time. Documenting benefits to the community such as garden development, harvests, and data on the people and families served will help show progress to gardeners, coordinators, and the overall community and can help justify the garden during times of hardship.

Community garden governance and organizations vary among groups with some gardens having a director and others functioning with a board of volunteers or committees. Umbrella nonprofit organizations exist in many parts of the United States and guide and manage smaller community gardens throughout regions.

Community gardens are often established on private or government property. Since the property is not owned by the garden, the land can become threatened by development. A legal agreement written at the beginning of the project can help prevent these situations. Land trusts set up by private or nonprofit groups are a means to secure the long-term preservation of community gardens.

Funding for community gardens is often through a variety of sources including fees, foundations, grants, donations, and crowd funding. Gardens find other resources through recycling and scavenging as well.

In this time of increased urbanization, densely populated cities, technological advances, and social media, it is ironic that people feel more isolated than ever. Community gardens have shown that on small parcels of land and with little investment from outside entities, gardeners can build something that affects positive change and greatly influences the quality of life for those in the community.

REVIEW QUESTIONS

1. Discuss the similarities and differences of today's community gardeners' motivations for gardening with those of community gardeners from the past.
2. Compare and contrast typical community gardeners to efforts of guerilla gardeners.
3. Describe the similarities and differences between the Victory Gardens of WWII and the Anti-Inflation Gardens of the 1970s.
4. Name three benefits of community gardens discussed in the chapter. Based on your own research and experience, suggest three more benefits which may still need to be researched and documented.
5. How do community gardens potentially help to alleviate "broken window syndrome" in cities?
6. In what ways could you see community gardens as being a place to build social capital amongst neighbors?

7. Who are some typical stakeholders in a community garden situation? Name some potential specific stakeholders for a community garden established in your own community.
8. Describe three potential opportunities for collective efficacy within a community garden situation.
9. Name three sources of funding for community gardens. Based on your own experience and research, are there also other opportunities which could be explored?

ENRICHMENT ACTIVITIES

1. Develop a mission statement for a university or city-based community garden for your own community.
2. Develop a list of stakeholders for a university or city-based community garden for your own community.
3. Locate grant opportunities on the Internet other than those listed here which would help support a neighborhood community garden. Email the list to a local community garden manager.
4. Identify a site locally or in your hometown where you could implement a potential guerilla gardening project. Identify appropriate plants for the site.
5. Divide the class into groups and plant multiple guerilla gardening sites around town. Host a guerilla garden walk with donations accepted to support a local community gardening effort.
6. Locate an established community garden or umbrella organization in your community. Interview one of the active gardeners about why he or she gardens at the community garden and the benefits he or she perceives from his or her participation.
7. Scavenge a novel container from the trash to produce a guerilla garden container design. Plant the container and display where others can witness beauty in unexpected places.
8. Organize a city-wide community garden open house bike tour day. Tour guides will take bike visitors to each garden on a rotation with donations being accepted to support the gardens.

REFERENCES

Adams, K. 2014. Community garden and farmer's market manager. Personal communication. June 14, 2014.

American Community Garden Association. 2004. Retrieved on August 8, 2014 from http://www.communitygarden.org/

American Community Gardening Association (ACGA). 2014. http://www.communitygarden.org/learn (accessed December 20, 2014).

Armstrong, D. 2000. A survey of community gardens in upstate New York: Implications for health promotion and community development. *Health and Place*, 6(4), 319–327.

Bacigalupi, R. (Producer). 2002. *A Lot in Common*. Oley, PA: Bullfrog Films.

Bassett, T. 1979. Vacant lot cultivation: Community gardening in America 1893–1978 MA thesis. University of California, Berkeley.

Bolitzer, B. and N.R. Netusil. 2000. The impact of open spaces on property values in Portland, Oregon. *Journal of Environmental Management*, 59, 185–193.

Boston Natural Areas Network (BNAN). 2014. Boston Community Garden Council. Retrieved on June 9, 2014 from http://www.bostonnatural.org/cgcouncil.htm.

Bremer, A., K. Jenkins, and D. Kanter. 2003. Community Gardens in Milwaukee: Procedures for their long-term stability and their import to the city. Milwaukee: University of Wisconsin, Department of Urban Planning.

Bureau of Labor Statistics, US Department of Labor. *Occupational Outlook Handbook*, 2014–15 Edition, Farmers, Ranchers, and Other Agricultural Managers, on the Internet at http://www.bls.gov/ooh/management/farmers-ranchers-and-other-agricultural-managers.htm (visited March 08, 2015).

Burger, C. 1975. Anti-inflation garden plan leads to jar/lid problem. Reading Eagle. Retrieved on August 19, 2015 from https://news.google.com/newspapers?nid=1955&dat=19750219&id=hg5XAAAAIBAJ&sjid=K0MNAAAAIBAJ&pg=5770,2397717&hl=en.

Carney, P.A., J.L. Hamada, R. Rdesinski, L. Sprager, K.R. Nichols, B. Liu, J. Pelayo, M. Sanchez and J. Shannon. 2011/2012. Impact of a community gardening project on vegetable intake, food security and family relationships: A community-based participatory research study. *Journal of Community Health*, 27:874–881.

City of Austin. 2011. Chapter 14-7. Sustainable Urban Agriculture. City Code. Retrieved on August 12, 2015 from http://www.amlegal.com/nxt/gateway.dll/Texas/austin/title-14useofstreetsandpublicproperty/chapter14-7sustainableurbanagriculture?f=templates$fn=default.htm$3.0$vid=amlegal:austin_tx.

City Farmer. 2001. Urban Agriculture in Philadelphia. Canada's Office of Urban Agriculture. Retrieved on June 11, 2014 from http://www.cityfarmer.org/Phillyurbag9.html.

Clinton Community Garden. n.d. Retrieved on June 11, 2014 from http://clintongarden.org/about/history-2/.

Danforth, P., T.M. Waliczek, S.M. Macey, and J.M. Zajicek. 2008. The effect of the National Wildlife Federation's Schoolyard Habitat Program on fourth grade students' standardized test scores. *HortTechnology*, 18(3), 356–360.

Dentine, Margaret. 2014. Registrar for the Eagle Heights Community Garden and Professor Emerita of University of Wisconsin. Personal communication. July 15, 2014.

Environmental Defense. 2002. Environmental Defense, Mayor Hahn and Verde Coalition Plan to Create Network of Green Spaces In Urban Core of L.A. Retrieved on August 22, 2015 from http://www.environmentaldefense.org/pressrelease.cfm?contentid=2259

Environmental Design Research Association. 1985. The Garden of Eden: An environmental "radical transformation." Retrieved on July 9, 2014 from http://www.zentences.com/edra.html.

Espey, M. and K. Owusu-Edusei. 2001. Neighborhood parks and residential property values in Greenville, South Carolina. *Journal of Agricultural and Applied Economics*, 33, 487–492.

Fields, Suzi. 2014. Community garden manager and former President of Edible San Marcos and owner of Suzi's Naturals. Personal communication. August 4, 2014.

Firth, C., D. Maye, and D. Pearson. 2011. Developing "community" in community gardens. *Local Environment*, 16(6), 555–568.

Gibson, L. 2014. Garden Manager and Environmental Consultant. Personal communication. July 18, 2014.

Glover, T. 2004. Social capital and the lived experiences of community gardeners. *Leisure Sciences*, 26(2), 143–162.

Gordon, B.H.J. and J.C. Dotter. 1996. City community gardens in partnership with the San Jose state university interdisciplinary student garden project for education, outreach, and sustainable agriculture research, a model for replication. In P. Williams and

J. Zajicek (eds.), *People-Plant Interactions in Urban Areas: Proceedings of a Research and Education Symposium.* Applying People-Plant Research in Classroom Education. Texas: San Antonio, pp. 215–218.
Gorham, M., T.M. Waliczek, A. Snelgrove, and J.M. Zajicek. 2009. The effect of community gardens on incidence of crime in Houston. *HortTechnology*, 19(2), 291–296.
Green Guerillas. 2002. Retrieved on June 9, 2006 from http://www.greenguerillas.org.
Harris, M. 2011. Gutted south central farm remains vacant. NBCUniversal Media, LLC. http://www.nbclosangeles.com/news/local/South-LA-Urban-Farm-Lot-123050788.html (accessed January 27, 2015).
Hobden, D.W., G.E. Laughton, and K.E. Morgan. 2004. Green space borders—A tangible benefit? Evidence from four neighbourhoods in Surrey, British Columbia 1980–2001. *Land Use Policy*, 21, 129–138.
Hynes, P.H. 1996. *A Patch of Eden: America's Inner City Gardens.* White River Junction, Vermont: Chelsea Green Publishing Co.
Kuo, F.E. 2001. Coping with poverty: Impacts of environment and attention in the inner city. *Environment and Behavior*, 33(1), 5–34.
Kuo, F.E. and W.C. Sullivan. 2001a. Aggression and violence in the inner city: Effects of environment via mental fatigue. *Environment and Behavior*, 33(4), 543–571.
Kuo, F.E. and W.C. Sullivan. 2001b. Environment and crime in the inner city: Does vegetation reduce crime? *Environment and Behavior*, 33(3), 343–367.
Langellotto, G. and A. Gupta. 2012. Gardening increases vegetable consumption in school-aged children: a meta-analytical synthesis. *HortTechnology*, 22(4), 430–445.
Lawson, L. 2005. *City Bountiful: A Century of Community Gardening in America.* Berkeley, CA: University of California Press.
Lee, J.H. 2011. Community gardening and place identity, Texas A & M University unpublished manuscript. Retrieved on March 8, 2015 from files.figshare.com/1406402/Place_Identity__Place_Attachment.pdf.
Lutzenhiser, M. and N. Netusil. 2001. The effect of open spaces on a home's sale price. *Contemporary Economic Policy*, 19, 291–298.
Maslow, A.H. 1943. A theory of human motivation. *Psychological Review*, 50(4), 370–396.
Mast, G.S. 2012. The geography of motivation and participation among community gardeners in Austin, TX. Master of Science thesis. Texas State University, San Marcos, TX.
McPherson, E.G., J.R. Simpson, P.J. Peper, S.E. Maco, S.L. Gardner, S.K. Cozad, and Q. Xiao. 2005. Midwest community tree guide: Benefits, costs, and strategic planting. NA-TP-05–05. Newtown Square, PA: US Department of Agriculture, Forest Service, Northeastern Area State and Private Forestry.
Miceli, R., M. Roccato, and R. Rosato. 2004. Fear of crime in Italy. *Environment and Behavior*, 36(6), 776–789.
National Gardening Association. 2015. Grants and fundraising. Retrieved on August 22, 2015 from http://grants.kidsgardening.org/
Newman, M. 1997. In search of food security. *Seasonal Chef.* Retrieved on December 21, 2014 from http://www.seasonalchef.com/security.htm.
New York City Department of Parks and Recreation. 2014. The Community Garden Movement: Green Guerillas gain ground. http://www.nycgovparks.org/about/history/community-gardens/movement.
Not Bored. (n.d.). Retrieved on June 8, 2014 from http://www.notbored.org/gardens.html.
Pagano, M.A. and A. O'M. Bowman. 2000. Vacant land in cities: An urban resource. Center on Urban and Metropolitan Policy. The Brookings Institution Survey Series. Washington, D.C. 1–9.
Park, S.-A., A.-Y. Lee, K.-S. Lee, and K.-C. Son. 2014. Comparison of the metabolic costs of gardening and common physical activities in children. *Korean Journal of Horticulture Science and Technology*, 32(1), 123–128.

Patel, I.C. 1996. Rutgers urban gardening: A case study in urban agriculture. *Journal of Agricultural and Food Information*, 3(3), 35–46.

Pennsylvania Horticultural Society. 2014. Philadelphia Green. Retrieved on August 19, 2014 from http://phsonline.org/greening/neighborhood-gardens-association.

Pincetl, S., J. Wolch, J. Wilson, and T. Longcore. 2003. *Toward A Sustainable Los Angeles: A "Nature's Services" Approach*. Los Angeles: USC Center for Sustainable Cities.

Poole, S. 2006. *The Allotment Chronicles: A Social History of Allotment Gardening*. Kettering, UK: Silver Link Publishing.

Pudup, M. 2008. It takes a garden: Cultivating citizen-subjects in organized garden projects. *Geoforum*, 39(3), 1228–1240.

Putnam, R.D. 2000. *Bowling Alone: The Collapse and Revival of American Community*. New York: Simon and Schuster.

Reid, D. 1996. Community garden workshop. In P. Williams and J. Zajicek (eds.), *People-Plant Interactions in Urban Areas: Proceedings of a Research and Education Symposium*. Workshops Texas: San Antonio, pp. 223–226.

Reynolds, R. 2008. *On Guerrilla Gardening: A Handbook For Gardening Without Boundaries*. New York: Bloomsbury Publishing.

Robertson, B. 2015. Garden Manager. Personal communication. June 18, 2014.

Sampson, R.J., S.W. Randenbush, and F. Earls. 1998. Reducing violence through neighborhood collective efficacy. *Alternatives to Incarceration*, 4(4), 18–19.

Schweitzer, J.H., J.W. Kim, J.R. Mackin. 1999. The impact of the built environment on crime and fear of crime in urban neighborhoods. *Journal of Urban Technology*, 6(3), 59–73.

Sherer, P.M. 2006. The benefits of parks: Why America needs more city parks and open space. Retrieved on March 8, 2015 from http://www.tpl.org.

Skogan, W.G. 1990. *Disorder and Decline: Crime and the Spiral of Decay in American Neighborhoods*. New York: The Free Press.

Small, V. 2002. Gardens that build community. *Fine Gardening*, November/December 64–67.

Snelgrove, A.G., J.H. Michael, T.M. Waliczek, and J.M. Zajicek. 2004. Urban greening and criminal behavior: A geographic information system perspective. *HortTechnology*, 14(1), 48–51.

Taylor, A.F., F.E. Kuo, and W.C. Sullivan. 2002. Views of nature and self-discipline: Evidence from inner city children. *Journal of Environmental Psychology* 22, 1–2, 49–63.

Tranel, M. and L.B. Handlin, Jr. 2006. Metromorphosis: Documenting change. *Journal of Urban Affairs*, 28, 151–167.

The Lampert Report. 2014. Aging farmer population threatens US Food Supply. Retrieved on March 8, 2015 from http://www.manufacturing.net/news/2014/12/aging-farmer-population-threatens-us-food-supply.

The Pennsylvania Horticultural Society. 2014. Philadelphia Green. Retrieved on August 19, 2014 from http://phsonline.org/greening/neighborhood-gardens-association.

The Trust for Public Land. 2007. The economic benefits of parks and open space (Chapter 2 Attracting Investment). Retrieved on December 16, 2014 from http://slco.org/openSpace/doctopdf/econBenefitsLandConservTPL.pdf.

United States Bureau of Soils. 2014. *Soil Survey*, Volume 2. Memphis, TN: General Books, LLC.

United States Department of Agriculture. 2006. Cooperative State Research, Education, and Extension Service. Retrieved on September 20, 2014 from http://www.csrees.usda.gov/nea/food/in_focus/hunger_if_competitive.html.

Urban Harvest. (n.d.). Retrieved on June 9, 2014 from http://www.urbanharvest.org/community/community.html.

Vallianatos, M., R. Gottlieb, and M.A. Haase. 2004. Farm-to-school: Strategies for urban health, combating sprawl, and establishing a community food systems approach. *Journal of Planning Education and Research*, 23, 414–423.

Victory Seeds. 2015. A Victory Garden. Retrieved on August 27, 2015 from http://www.victorygardening.com/page2.html

Voicu, I. and V. Been. 2008. The effect of community gardens and neighboring property values. *Real Estate Economics*, 36(2), 241–283.

Waliczek, T.M., R.H. Mattson, and J.M. Zajicek. 1996. Psychological benefits of community gardening. *Journal of Environmental Horticulture*, 14(4), 204–209.

Waliczek, T.M., J.M. Zajicek, and R.D. Lineberger. 2005. The influence of gardening activities on consumer perceptions of life satisfaction. *HortScience*, 40(5), 1360–1365.

Wilson, J. and G.L. Kelling. March 1982. Broken Windows: The police and neighborhood safety. *The Atlantic*, retrieved August 19, 2015. http://www.theatlantic.com/magazine/archive/1982/03/broken-windows/304465/

Whitmire Study. 2008. Gateway Greening Public Policy Research Center at the University of Missouri, St. Louis, MO. Retrieved on July 14, 2014 from http://actrees.org/files/Research/gateway_greening_whitmire.pdf.

Zukin, S. 2010. *Naked City: The Death and Life of Authentic Urban Places*. New York: Oxford University Press.

FURTHER READING

Baker, L. 2004. Tending cultural landscapes and food citizenship in Toronto's community gardens. *The Geographical Review*, 94(3), 305–325.

Beilin, R. and A. Hunter. 2011. Co-constructing the sustainable city: How indicators help us "grow" more than just food in community gardens. *Local Environment*, 16(6), 523–538.

Bremer, A., K. Jenkins, and D. Kanter. 2003. Community Gardens in Milwaukee: Procedures for their long-term stability and their import to the city. Milwaukee: University of Wisconsin, Department of Urban Planning.

Corrigan, M. 2011. Growing what you eat: Developing community gardens in Baltimore, Maryland. *Applied Geography*, 31(4), 1232–1241.

Evers, A. and N. Hodgson. 2011. Food choices and local food access among Perth's community gardeners. *Local Environment*, 16(6), 585–602.

Ferguson, S. A brief history of grassroots greening in NYC. Retrieved on March 8, 2015 from http://skillshares.interactivist.net/gardens/h_1.html.

Flachs, A. 2010. Food for thought: The social impact of community gardens in the greater Cleveland area. *Electronic Green Journal*, 1(30), 1–9.

Goddard, M.A., A.J. Dougill, and T.G. Benton. 2009. Scaling up from gardens: Biodiversity conservation in urban environments. *Trends in Ecology and Evolution*, 25(2), 90–96.

Hale, J., C. Knapp, L. Bardwell, M. Buchenau, J. Marshall, F. Sancar, and J. Litt. 2011. Connecting food environments and health through the relational nature of aesthetics: Gaining insight through the community gardening experience. *Social Science and Medicine*, 72, 1853–1862.

Hassan, B. 1990. Horticultural experience increases garden size and economic investment within a community garden program. Master of Science thesis. Kansas State University, Manhattan, KS.

Litt, J., M. Soobader, M. Turbin, J. Hale, M. Buchenau, and J. Marshall. 2011. The influence of social involvement, neighborhood aesthetics, and community garden participation on fruit and vegetable consumption. *American Journal of Public Health*, 101(8), 1466–1473.

National Recreation and Park Association (NRPA). 2014. Grow your park initiative. Building a community garden in your park. http://www.nrpa.org/uploadedFiles/nrpaorg/Grants_and_Partners/Parks_and_Conservation/Resources/Community-Garden-Handbook.pdf (accessed June 14, 2014).

Reynolds, R. 2004. Video of famous Elephant and Castle district of London and the guerilla gardening efforts there. http://www.guerillagardening.org/ (accessed June 16, 2014).

Russell, S.E. and C.P. Heidkamp. 2011. "Food desertification": The loss of a major supermarket in New Haven, Connecticut. *Applied Geography*, 31, 1197–1209.

Saldivar-Tanaka, L. and M.E. Krasny. 2004. Culturing community development, neighborhood open space, and civic agriculture: The case of Latino community gardens in New York City. *Agriculture and Human Values*, 21(4), 399–412.

Smith, C.M. and H.E. Kurtz. 2003. Community gardens and politics of scale in New York City. *Geographical Review*, 93(2), 193–212.

Stan Herd, an artist, initiated a project in Rio de Janiero, taking trash strewn areas in urban favelas and transforming it into the largest community green strip in the country. http://stan-herd-art.com/?p=152.

Teig, E., J. Amulya, L. Bardwell, M. Buchenau, J.A. Marshall, and J.S. Litt. 2009. Collective efficacy in Denver, Colorado: Strengthening neighborhoods and health through community gardens. *Health and Place*, 15(4), 1115–1122.

Turner, B. 2011. Embodied connections: Sustainability, food systems and community gardens. *Local Environment*, 16(6), 509–522.

Turner, B., J. Henryks, and D. Pearson. 2011. Community gardens: Sustainability, health and inclusion in the city. *Local Environment*, 16(6), 489–492.

University of Wisconsin—Madison. 2010. Eagle Heights Community Garden. http://www.eagleheightsgardens.org/index.shtml.

Zanetti, O. 2007. Guerilla gardening. Geographers and gardeners, actors and networks: Reconsidering urban public space. http://www.guerillagardening.org/books/ZanettiGG.pdf (accessed June 16, 2014).

4 Public Gardens and Human Well-Being

Sonja M. Skelly

CONTENTS

OBJECTIVES

Upon completion of this chapter, the reader should be able to

- Define what a public garden is.
- List the ways public gardens are governed and funded.
- Describe the development of public gardens throughout history.
- Identify the types and characteristics of public gardens.
- Identify the major roles of public gardens.
- List and describe the numerous benefits of public gardens to individuals, their communities, and the world.

KEY TERMS

- Accession
- Accessioned collection
- Advisory board
- Aesthetic arrangement
- All-American selections
- Apothecary
- Arboreta/arboretum
- Biodiversity
- Biodiversity "hot spot" regions
- Bioswale
- Board of trustees
- Botanical garden
- Botany
- Certificate programs
- Children's garden
- College and University garden
- Conservatory
- Continuing education
- Continuing Education Units
- Curate
- Curriculum standards
- Demonstration garden
- Display garden
- Earned income
- Ecology
- Economic botany
- Endowment
- Environmental degradation
- Espalier
- Estate garden
- Ethnobotany
- *Ex situ* conservation

- Floristics
- Formal education
- Governance
- Grants
- Historic garden
- Historical landscape
- *In situ* conservation
- Informal education
- Interdisciplinary
- Internship
- Interpretation
- Invasive species
- Landscape for Life
- LEED
- Membership
- Mission
- Mulch
- Native plants
- Ornamental plant science
- Outreach program
- Phenology
- Phylogenetics
- Physic garden
- Physiology
- Plant conservation
- Plant genetics
- Plant morphology
- Plant systematics
- Plant taxonomy
- Population biology
- Private foundation
- Professional training and education
- Public garden
- Raised bed garden
- Research display
- SITES
- Social inequity
- Social justice
- Sponsorship
- Taxonomic arrangement
- Thematic displays
- Tourism industry
- Trial garden
- Universal design
- Water conservation garden
- Wetland garden

Public gardens are places that provide many benefits for individuals and communities. They are places of beauty, respite, and inspiration. Public gardens are varied, each with its own unique set of plants, gardens, programs, and goals for improving the lives of those who visit and those with whom they work. These goals, programs, plants, and gardens provide opportunities for people of all ages to find relief from stress, gain knowledge about the plant and natural world, connect with nature, help conserve the natural world, develop leadership skills, connect with their community and improve social relationships, improve healthcare and nutrition, and alleviate financial poverty (Waylen 2006).

WHAT IS A PUBLIC GARDEN?

A *public garden* is an organization, open to the public, that holds a collection of plants and uses that collection to further its mission. A *mission* is an organization's reason for existing. The mission of many public gardens includes the cultivation and display of a wide range of plants—its collection—to address some purpose such as education, conservation, or aesthetic display (Rakow 2011).

Mission statement examples are as follows:

Brooklyn Botanic Garden—Brooklyn Botanic Garden (BBG) is an urban botanic garden that connects people to the world of plants, fostering delight and curiosity while inspiring an appreciation and sense of stewardship of the environment.

Cornell Plantations—To preserve and enhance diverse horticultural collections and natural areas for the enrichment and education of academic and public audiences, and in support of scientific research.

The Fells—To stimulate appreciation of the environment, horticulture, and the significance of the past by preserving and sharing the Hay family's historic lakeside summer home.

At the heart of each of these public gardens is an *accessioned collection* or cataloged collection of plants. This collection is maintained for conservation, research, education, teaching, recreation, and display. Public gardens *accession*, or keep detailed records, of their plants. This is done by a paid staff who may be a plant records specialist, a curator, and a gardener with botanical expertise to ensure botanical accuracy. This person(s) may also be responsible for research and interpretation related to the collection. The plant collections are *curated* in a similar fashion to the way a museum curates its art. The garden is responsible for researching, organizing, and maintaining the collection of plants. Under these criteria parks, resorts, and amusement parks landscaped with beautiful gardens are not public gardens unless they have a mission and a curated collection of plants (Michener 2011).

There are three main types of public garden *governance*, the way in which the garden is governed or overseen by a governing body: nonprofit board of trustees, advisory boards, or a government unit. A nonprofit *board of trustees*, or board of directors, is a group of individuals responsible for overseeing the management of the garden, the financial and legal operations of the garden, long-range planning, fund-raising, and the management of the garden director. *Advisory boards* have many of the same roles and responsibilities as a board of trustees except that the financial and legal operations are overseen by a larger governing unit, such as a college or university. For public gardens that are owned and operated through a government office at the national, state, county, or city level the government unit itself, such as a city council, oversees the gardens' operations (Matheson 2011).

The very definition of a public garden is that it is one that is open to the public. There are an estimated 2500 public gardens worldwide (Waylen 2006). In the United States an estimated 30 million people visited a public garden in 2003 (Benfield 2006). In a 2010 national study of 40,000 museum-goers, 60% (23,923) of those surveyed indicated that they enjoy visiting botanical gardens and arboreta (Wilkening and Chung 2011).

FUNDING A PUBLIC GARDEN

Public gardens derive their funding from a variety of sources depending on their governance and history; all need a variety of funding streams. Some gardens have an *endowment*, an income typically from the investment of a donation, to support their organization. Other gardens are privately funded, while many receive support from their governing organizations such as a college or university or government units. In most cases, these funds are not enough to sustain a garden so other funding streams are sought and typically include fund raising, corporate and foundation support, grants, and earned income.

Fund raising programs at public gardens vary with the size and maturity of a garden, but activities often include a membership program, individual gift giving, and capital campaigns. *Membership* programs help gardens earn *unrestricted money*, money not restricted to a specific purpose, as well as help them to identify individuals with interest and perhaps greater capacities to donate. Individual gift giving is an important source of funding. Garden staff often cultivate potential and current donors for support by learning of their interests in the garden, engaging with them on visits to the garden, communicating with them, and sometimes visiting them at their home to develop long-standing relationships that can result in significant monetary gifts. Support from corporations, often in the form of *sponsorship* provides funding for specific gardens, programs, and events. *Private foundations* such as family trusts and foundations can provide funding for gardens especially as the garden's mission and the request for funding align with the foundation's mission. The Stanley Smith Horticultural Trust is a private foundation that provides support to public gardens for education and research in ornamental horticulture. The Trust grants awards to organizations who advance the research of ornamental horticulture, who create, develop, preserve, and maintain gardens for educational purposes, and promote environmentally responsible cultivation of plants with ornamental horticultural value (Stanley Smith Horticultural Trust 2014).

Grants can provide a significant level of revenue to a garden. Grants often support educational programming, research, and capital projects. Grants come from a variety of sources, though federally offered grants through the Institute of Museum and Library Services, the National Science Foundation, and the National Endowment for the Humanities are often sought after by public gardens.

An increasing revenue stream for public gardens is from earned income. *Earned income* is revenue derived from a fee for a service or product. These services and products include admission to the garden, facility rental (such as use of the garden or garden building for a wedding), retail (gift shop), food services, plant sales, special events (such as a concert or art exhibit), and education program income (fee for classes and workshops) (Daley 2011).

HISTORY OF PUBLIC GARDENS

The first public gardens were known as *physic gardens*, or gardens growing herbs for medicinal purposes, and were found in Italy in the fifteenth century. These gardens, found at Universities in Padova, Firenze, and Bologna, Italy were for the academic study of medicinal plants. Such medicinal gardens were adopted by universities and *apothecaries*, practitioners who dispensed medicines throughout Europe in the sixteenth and seventeenth centuries.

During the sixteenth and seventeenth centuries, exploration and trade expansion led to the creation of places such as the Royal Botanic Gardens, Kew and the Royal Botanic Garden, Madrid where new species of plants from around the world were displayed and cultivated. Botanical gardens were often set up to display new plants from European colonies and newly visited lands as trophies (Botanic Gardens Conservation International 2014). Beyond their novelty, many plants were also of interest for their economic importance. In the eighteenth century, many garden were where *botany*, the study of plants was fostered along with the development of *plant taxonomy*, the science that identifies, describes, classifies, and names plants. Prior to taxonomic classification, gardens often displayed their plants in four quadrants, representing the four corners of the earth or the known continents of the time—Europe, Asia, Africa, and America. With the establishment of plant taxonomy, plants began to be placed in displays based on their relationships with other plants (Hobhouse 1997) (Figure 4.1).

John Bartram's nursery, established in 1728 in Philadelphia, Pennsylvania, is considered to be the earliest public garden in North America. His collection of plants native to America, which he supplied to European nurseries, was visited by notable figures such as Benjamin Franklin and Thomas Jefferson. The mid-1800s saw the growth of many parks and public gardens including Boston's public garden in 1837, The United State Botanic Garden in 1850, the Missouri Botanical Garden in 1859, the Arnold Arboretum of Harvard University in 1872, and the New York Botanical Garden in 1891 (Flanagan 2011). Establishing these gardens was seen by early leaders as important to the development of America and its cities. Frederick Law Olmsted, the father of landscape architecture, embraced this vision by designing more public parks than any other designer. His best known work, Central Park in New York City, was "designed to be a place where city workers with no chance to travel to the country, could enjoy God's handiwork" (Hobhouse 1997, p. 271).

FIGURE 4.1 The Garden of Padua in Italy in 1545 showing the four quadrant representing the four corners of the earth. (https://commons.wikimedia.org/wiki/File:Orto_dei_semplici_PD_01.jpg.)

TYPES OF PUBLIC GARDENS

There are many different types of public gardens. Commonly referred to as botanical gardens, public gardens can exist in urban centers, at universities, as estate gardens, arboreta, and conservatories.

The American Public Garden Association's (2014a) categories of public gardens include

- Arboretum
- Aviary
- Botanical garden
- Cemetery
- Children's Garden
- College and University Garden
- Community Garden
- Conservatory
- Cultural Center
- Display
- Experimental
- Farm Garden
- Historical Landscape and Site
- Japanese Garden
- Museum

- Native Wildlife Habitats
- Nature Garden
- Park
- Park for the Arts
- Public Park
- Research
- Sculpture Garden
- Special Collection
- Tropical
- Urban Garden
- Zoo

What distinguishes public gardens from one another generally stems from the garden's focus. Traditionally, *arboreta* were public gardens whose sole focus was on the curation of woody plant material—trees and shrubs, whereas *botanical gardens* tended to focus on herbaceous plants as well as trees and shrubs. *Historical landscapes* and *estate gardens* are public gardens that illustrate, document, and attempt to preserve a landscape and garden in a particular time period. For estate gardens, the garden is often preserved as it was in when it was an active estate for display and interpretation to the public. Some gardens' mission is to focus on aesthetically pleasing horticultural displays. In these *display gardens*, the display of the plants is paramount and although plant records are kept, plants are not often labeled though education and interpretations are provided. A *college and university garden* is, as its name implies, a public garden at a college or university. These gardens' missions are often to serve the educational and research purposes of the college or university with which it is associated. In some cases, the college and university garden is comprised of the university grounds and accessions of all of the plant material in the campus while other college and university gardens are adjacent to and sometimes removed from the campus (Figure 4.2).

An emerging type of public garden are *children's gardens*, gardens designed specifically for children. Some children's gardens are stand-alone gardens while the majority are housed within a bigger botanical garden. Whether stand-alone or at a public garden, the mission and intent of children's gardens are to connect children with plants, gardening, and nature (Rakow 2011).

Examples of public gardens by type are as follows:

Arboreta

- Morris Arboretum of the University of Pennsylvania, Philadelphia, Pennsylvania
- Morton Arboretum, Lisle, Illinois

Botanical gardens

- Brooklyn Botanic Garden (see Case Study #1), Brooklyn, New York
- Missouri Botanical Garden, St. Louis, Missouri

FIGURE 4.2 University faculty use the plant collections at a University public garden to enhance their courses. (Photo courtesy of Cornell Plantations, photo taken by Cornell University Photography member.)

Estate gardens

- Filoli, Woodside, California
- Biltmore estate, Asheville, North Carolina

Display gardens

- Chanticleer, Wayne, Pennsylvania
- Longwood Gardens, Kennett Square, Pennsylvania

College and University garden

- Cornell Plantations (see Case Study No.#2), Ithaca, New York
- Scott Arboretum of Swarthmore College, Swarthmore, Pennsylvania

Children's garden

- Ithaca Children's Garden, Ithaca, New York
- Rory Meyers Children's Adventure Garden at the Dallas Arboretum, Dallas, Texas

ROLES OF PUBLIC GARDENS

Plant Display

The display of plants, whether in taxonomically ordered beds or in an aesthetically displayed manner, has been the hallmark of public gardens since their inception. A garden's collection of plants goes to the heart of the garden's mission and influences the garden design and even the programming that occurs at the garden. These collections can be arranged in a multitude of ways. A *taxanomic arrangement* places the plants into groups based on how they are related by plant classification systems (Michener 2011). *Plant classification systems* place plants into groups or categories to show relationship based on the standard hierarchy for biology: Kingdom, Phylum, Class, Series, Family, Genus, and Species. For example, oak trees belong to the family Fagaceae, the genus *Quercus,* and have a wide range of species such as *Quercus alba* (white oak), *Quercus rubra* (red oak), and *Quercus stellate* (post oak). In a taxonomic arrangement, all oak trees are placed together because they share the same family and genus group but may differ in their species designation. This traditional manner of display is a useful way to understand which plants are related to one another. At the Arnold Arboretum of Harvard University, plants are grouped by family for easy comparison.

An *aesthetic arrangement* of plants is a grouping of plants together in an aesthetically pleasing manner. Gardens often employ landscape architects to design such displays. These can be new and innovative displays of plants such as the work of renowned landscape architect, Piet Oudolf, whose displays of bold drifts of herbaceous perennials and ornamental grasses grace the Highline in New York City and the Millennium Park's Lurie Garden in Chicago. Gardens like Chanticleer and Longwood Gardens, both in Philadelphia, are renowned for their artistic displays of plants (Figure 4.3).

Other types of displays may include *thematic displays*, those developed around concepts such as garden styles, specific group of plants, historical periods, ecological settings and services, and human uses of plants and gardens (Michener 2011). Two examples include St. Louis' Missouri Botanic Garden's Japanese Garden and Cornell Plantations' Herb Garden in Ithaca, New York (Figure 4.4).

Missouri Botanic Garden's Japanese Garden is a 14-acre garden representing centuries of Japanese cultural influences. Cornell Plantations' Herb Garden displays

FIGURE 4.3 Chanticleer Garden in Philadelphia, PA is a premier display garden. (Photo courtesy of Chanticleer, taken by Lisa Roper.)

FIGURE 4.4 Cornell Plantations Robison York State Garden in Ithaca, NY. (Photo courtesy of Cornell Plantations, taken by F. Robert Wesley.)

380 different types of herbs displayed in 17 different themed beds such as herbs of the ancients, medicinal herbs, culinary herbs, herbs of the Native Americans, and tea herbs. Displays showing ecological settings such as a *wetland garden*, one that includes water and land, or ecological services such as *bioswales*, ditches planted as a garden to cleanse storm water runoff, and gardens of *native plants*, plants that occur naturally and are indigenous to an area, are becoming increasingly popular at public gardens (Figure 4.5). The Ladybird Johnson Wildflower Center in Austin, Texas, has become a model for showcasing the use of plants native to Texas in their display gardens as well as in their landscape gardens which include examples of home gardens in traditional, formal, and naturalistic styles.

Demonstration, trial, and research gardens are other popular types of displays. *Demonstration gardens* showcase something in particular that is often of interest to the garden visitor. These can include specific types of gardens such as container, backyard vegetable, and butterfly gardens as well as specific gardening techniques such as *espalier*, the training of fruit trees along a wall. Another popular technique is to demonstrate a *raised bed garden*, one in which a frame is built on top of the ground and soil is added and plants grown. In *water conservation gardens* demonstrations can include drought tolerant plants, different methods of irrigation, or ways to water plants, and examples of *mulch*, a layer of material applied on top of the soil and around plants; the intent with these gardens is to show visitors how they can conserve water in a garden. In *trial gardens*, public gardens often partner with a nursery or grower to trial or try out and evaluate new plants. One popular program is the "*All-American Selections*" (2014) plant program, a nonprofit organization that tests new varieties of plants and only introduces the best performers based on criteria such as new colors, flavors, and forms, improved pest and disease resistance,

FIGURE 4.5 A bioswale garden designed to cleanse storm water runoff. (Photo courtesy of Cornell Plantations, photo taken by Sonja Skelly.)

bloom time, total yield, and length of flower. Gardens display these new plants for visitors and professionals to observe and garden staff evaluate the plants and provide feedback to the growers on how the plants perform in that garden's region. *Research displays*, often at university public gardens, but not exclusive to them, offer researchers the chance to assemble a collection of plants for some research purpose (Michener 2011).

Many public gardens have *conservatories*, glass buildings built for the display of plants. Conservatories such as those found at Longwood Gardens in Philadelphia, the Phipps Conservatory in Pittsburgh and the Conservatory of Flowers in San Francisco's Golden Gate Park allow gardens in temperate climates to display tropical and nonhardy plants. Conservatories allow gardens to show plants found in the tropics and to tell their conservation stories; they also give gardens the ability to display plants year-round. While tropical plant displays are the most popular collections found within conservatories, gardens such as Longwood use their conservatories for themed plant exhibits such as their orchid and desert plant displays, their chrysanthemum festival, and their extravagant and beautiful holiday displays (Figure 4.6).

Plant Conservation

With an estimated loss of two-thirds of the known 300,000 plant species by the end of the century, public gardens are increasingly at the forefront of plant conservation efforts. It is estimated that 100,000 species of plants are grown in public gardens and some of these gardens maintain 250,000 seeds in seed banks (Wyse Jackson 1999). As plant species continue to decline due to habitat degradation and loss, pests and disease, and competition from *invasive species*, pests and plants that out compete

FIGURE 4.6 The East Conservatory at Longwood Gardens. (Photo courtesy of Cornell Plantations, taken by Sonja Skelly.)

or attack native plants, gardens are working to prevent plants from going extinct. This work includes *in situ* and *ex situ* conservation, invasive species control efforts, working with conservation agencies, and educating garden visitors (Reichard 2011).

In situ conservation is work being done by many gardens to conserve plant species on-site in their native ecological habitats. An estimated 62,000 acres of natural areas are managed by public gardens in the United States and Canada. Within these natural areas, plant species that are rare or endangered are protected and cared for by garden staff. These efforts are typically done in concert with international and federal agencies such as the Center for Plant Conservation and the U.S. Fish and Wildlife Service. Worldwide, botanical gardens have conservation programs in *biodiversity "hot spot" regions*, areas that are rich in the diversity of life, that work directly with the local community and governments to protect critical plants and ecosystems (Reichard 2011) (Figure 4.7).

Ex situ conservation, an effort to conserve plant species that occur away from the site of wild origin is also practiced by public gardens. Collecting plants in the wild, documenting their source and ecological setting, and propagating them for display is one form of *ex situ* conservation. Collecting seeds for seed banking, storing seeds for a long duration, is another common tactic used by public gardens. Seeds are carefully collected and conservationists are trained to know how to select seeds, how many to collect and how many to leave for regeneration, to document, and ultimately store seeds (Byrne and Olwell 2008; Reichard 2011).

Invasive species control and education is a focus of many public gardens. Invasive plant and pest species are those that compete or attack native plant species and cause their decline. Invasive plants such as Kudzu (*Pueraria Montana* var *lobata*), a vine that engulfs natural forests in the Southeastern United States, and the Emerald Ash Borer (*Agrilus planipennis*) aninvasive pest which has killed millions

FIGURE 4.7 A Cornell Plantations' natural area managed for plant conservation. (Photo courtesy of Cornell Plantations, taken by Cornell University Photography member.)

of ash trees in North America, can devastate native plant populations. Staff at public gardens actively work to control the spread of invasive plant and pest species in natural areas as well as within display garden areas. This is done through removal, protection, and ongoing research to develop measures of control of the plant or pest. Public gardens often have invasive species policies that follow *The St. Louis Declaration: Codes of Conduct for Public Gardens* that dictates control measures that keeps invasive species from being introduced into the garden and bordering landscapes, considers removal of invasive species from display collections/gardens, seeks to control invasive species, increases public awareness of invasive species, participates in local, regional, and national efforts to control invasive species; and follows all laws on the importation, exportation, and distribution of plant materials across political boundaries (Reichard 2011).

Public gardens find it necessary and helpful to work with plant conservation agencies and organizations to further their conservation efforts. Among the groups that public gardens work with are Botanic Gardens Conservation International (BGCI), the Center for Plant Conservation, the Convention on International Trade in Endangered Species, the Convention on Biological Diversity, the Global Strategy for Plant Conservation, and the Gran Canaria Declaration on Climate Change and Plant Conservation II (Reichard 2011).

Many gardens, even those not engaged directly in *in situ* or *ex situ* conservation efforts, can use their collections to educate their visitors about the importance and necessity of plant conservation.

Education

Education has always been a hallmark role of most public gardens. From the earliest physic gardens which were set up to teach the medicinal qualities of plants to present-day gardens that are educating about plant conservation and the cultural connections of plants, education is at the heart of the mission for many public gardens. Education programs at public gardens generally fall within the following types of offerings (the type, size, and extent of programs vary from garden to garden).

Formal education comprises the structured programming offered by public gardens. In formal education programs, a teacher or program leader is present and whether the program is occurring inside a classroom or outside, there are clear learning objectives for the program. Programs offered by public gardens for students in pre-K-12th grades have increased over the years. Garden educators may take plant- and nature-based learning activities and programs directly to the schools. Many gardens also offer field trip programs that support a school's *curriculum standards* the educational expectations of each state for teaching and learning by grade level. Garden educators often partner with teachers and school administrators to develop and link garden-based programs to the curriculum to enhance the learning outcomes as well as the likelihood that schools will participate (Benveniste and Schwarz-Ballard 2011). Other formal education programs include public garden/school partnerships such as the BBG's collaboration with the New York public schools to run the Brooklyn Academy of Science and the Environment. In Ithaca, New York, Cornell Plantations' Wildflower Explorations program provides every third grade class in the

Ithaca City School District with in-class lessons about the life cycles of the region's native wildflowers. This visit is followed by a class field trip to Plantations' wildflower garden where students see and learn more about the wildflowers. Both programs are led by skilled educators in a formal setting and manner and the program's coordinator closely collaborates with teachers and school administrators to ensure the program's success for the students.

Other formal education programs include academic programs at college and university gardens which provide opportunities for undergraduate and graduate students to learn more through lessons taught at the garden or by garden staff.

Most public gardens offer *continuing education* programs such as classes, workshops, lectures, and demonstrations for audiences interested in learning for personal enrichment. These programs are often aimed at the adult, lifelong learning, but increasingly gardens are developing family (parent, grandparent, and child) programs as well. As gardens vary, so do their programs though most offerings include classes in the botanical arts, horticulture and gardening, landscape design, floral design, botany, and lectures on similar topics (Figure 4.8).

Professional training and education is another form of programming offered by gardens for individuals who seek knowledge related to or working toward a profession, typically in horticulture, landscape design, arboriculture, botanical arts, and plant conservation. Both Longwood Gardens (Philadelphia, Pennsylvania) and the New York Botanical Garden (Bronx, New York) offer professional gardening programs which prepare individuals for work in the horticulture sector. *Certificate Programs* provide in-depth instruction on particular topics and can be quite rigorous if aimed at the professional audience and less intense for the lifelong learner audience (DeBuhr 2011). The New York Botanical Garden offers Certificate Programs

FIGURE 4.8 Botanical art classes are a common class offering at public gardens. (Photo courtesy of Cornell Plantations, taken by Cornell University Photography member.)

in seven disciplines: Botanical Art and Illustration, Landscape Design, Horticultural Therapy, Botany, Gardening, Horticulture, and Floral Design. Each program has a set of core requirements, a selection of electives, and in some cases a required portfolio or internship. Students in NYBG's Certificate Programs generally take a year to complete the program, though they now offer intensive summer programs that help interested participants complete the program requirements faster (New York Botanical Garden 2014a). Some public gardens offer *Continuing Education Units* (CEU) for their programs. A CEU is a measure used/required by those in licensed professions. Certificate programs can provide the necessary professional education accreditation recommended and required by many professional associations. Professionals such as landscape designers, arborists, and golf superintendents are required to continue their professional education and gain CEU credits usually after 10 hours of participation in an accredited program. Public gardens wishing to fulfill the CEU through their programs meet standards outlined by the International Association for Continuing Education and Training (DeBuhr 2011).

Internships, offered by many public gardens, are temporary positions that provide practical hands-on learning opportunities. The majority of internships are completed by undergraduate or graduate students who are required to complete an internship as part of their course of study. For an internship to be truly educational, the program should offer learning opportunities, mentorship by garden staff, and practical guided experiences. Depending on a garden's capacity, internships can be offered in almost every department at a garden though the majority of internships tend to be in horticulture and gardening (DeBuhr 2011). The Morris Arboretum of the University of Pennsylvania offers a notable public garden internship program that provides full-time internships for 1 year. Internships are in areas of arboriculture, education, horticulture, natural lands management, plant protection, propagation, rose and flower cultivation, urban forestry, and in the study of the flora of Pennsylvania. In addition to their work experience, students participate in weekly seminars, field trips, take and receive university course credit, and staff the Arboretum's plant clinic (Morris Arboretum 2014) (Figure 4.9).

Informal education is "the lifelong process whereby every individual acquires attitudes, values, skills, and knowledge from daily experience and the educative influences and resources in his or her environment" (Connolly 2011, p. 219). Since many visitors to public gardens visit for relaxation, recreation, to enjoy and see plants, informal education can create a bridge between the educational goals of the garden and a visitor's willingness to learn and be engaged in learning. The most common type of informal education at public gardens is through interpretation. *Interpretation* is a mission-based communication process that forges emotional and intellectual connections between the interests of the audience and meanings inherent in resources (National Association for Interpretation 2014). Interpretation often starts with an overarching theme and subsequent messaging that gets conveyed through signs, brochures, plant labels, exhibits, mobile guides, tours, websites, and mobile phone applications (Figure 4.10).

Educational *outreach programs* offered by public gardens respond to the garden community's needs and usually occur in the community instead of at the garden. Outreach programs occur as a way to reach audiences where they are, overcome

FIGURE 4.9 Internships provide hands-on learning opportunities. (Photo courtesy of Cornell Plantations, taken by Jay Potter.)

barriers that prevent people from being engaged with a garden, and for social justice purposes. Such programs help a garden realize its mission by serving greater numbers of people than would otherwise visit (Lacerte 2011). Great examples of community outreach include BBG's Greenbridge program (see Case Study: Brooklyn Botanic Garden), which promotes urban greening through education, conservation, and creative partnerships and the Queen's Botanical Garden Cultural Research Program. Located in one of the most ethnically diverse counties in the United States (Queen County, New York), the Queen's Botanical Garden is looking to understand the ways in which people use plants in various cultures. They are working with the communities surrounding the garden to understand and document these plant and cultural connections. Examples of these cultural connection stories include "Hot and Cold: The Art of Traditional Chinese Medicine," "Persimmons and Papayas: The Ethnic Markets of Queens," and "Green in Memory: The Food Traditions of Greece" (Queens Botanical Garden 2014).

Research

The earliest botanical gardens were developed to support research in identifying and using plants for medical purposes. Presently, public gardens are engaged in research, either by having their own researchers and research departments or by partnering with research institutions and researchers, often at Universities. Public Gardens do support and often partner with researchers conducting a wide range of studies beyond plant science that explore the effects of gardens, plants, nature, and garden-based education on people.

One good example of a public garden research program can be found at The Chicago Botanic Garden (CBG). CBG has a robust research program both onsite

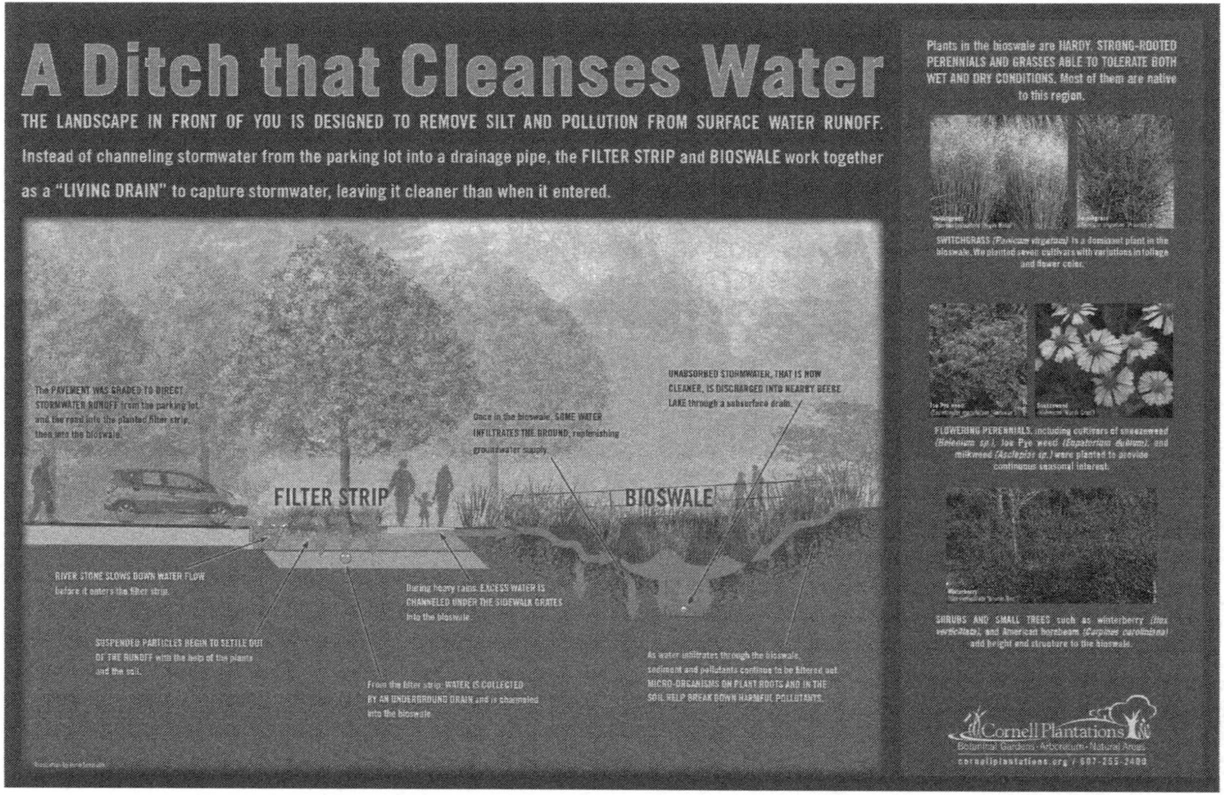

FIGURE 4.10 This graphic is the interpretive sign that was developed for Cornell Plantations' Bioswale Garden. (Photo courtesy of Cornell Plantations.)

in its Daniel F. and Ada L. Rice Plant Conservation Science Center and through partnerships with faculty at the University of Illinois at Chicago, Yale University, Niigata University in Japan, and Northwestern University. The garden is conducting or working with researchers to conduct plant science studies in

- *Biodiversity*—the study of variety of living organisms
- *Plant genetics*—the study of genes, gene variation, and heredity in plants; genes are the fundamental units of heredity in living organisms
- *Ecology*—the study of the relationship and interactions among organisms and their environment
- *Population biology*—the study of a population of organisms, how they work together, their health, growth, and development
- *Plant conservation*—conservation studies seek to understand how to protect and prevent the extinction of plants
- *Economic botany*—studies focus on the use of plants for commercial use and economic benefit
- *Ornamental plant science*—studies dedicated to improving the ornamental or aesthetic properties of plants (Chicago Botanic Garden 2014a)

Advancing knowledge of the plant kingdom through research is a central tenant of the mission of The Arnold Arboretum of Harvard University (2014). In addition to having staff scientists, the Arnold Arboretum facilitates research of its living collection of 15,000 well-documented plants representing 4000 species by an extensive network of scientists. Research includes

- *Plant morphology*—studies explore plants' physical form and external structures to compare plants
- *Phylogenetics*—studies that focus on the evolutionary relationships of plants
- *Physiology*—studies that explore how a plant functions
- Ecology (see above definition)
- Biodiversity (see above definition)

Research being conducted by public gardens tends to focus on plant science with a present day emphasis on plant conservation. Gardens such as The Missouri Botanical Garden and the Royal Botanic Gardens, Kew conduct research related to plant taxonomy, *plant systematics* (the classification of plants) and *floristics* (how plants are distributed over geography); other gardens often specialize in particular groups of plants. The most common research being conducted and supported by public gardens includes plant conservation, ecology, seed biology, plant population biology and genetics, soil science, *phenology* (the timing of natural events such as bloom time), *ethnobotany* (the study of how people of particular cultures and regions make use of plants), horticulture, climate change, invasive species, land management, and sustainability (Cook 2006; Havens 2011). Public gardens also support research, often through partnership with University researchers, that looks at the benefits of public gardens to people. The results of some of these studies are addressed in the Section "Benefits of Public Gardens."

Historic Preservation

Another role of some public gardens is in the historic preservation of a *historic garden*, one that has historical or artistic significance. A historic public garden is a place where the public can visit a historic cultural landscape that is being preserved for such historical relevance. At these gardens there is usually some defining feature, character, and style associated with the historic site. The U.S. Interior Department keeps the National Registrar of Historic Places and gardens that seek this designation adhere to professional preservation standards outlined by the *Secretary of the Interior's Standards for the Treatment of Historic Properties*. Historic gardens can include, but are not limited to, estate gardens (Barnes 2008). Examples of historic and estate gardens include notable places such as Monticello in Charlottesville, Virginia, The Fells in Newbury, New Hampshire, and Bartram's Garden in Philadelphia, Pennsylvania.

Monticello is the home and gardens of Thomas Jefferson located in Charlottesville, Virginia. Both the home and the gardens are open to the public. Monticello is famed for its display of Jefferson's botanical research into the ornamental and usefulness of flowers, the over 330 vegetable varieties planted in the Vegetable Garden, the fruit varieties, and the unique landscape features leading up to and surrounding Jefferson's home (Monticello 2014).

The Fells, in Lake Sunapee, New Hampshire, is the former summer estate of John Milton Hay, a private secretary to Abraham Lincoln during the Civil War. The property includes historic structures, historic gardens, and a surrounding forest, all of which are part of a larger 164-acre National Wildlife Refuge. The Fells' notable garden features include a walled garden, rock garden, perennial border, rose terrace, lawns, and a pebble court all associated with the development of artistic social summer colonies in New England (Barnes 2008).

John Bartram, America's first botanist, established Batram's Garden, a 45-acre garden in 1728 in Philadelphia, Pennsylvania. Bartram systematically began gathering and growing the most varied collection of North American plants in the world on the property. He began selling seeds and plants around the world and his home and garden was visited by Thomas Jefferson and George Washington among other notable figures. Today the garden, including the living collections, parkland, wildlife habitats, tidal wetlands, and a reclaimed meadow, is being actively restored to showcase the collection of plants collected, grown, and studied by the Bartram family from 1728 to 1850 (Bartram's Garden 2014).

BENEFITS OF PUBLIC GARDENS

The benefits public gardens provide have greatly expanded from their earliest days of providing a learning laboratory of medicinal plants for clergy. Today they are centers of global plant conservation and environmental stewardship, they advance knowledge of the plant world and its many associated disciplines, they promote social justice, enhance the well-being of their communities and the individuals who visit them, provide engines for economic development, and provide places of beauty.

Plant Conservation

Easily one of the greatest benefits public gardens provide is their work protecting plants around the world. Through work at their home institutions, in locations around the world, and in concert with plant conservation agencies and organizations, public gardens are seeking to conserve the estimated 300,000 plants known in the world. BGCI supports a view that plant conservation is directly linked to human well-being and that to conserve plants and biodiversity is to improve human well-being. They contend that by conserving plants, public gardens can improve nutrition, improve health care, help alleviate of poverty through income-generating opportunities, and provide a wealth of social and community benefits (Waylen 2006).

Improving Nutrition

Plants provide food and human well-being depends on a reliable food supply. According to the Food and Agriculture Organization of the United Nations, there are links between sustainable management of natural resources, plants, and ecosystems and the ability to reduce hunger. Plant diversity provides the genetic resources such as disease resistance, improved yields, and adaptations to climate conditions that can improve food crops (Waylen 2006).

Worldwide, especially in developing countries, public gardens are teaching people and communities how to grow their own food at home or in community gardens. In addition to teaching proper cultivation techniques, public gardens provide ongoing advice and support. Gardens' research and education programs provide residents with information about food plants that grow well in local regions along with specific cultivation techniques to improve yield and quality. Programs also share information about proper nutrition and healthy eating.

The Missouri Botanic Garden's Center for Conservation and Sustainable Development has collaborated with indigenous groups in Peru to help them establish vegetable and fruit tree gardens that school teachers and children manage. They also experiment with vegetables and fruit trees not previously grown by the groups (Waylen 2006). In Richmond, Virginia the Lewis Ginter Botanical Garden partnered with the Central Virginia Food Bank, Richmond's Community Kitchen and Meals on Wheels organizations to create the FeedMore program which provides fresh home-cooked meals for at-risk youth and homebound adults (Gough and Accordino 2012). In 2012, the Lewis Ginter Botanical Garden harvested and donated 12,454 lbs of vegetables to FeedMore's Community Kitchen which produced ~9580 meals (FeedMore 2014). The Rio Grande Botanic Garden in Albuquerque, New Mexico, runs a 10-acre Heritage Farm which recreates a 1930s New Mexico farm and is used to grow food and help the Garden promote the importance of growing produce. Heritage Farm showcases sustainable gardening practices and extols the virtues of supporting local agricultural economies as well as the benefits and enjoyment of fresh produce (Gough and Accordino 2012).

Improving Healthcare

In many parts of the world, people depend on plants as their primary sources of medicine. In Africa, an estimated 80% of the population uses medicines that incorporate plants. Herbal preparations account for 30%–50% of the medicines consumed in China and in Europe. In North America and other industrialized countries, over 50% of the population have used plant-based medicinal products (WHO 2003). Public Gardens often offer programming, demonstrations, and community-based plant conservation programs aimed at the use of plants for healthcare. For example, The Nature Palace Botanic Garden in Uganda works with a community group to teach participants about the cultivation and conservation of medicinal plants through workshops, field experiences, demonstration gardens, and handouts. The Royal Botanic Gardens, Kew is also working in partnership with organizations in Uganda to develop the use of plants to treat HIV/AIDS. The Calicut University Botanic Garden in South India is working to conserve Indian Zingiberaceae, Indian ginger, which is one of the most popular herbal medicines. The Garden is educating residents about sustainable harvesting so that not all of the plants are taken at harvest, establishing home gardens to grow ginger, and the development of nurseries for residents to sustainably grow and then sell the plant for profit (Waylen 2006).

Income Generation

Millions of people around the world depend on plants for their incomes. Many public gardens' plant conservation efforts are simultaneously working on economic improvement projects as plant conservation and the alleviation of poverty are linked. Programs that emphasize the use of plants for supporting livelihoods are an important benefit provided by public gardens. Educating and empowering people to grow and sell plants and plant products is the most common avenue for income generation. Earth Botanic Garden at Earth University in Costa Rica offers workshops on the development of products from plants such soaps, teas, creams, ointments, and agricultural insecticides which students sell for a profit (Waylen 2006). The Chicago Botanic Garden's Windy City Harvest Apprenticeship Program is a hands-on training program in organic vegetable production designed to prepare students with the skills needed to earn jobs in urban agriculture and green horticulture industries. The program typically serves people who are changing careers, young adults previously incarcerated, and people who have some type of barrier to employment. Nearly 90% of the program's graduates are employed full-time (Ciaccio 2013).

Social Justice

Increasingly public gardens are working on efforts to address social justice and affect social change. *Social justice* is the view that everyone deserves equal economic, political and social rights, and opportunities. Very often *environmental degradation*, the depletion and destruction of the environment, and *social inequity*, the unequal opportunities for people with different social positions and wealth are linked. Public gardens, which seek to safeguard the environment, are increasing their role in addressing pressing social issues. They are reaching out to under-represented and disenfranchised populations and engaging them at their gardens or in activities in their communities (Waylen 2006).

The Great Day Out program offered by The Eden Project, a public garden in Cornwall, England, provides an opportunity for people who are often excluded from society—young offenders, homeless people, those suffering from physical or mental disabilities, and refugees—a chance to visit the garden. Participants are given a personalized tour of the Garden; they go behind the scenes to see how the garden operates and meet employees, and they participate in a hands-on activity. Eden staff continue to work with participants through partnerships to support projects that result from the visit such as garden projects, exhibits, and storytelling workshops in shelters. Participants indicate that the visit was enjoyable to them, that they wish to become more involved with Eden, that they intend to take positive action, and that they felt happy, inspired, and normal (Eden Project 2014a).

The Green Youth Farm is one of Chicago Botanic Garden's (CBG) programs aimed at serving high-risk youth as one means to help address the well-being of communities in Chicago struggling with poverty. The program engages high-risk students ages 13 to 19, connects them with caring adult mentors, and provides them with knowledge and skills about organic agriculture, community service, healthy eating habits, and entrepreneurship. With these skills, students are empowered to give back to their communities by providing local, fresh produce as well as local green spaces for residents to enjoy. CBG has four Green Youth Farms throughout Chicago and works with ~80 to 90 youth from low income communities. The staff and community mentors of the Green Youth Farm program provide students with respect, skills for conflict management and problem solving, and team building. The goals for this program are to "operate a social enterprise that creates training and job opportunities for hard-to-employ youth adults in an economy that is putting increasing emphasis on local production, remedies for food insecurity and poor diet, and models that catalyze long-term economic development for poor communities" (Benveniste 2010, p. 10; Chicago Botanic Garden 2014b).

Horticultural Therapy (addressed in Chapter 7 Horticulture Therapy—Gardens for Special Populations) is offered by public gardens as a way to make their gardens and programs accessible to people with disabilities. Garden staff, recognizing the therapeutic benefit of working with plants, develop programs that provide opportunities for disabled visitors and program participants to interact with plants. The Tucson Botanical Garden (TBG) in Arizona started their horticultural therapy program in 1982 when docents began leading tours for people with disabilities, mainly those who were visually impaired and for residents of local nursing homes. Through this effort, the garden further developed their programs by taking horticultural activities to the nursing homes for residents who were unable to leave the facility. In the late 1990s, TBG hired a professional horticultural therapist to develop programs that included the ongoing outreach and tour programs as well as volunteer programs for high school students with developmental disabilities, brain injury, and mental illness and a horticulture therapy training program for garden staff, healthcare facilities, and human service agencies (Niehaus and Hassler 2005).

Other public gardens offer similar programs. The Chicago Botanic Garden is recognized as a world leader in horticultural therapy at its garden and for facilities in

Chicago. They provide direct services at off-site facilities, using horticulture therapy to help improve the health of clients through active engagement with plants. In their Buehler Enabling Garden, a fully accessible garden, they provide activity sessions for visiting groups and have programs for students with special needs (Chicago Botanic Garden 2014c). The Enabling Garden is also a teaching garden that promotes *universal design*, the design of products and spaces, so that they can be used by the widest range of people possible (Universal Design 2014), and provides tools, equipment, and techniques to engage people of all abilities. They also offer a Horticultural Therapy Certificate Program to train individuals, often healthcare professionals, interested in learning how to integrate horticultural therapy into a therapy practice (Chicago Botanic Garden 2014c). Other public gardens that offer horticultural therapy programs include the North Carolina Botanical Garden (Chapel Hill, North Carolina), the Denver Botanical Garden (Denver, Colorado), the Missouri Botanical Garden (St. Louis, Missouri), the Birmingham Botanical Gardens (Birmingham, Alabama), the Cheyenne Botanic Gardens (Cheyenne, Wyoming), Naples Botanical Garden (Naples, Florida), the Dallas Arboretum (Dallas, Texas), The Frelinghuysen Arboretum (Morristown, New Jersey), Guelph Enabling Garden (Guelph, Ontario), Minnesota Landscape Arboretum (Chaska, Minnesota), The New York Botanical Garden (Bronx, New York), The Royal Botanical Gardens, Ontario, and the Coastal Maine Botanical Gardens (Boothbay, Maine) (American Horticulture Therapy Association 2014).

Knowledge

Public gardens provide a myriad of opportunities for people to learn about the plant and natural world. Education is usually a hallmark of most public gardens' mission and though the size, scope, and style of program varies dramatically, education is offered in some capacity. In a study of museum goers who enjoy visiting public gardens, nearly half of those surveyed indicated that they were curious and loved to learn (Wilkening and Chung 2011). Whether through an off-site formal program such as a school visit by a public garden educator to teach third graders about wildflowers, an on-site interpretive brochure explaining a garden's features, or a community outreach program teaching organic gardening techniques, gardens are sharing their knowledge of the plant and natural world in ways that can have significant impact.

Some of the most significant impacts public gardens are having on increasing knowledge is through their youth outreach programs. A majority of gardens offer some programming aimed at young children (Figure 4.11). Increasingly, gardens are also developing programs for the youth and families. With evidence strongly supporting the positive impacts of school gardening programs (see Chapter 2: Children and Nature), many gardens are working directly with schools to help teachers plan and install school gardens as well as helping them connect the garden to their curriculum standards. BBG's Brooklyn Urban Gardener volunteers have strong ties to local schools and often help to develop school gardens (Gough and Accordino 2012). Beyond school gardens, public gardens often have programs on-site and off-site at schools that support schools' curriculum. An excellent example of such programs

FIGURE 4.11 Most public gardens offer programs designed to teach youth about plants and the natural world. (Photo courtesy of Cornell Plantations, taken by Sonja Skelly.)

can be found at the Dallas Arboretum in Texas, which serves 100,000 school-aged children every year. Recognizing that it was becoming increasingly expensive for schools to travel to the Arboretum for programs, staff at the garden developed a robust offering of programs they take to the schools all of which are tied to the Texas Essential Knowledge and Skills (TEKS) curriculum standards. With the "Nature Naturally" pre-K-6th grade program, Arboretum staff bring nature into the classroom with lessons about seeds, trees, frogs, plants, insects, earth systems, and ecosystems. Their "Science on the Grow" program also focuses on life science lessons on topics such as seeds, trees, butterflies, bees, ecosystems, and life adaptations. This program is a precursor to class field trip to the Arboretum as part of the Arboretum's "Adventure Backpack" program. Classes visiting the garden receive a backpack that is full of materials (hand lenses, prisms, journals, and a teacher's manual) for 30 students to have a fun, interactive, and educational visit (Dallas Arboretum 2014).

Increasingly, public gardens are developing programs to better serve teenage audiences primarily through a range of authentic and engaging environmental education programs (Klemmer and Skelly 2006). One of the most notable of these programs is the Fairchild Challenge Program offered by the Fairchild Botanical Garden in Miami, Florida. This annual free program addressing Florida's school standards for middle and high school students aims to "foster interest in the environment by encouraging young people to appreciate the beauty and value of nature, develop critical thinking skills, understand the need for biodiversity and conservation, tap community resources, become actively engaged citizens, and recognize that individuals do make a difference" (Lewis 2011, p. 202). The program offers a multitude of challenges that students and schools can choose from to accrue points, schools that surpass 800 (middle school) or 1000 (high school) points receive special recognition. The challenge options are *interdisciplinary* (combining one or more subject) including things such as creative writing, writing and performing songs and poetry, designing solar devices, clothing design, video production, debates, dance, and biodiversity inventories of backyards. The program has had enormous success in Miami and has been adopted by public gardens in Illinois, California, Utah, Pennsylvania, and Costa Rica (Lewis 2011). Survey results of Miami Fairchild Challenge participants indicate that more than 70% of participating middle and high school students are able to discuss the importance of plants and agree that they know of ways to help and are willing to preserve the environment (Fairchild Tropical Botanic Garden 2014).

Providing opportunities for lifelong learning through continuing and professional education programs offers a range of benefits from enrichment to career advancement. Participation in public gardens' continuing education programs is one of the most popular ways that gardens' adult audiences gain enrichment. This occurs through programs aimed at helping them learn how to better care for their home gardens or through the attainment of a new skill such as outdoor photography. For many participants, these programs offer more than just knowledge, they are also ways to have social interaction and to improve their lives (DeBuhr 2011) (Figure 4.12).

Stress Reduction

Many studies have shown that nature has an impact on psychological well-being (Wells and Evans 2003). Public gardens are places that invite visitors to be in nature,

FIGURE 4.12 Tours are a key component of many public gardens' continuing education programs. (Photo courtesy of Cornell Plantations, taken by Kevin Moss.)

interact with nature, learn about nature in all its many forms. In a study of visitors to the Brooklyn Botanic Garden and the New York Botanical Garden, visitors indicated that their most important reasons for visiting the garden were for relaxation, stress reduction, and inspiration. This same study showed that visitors had perceived decrease in their stress after their visit (Bennett 1995). A study of visitors' blood pressure before and after a visit to The Wichita Gardens (Wichita, Kansas) showed a significant decrease in visitors' blood pressure indicating that the visit positively affected physical stress levels (Owen 1994). A study of the components of stress as they related to a public garden visit showed that a visit to a public garden was an effective *coping strategy* for stress or a way of managing the effects of stress. Those visitors who most needed a coping strategy for stress benefited the most from the public garden visit (Kohlleppel et al. 2002).

Improved Physical Health

Public gardens also provide opportunities for improving individuals' physical health. Public gardens, especially those in urban environments, are places for visitors to take walks or run for exercise. Recognizing this benefit, public gardens often promote the use of their paths and trails for the express purpose of exercise. Some will post routes and associated mileage on signs or maps, while other gardens offer running tours of their grounds. Staff and docents of Cornell Plantations offer a series of guided 5K running tours each year. The tours are well attended and participants remark on how rewarding it is to exercise in a beautiful location while also learning something new. Gardens also offer programs aimed at improving physical health such as Tai Chi and Yoga practice in the garden (Kennedy 2005). Children's Gardens, whether stand alone or part of a larger garden, provide space where children can be physically active. The Eden Project hosts a weekly walking program for people with the breathing ailment, Chronic Obstructive Pulmonary Disease. The program is a collaborative effort between the Eden Project and local healthcare providers to help improve the health of participants by providing a safe but challenging environment as well as a means for social interaction (Eden Project 2014b).

Many public gardens in the United States participated in First Lady Michelle Obama's "Let's Move" campaign to combat the epidemic of childhood obesity (Let's Move 2014). Programs include Kansas City's Powell Gardens' Growing Healthy Snacks program in which students learn how to make smart food choices and experience the benefits of outdoor exercise. Participants in Cornell Plantations' Let's Move Family Hike tallied 325,156 steps or the equivalent of 149 miles as they collectively hiked through the garden's natural areas (American Public Garden Association 2014b) (Figure 4.13).

Healing

Public gardens provide places for quiet solitude, reflection, and contemplation. In urban settings, public gardens provide respite from the urban stressors of noise, visual stimulation, and lack of nature. Public gardens are often places for spiritual reflection and healing as well as remembrance of loved ones (Heffernan 2006). After

FIGURE 4.13 Families participating in a "Let's Move!" program designed to encourage physical activity at a public garden. (Photo courtesy of Cornell Plantations, taken by Cornell University Photography member.)

the September 11, 2001 attacks in New York City and Washington, DC, many of the public gardens in those cities saw a dramatic increase in attendance as people sought out solace. These gardens often waived their admission fees and extended their hours in order to accommodate the need of people to seek out nature, beauty, and to grieve during the national tragedy. The Meadowlark Botanical Gardens in Vienna, Virginia (just outside of Washington, DC), where many of the federal workers reside, opened on September 12 and the days that followed and had a steady stream of visitors to the garden. The director, Keith Tomlinson, believes that the collective mourning that occurred in the garden helped residents cope with the traumatic events of September 11 (Tomlinson 2003).

To help local residents dealing with the loss of babies and children, the Ithaca (New York) Perinatal Loss support group partnered with the Ithaca Children's Garden to build a Bulb Labyrinth Memorial Garden. Since labyrinths have been used for centuries for meditation and to help foster healing, the group along with garden staff chose to plant a bulb labyrinth since bulbs symbolize the cycle of life. The goal for the garden is to provide a beautiful space where families suffering from the loss of a child can remember their child while celebrating lives that carry on (Ithaca Children's Garden 2014) (Figure 4.14).

Improving Urban Environments

Public gardens, especially those in urban areas, provide oases of gardens, nature, and green space nestled among dense buildings, concrete roads, and the hum of city life. In addition to providing respite from these urban environments, public gardens provide important *ecological services*, those benefits society gains from nature.

FIGURE 4.14 The Bulb Labyrinth Memorial Garden at Ithaca Children's Garden is a place for grieving families to remember their child. (Photo courtesy of Ithaca Children's Garden, taken by Erin Marteal.)

These services include providing food and water, regulation of flooding, disease, and climate, support of pollinators, soil health, decomposition and nutrient recycling, as well as cultural and recreational benefits (Millennium Ecosystem Assessment 2005). But public garden leaders recognize the importance of taking their services beyond their garden's gates and into their communities for urban greening and revitalization.

The New York Botanical Garden's "Bronx Green-Up" initiative provides the residents of the Bronx with tools to transform vacant lots into community gardens. These gardens not only provide natural green spaces and places of beauty, but they also help to stabilize the neighborhoods they are in by bringing residents together to work and socialize together. The community gardens provide places for residents to grow food and flowers for themselves, their neighbors, or to sell at community markets. They are places where residents of all ages come together to learn from one another. The NYBG contributes landscape design assistance, gardening advice, plants to grow, hands-on workshops, and a Harvest Festival which brings gardeners together to learn, share, and socialize (New York Botanical Garden 2014b) (see Chapter 3: Community Gardening for more).

Similar community greening efforts are occurring in cities across the United States. The Franklin Park Conservancy in Columbus, Ohio, is helping to create innumerable community gardens, school gardens, and community beautification projects throughout the city. This effort, part of the Conservancy's "Growing to Green" program has assisted in the start up of more than 250 gardens since the program started in 2000. In addition to assisting and facilitating the conversion of vacant lots into community gardens, FPC has developed "The Community Gardening Resource Guide" a comprehensive collection of information about community gardening for communities. They have also developed a train-the-trainer program designed to teach community residents leadership skills in order to sustain community involvement in the gardens from year to year. These trained community leaders learn about gardening, community engagement, leadership, nutrition, and wellness; school administrators and teachers who oversee school gardens also participate in the program (Dawson 2010; Franklin Park Conservancy 2014).

Environmental Stewardship

Public gardens are also becoming models of environmental stewardship. Many gardens have display gardens and educational programs aimed at showing visitors how to be better stewards of the environment. To showcase stewardship tactics, gardens are embracing the sustainability movement by constructing *LEED* (Leadership in Energy and Environmental Design) buildings. Such buildings achieve sustainable design by aiming for efficiencies in energy and water use, selection of sustainable products, promotion of waste reduction, smart site use, health of users, and design innovation (United States Green Building Council 2014). When erecting new buildings, many public gardens are opting to build LEED facilities and to share information about their sustainable design with their visitors. The Phipps Conservatory in Pittsburgh has one of the most notable LEED buildings among public gardens, achieving a Net Zero Energy Building and LEED Platinum Certification, the highest possible LEED rating with the construction of their Center for Sustainable Landscapes. This building that generates 100% of its own energy, treats and reuses all the water on-site, and is surrounded by sustainable landscapes is meant to inspire and teach visitors about sustainable building and landscape design (Phipps Conservatory 2014).

Public gardens are contributing significantly to the sustainability movement through partnerships with the American Society of Landscape Architects to develop guidelines and benchmarks for the Sustainable Sites Initiative (SITES). *SITES* is similar to the LEED program for buildings but instead focuses on sustainable landscape design. Sustainable landscapes reduce the demand for water, reduce stormwater runoff, provide habitat for wildlife, reduce energy consumption, improve air quality, improve human health, and increase outdoor recreation opportunities (Sustainable Sites 2014). The United States Botanical Garden in Washington, DC along with the Ladybird Johnson Wildflower Center at the University of Texas in Austin have developed *Landscape for a Life*, a program that teaches about sustainable home gardening. The main effort of the program is the train-the-trainer program, a program to teach educators how to teach the principles of the SITES to homeowners (Landscape for Life 2014).

Economic Development

Public gardens are sources of economic growth for the cities and regions where they exist. They do so by attracting tourists, contributing to a region's *tourism industry*, out-of-town visitors spending money at local establishments such as restaurants, hotels, and stores. Gardens elevate a city or region's cultural standing by being places to see art, enjoy high quality gardens, and participate in educational programs. As discussed previously, gardens often work within their communities on beautification efforts which often lead to improved economic conditions. Lauritzen Gardens set out to be a driver of economic development in Omaha, Nebraska when it was established in 1999. To achieve this goal they focused their earliest efforts on developing high-quality visitor experiences. Their gardens were designed to showcase regionally appropriate gardening practices while maintaining a high standard of excellence. They developed robust educational opportunities for all visitors with a special focus on family programming to attract younger audiences. The Garden credits itself with Omaha's economic boom, which includes attracting new people to the city as well as keeping their own talented community members in the area (Crews 2006). The Coastal Maine Botanical Garden helped to develop the local economy of Boothbay, Maine. As one of the few waterfront public gardens in the United States, the garden has increased the tourism to the region (Heffernan 2006). Many of the community gardening programs supported by public gardens (discussed previously) also provide means for a community's economic development (Gough and Accordino 2012).

Beauty

Beauty, respite, and personal enjoyment are among some of the common responses visitors give when asked why they visit public gardens. Public gardens are able to produce beautiful gardens on a grand scale and provide places where visitors can enjoy and appreciate the beauty of the natural world. Such beautiful places allow visitors to feel elevated, dignified, and perhaps even noble. Beauty can provide a refuge for people from present day stressors such as an urban environment, media oversaturation, and environmental degradation. Public gardens are places where visitors can reliably see the beauty of the natural environment in every season and they are often the best places to showcase a region's natural beauty and cultural heritage (Heffernan 2006; Lewis 2006).

CASE STUDY 1 Brooklyn Botanic Garden: An Orban Oasis, Brooklyn, New York

The Brooklyn Botanic Garden (BBG) started in 1897 when New York legislation reserved 39 acres for a botanic garden; the Garden has since grown to encompass 52 acres. The Garden has become an urban oasis whose mission is to "connect people to the world of plants, fostering delight and curiosity while inspiring an appreciation and sense of stewardship of the environment. In the Garden, in its community, and well beyond, BBG inspires people of all ages through the conservation, display, and enjoyment of plants, with educational programs that emphasize learning by doing, and with research focused on understanding and conserving regional plants and plant communities" (Brooklyn Botanic Garden 2014).

Within the boundaries of the Garden are thirteen specialty gardens such as the Japanese, Herb, Rock, Rose, and Native Flora gardens as well as a Children's Garden. There is also a Conservatory which houses warm temperate, tropical, and desert plants and includes a Bonsai Museum. BBG has several notable plant collections such as their flowering cherries, lilacs, magnolias, orchids, and roses.

BBG's education program is very robust with offerings for adults, families, teachers, schools, and youth. They offer a series of classes and workshops, a certificate program in horticulture, internships, family discovery programs, school visits and field trips, teacher training, and a teen apprentice program. Additionally, they partner with the New York City school system to host the Brooklyn Academy of Science and the Environment High School. The garden's resources are used to support the School's academic goals by serving as a natural laboratory for project-based learning within the garden.

BBG is renowned for their community outreach programs which fall under the umbrella of their GreenBridge program. It promotes urban greening through education, conservation, and creative partnerships. Garden staff work with block associations, community gardens, and other service groups to build a network of people, places, and projects dedicated to making Brooklyn a greener place. Several programs comprise the GreenBridge program: the Greenest Block in Brooklyn Contest, the Brooklyn Urban Gardener Program, the Community Garden Alliance, its Street Tree Stewardship program, and the Making Brooklyn Bloom annual conference.

The Greenest Block in Brooklyn Contest is an annual free contest dedicated to community beautification and greening efforts. Its purpose is to promote street side gardening, tree stewardship, and community development. Block and merchant associations along with other community groups participate and winners receive cash prizes, plants, gardening tools, and community recognition. Categories include Greenest Storefront, Best Street Tree Beds, Best Window Box, and Best Community Garden Streetscape. BBG offers off-site gardening clinics as well as on-site workshops and tours to participating groups.

The Brooklyn Urban Gardener Certificate Program supports community greening efforts by offering a horticulture training program resulting in a certificate of urban gardening. Graduates complete 35 hours of intensive workshops that cover the essentials of urban gardening and community organizing principles, 30 hours of volunteer service working with Brooklyn-based organizations to support local greening efforts, required reading and assignments, and group presentations. Upon completion of the program, the Brooklyn Urban Garden (BUG) volunteers continue to support greening projects at schools, senior centers, block associations, community gardens, and other organizations.

The Greenbridge Community Garden Alliance is a network of grassroots gardeners which promotes sustainable gardening practices. Alliance members share gardening skills through gardener-led workshops and seasonal gatherings both on- and off-site. The Alliance was developed as a way for community gardeners to learn, communicate, and forge personal connections while gaining horticultural assistance. Each spring, BBG gives plants not sold at their plant sale to the members of the Alliance.

Given the importance of street trees in urban environments, BBG offers a Street Tree Stewardship program for residents who are interested in taking proper care of the trees in their neighborhood. Stewards receive free training, tools, and a permit from the New York City Parks Department to work on street tree beds. The goal of this program is to educate people about the proper care of trees for their beauty, health, and survival.

Each year BBG kicks off the spring gardening season with its Making Brooklyn Bloom event featuring exhibits, workshops, networking opportunities, and speakers. Community gardeners, BUG volunteers, BBG staff, and gardening experts provide hands-on workshops about organic and sustainable gardening practices. Attendees are surveyed to identify future relevant themes and workshop topics. Past themes have revolved around trees, soil health, growing food, small space garden design, and sustainability. Community greening organizations exhibit at the event to meet and develop relationships with community members.

The Greenbridge program is, according to BBG staff, mutually beneficial to the communities throughout Brooklyn as well as to BBG. It celebrates horticulture and promotes sustainable gardening and community building while also achieving the mission of the garden. It is a means of taking BBG into the community and sharing its resources while attracting new visitors to the Garden and gleaning knowledge from urban gardeners "on the front lines." Greenbridge staff remark how effective the program is at promoting sustainable gardening practices as well as teaching and helping people become leaders in their communities. The program empowers people of all ages, ethnicities, socioeconomic status, and gender found in the Brooklyn community. The BUG program, for example, teaches residents how to identify community assets (where participants can find soil, timber, plants, and technical expertise), develop their community network, practice peaceful coexistence, and conflict resolution and seek out funding for their gardens. By gardening and taking care of the plants, gardens, trees, and each other, Greenbridge is helping connect people to each other, to plants and nature, and to a deeper sense of place. It is also helping participants understand horticulture and ecology and how their actions have positive effect. When asked about the programs' effectiveness, BBG shared these quotes from the programs' participants, "Plants help me to breathe. Gardening increases property values. There is less trash. The trees keep us cool, help us commune with nature. Gardening is fun. It's therapeutic. It connects children with adults. It encourages respect and caring for our block. It is a spiritual endeavor. It helps us slow down. The Garden is a jewel and sanctuary. The Garden helps us get out of the city—provides relief." In an urban environment, BBG is truly an urban oasis, using its resources to connect people and plants within the Garden's walls and beyond into their community (Brooklyn Botanic Garden 2014; Browne, N. 2014, personal communication; O'Brien, M. 2014, personal communication) (Figures 4.15 through 4.18).

CASE STUDY 2 Cornell Plantations: A University's Living Laboratory, Ithaca, New York

Cornell University is home to an arboretum, botanical garden, and natural areas that collectively comprise Cornell Plantations (CP). This university public garden is located directly on the campus grounds and serves as a living laboratory

FIGURE 4.15 Community Garden Alliance members learn hands-on about integrated pest management. (Photo courtesy of Brooklyn Botanic Garden, taken by Gulshan Kirat.)

FIGURE 4.16 Brooklyn Urban Gardener students study intensive gardening using BBG's Children's Garden for inspiration. (Photo courtesy of Brooklyn Botanic Garden, taken by Nina Browne.)

FIGURE 4.17 Greenest Block contestants, young and old, organize to keep their blocks' plants well-watered. (Photo courtesy of Brooklyn Botanic Garden, taken by Robin Simmen.)

FIGURE 4.18 BUG students learn to tend compost at a local urban farm. (Photo courtesy of Brooklyn Botanic Garden, taken by Nina Browne.)

for many of Cornell's classes and research programs. Its mission is to preserve and enhance diverse horticulture collections and natural areas for the enrichment and education of academic audiences and in support of scientific research.

Contiguous with Cornell's campus, CP cultivates a 25 acre botanical garden including specialty gardens that display ornamental and practical herbs, flowers, heritage vegetables, rhododendrons, perennials, ornamental grasses, a bioswale or rain garden, specialty groundcovers, conifers, containers, plants with winter interest, and demonstration gardens for sustainability and climate change. Within the 100 acre arboretum are collections of maples, oaks, crabapples, conifers, dogwoods, urban trees, nut trees, lilacs, and pocket gardens showcasing ornamental shrubs and streamside plantings. CP also manages over 30 preserves of natural areas across 3000 acres that support habitats of rare wild plants and unique ecosystems.

As a public garden associated with a university, education is at the heart of CP's mission. Similar to other public gardens, CP provides many meaningful opportunities for people of all ages to learn about plants. A very important audience for CP is the academic community—faculty, students, and staff of Cornell University. Many of its core programs support the academic curriculum and offer multiple venues to support research and outreach. CP partners with faculty to provide a natural classroom for hands-on learning. Plant identification and landscape design students visit CP weekly to study its plants and landscapes, while students in biology and natural resources classes explore CP's campus natural areas to study ecology and insect behavior. CP's communications and education staff regularly provide lectures about interpretation and informal education techniques.

While CP does not have its own research program or facilities; it supports Cornell researchers. Studies conducted at CP have included investigations of plant and soil health, the effect of soil compaction on tree growth and vigor, climate change impacts on plants wildlife control techniques, and invasive species management.

Cornell University students engage with CP in numerous ways ranging from using the grounds for exercise, for quiet contemplation and study, for romantic strolls, for focused study, and for hands-on learning. Two key programs offered by CP for Cornell students are its summer internship program and Masters of Professional Studies program. The internship program, open only to Cornell students, is a 14-week intensive learning and working experience. Students are hired to work in the gardens, natural areas and in the education and communications departments. One day a week is set aside as a learning day for the interns to learn about skills and issues central to the work of public gardens. The remaining time is spent working alongside CP staff who mentor and teach interns about the worked they are engaged in. For individuals seeking an even more in-depth understanding of the public garden profession, CP, in conjunction with the Department of Horticulture, offers a Masters of Professional Studies in Public Garden Leadership. This graduate program provides students with learning and research opportunities focused on issues related specifically to public garden management and leadership development.

As a university public garden, CP is governed by university policies. It is considered a unit of the College of Agriculture and Life Sciences and the director of CP reports to the Dean of the College. The university association provides benefits such as financial investment of endowment funds, access to alumni donors, and benefit services such as retirement and health care. Access to faculty experts in a wide range of disciplines provide enormous opportunities for CP staff to develop high-quality education programs that serve all audiences, while also providing an avenue for faculty to participate in outreach which is often a requirement of their academic appointment or needed to fulfill a grant obligation (Cornell Plantations 2014).

SUMMARY

A public garden is an organization that is open to the public and that holds and cares for a collection of living plants for the purposes outlined in the organization's mission. The mission of public gardens often includes the use of their plant collections to educate, conserve species, and for the aesthetic display of those plants. Public gardens are governed by boards of trustees, advisory boards, or government units. They are supported by a variety of funding programs including endowments, membership programs, sponsorships, private foundations, grants, and earned income ventures. The main roles that public gardens serve are to display plants in a variety of ways to conserve plants, to provide educational programming and outreach, to conduct and support research, and to preserve historic landscapes. Public gardens provide numerous benefits to individuals, communities, and to the world such as plant conservation, improved nutrition, improved healthcare, income generation, social justice, knowledge, stress reduction, improved physical health, healing, improved urban environments, environmental stewardship, economic development, and beauty.

REVIEW QUESTIONS

1. What is a public garden? How is it like a museum?
2. Describe the three ways public gardens can be governed.
3. List the ways public gardens are funded.
4. What are the key events in the history of public gardens?
5. What are key types of public gardens?
6. What are the different ways gardens display their plant collections?
7. What is *in situ* and *ex situ* conservation? How are they different and how are they carried out at public gardens?
8. Beside *in situ* and *ex situ* conservation, describe two other ways public gardens are conserving plants.
9. Describe the difference between formal and informal education programs at public gardens.
10. Describe the similarities and differences between a continuing education program and a professional training program.
11. What are three reasons public gardens have outreach programs?
12. What are the main areas of research focused on by public gardens?

13. What is the defining feature of a historic garden and what government agency provides preservation standards?
14. Describe how public gardens benefit individuals.
15. Describe the ways public gardens benefit communities.
16. Describe the ways public gardens benefit the world.

ENRICHMENT ACTIVITIES

1. Identify the public gardens in your town, hometown, or a city you would like to visit or live in. Look at their website, read their mission, review their collections, educational programs, and events. What resonated with you, which gardens are you most excited about and why? What more would you like to learn about the garden?
2. Select a public garden and interview a staff member from each department to find out more about the garden. Ask them about their mission, their collections, their programs, what benefits they provide to their visitors and community, their fund raising strategies, and what makes them unique among other public gardens.
3. Plan a public garden. Pick a region of the world or United States and craft a proposal for developing a garden in that region. What would the mission and focus of the garden be and why? What unique plant collections would this garden have? What programs would they provide? Would they conduct research? and How would they fundraise?
4. Visit www.publicgardens.org, go to the Career Center and review the listing of current openings. Which job appeals to you? Why? How might you learn more about this position? What types of skills are needed? How might you gain those skills? Then review the listing of internship opportunities and see if any provide the skills that you would need for the job that interested you.

REFERENCES

All-American Selections. 2014. About Us. http://all-americaselections.org/about/index.cfm.

American Horticultural Therapy Association. 2014. *Botanic Gardens and Arboreta offering Therapeutic Horticulture Programs.* http://ahta.org/horticultural-therapy/helpful-links, accessed November 2014.

American Public Garden Association. 2014a. *Search Member Directory.* https://www.publicgardens.org/content/membership-central-0, accessed October 2014.

American Public Garden Association. 2014b. *Let's Move! Museums and Gardens.* https://www.publicgardens.org/content/lets-move-museums-and-gardens-0.

Arnold Arboretum of Harvard University. 2014. *Research.* http://arboretum.harvard.edu/research/.

Barnes, M. 2008. Historic garden survival: Strategically growing human, funding, physical site, and organizational resources. MPS thesis. Cornell University, Ithaca, NY.

Bartrams Garden. 2014. *Plants are Our Story.* http://www.bartramsgarden.org/plants-are-our-history/, accessed December 2014.

Benfield, R. 2006. Who are our visitors and what do they like? *Public Garden*, 21(2), 7.

Bennett, E. 1995. The psychological benefits of public gardens for urban residents. MS thesis. University of Delaware, Newark, NJ.
Benveniste, P. 2010. The city and the garden feeding the movement. *Public Garden*, 25(1), 9–10.
Benveniste, P. and J. Schwarz-Ballard. 2011. Formal education for students, teachers, and youth at public gardens. In D. Rakow and S. Lee (eds.), *Public Garden Management*. Hoboken, NJ: John Wiley & Sons, Inc., pp. 190–204.
Botanic Gardens Conservation International. 2014. *The History of Botanic Gardens*. http://www.bgci.org/resources/history/, accessed November 2014.
Brooklyn Botanic Garden, 2014. Mission and Reports. http://www.bbg.org/about/mission, December 2014.
Byren, M. and P. Olwell. 2008. Seeds of success: The national native seed collection program in the United States. *The Public Garden*, 23(3), 24–25.
Chicago Botanic Garden. 2014a. *Plant Science and Conservation*. http://www.chicagobotanic.org/research.
Chicago Botanic Garden. 2014b. *Windy City Harvest Youth Farm*. http://www.chicagobotanic.org/urbanagriculture/youthfarm.
Chicago Botanic Garden. 2014c. *Horticultural Therapy Services*. http://www.chicagobotanic.org/therapy.
Ciaccio, G. 2013. *Chicago Botanic Garden's Sixth Year of Windy City Harvest Program*. Chicago Botanic Garden website. http://www.chicagobotanic.org/pr/release/windy_city_harvest_six.
Connolly, K. 2011. Interpreting gardens to visitors. In D. Rakow and S. Lee (eds.), *Public Garden Management*. Hoboken, NJ: John Wiley & Sons, Inc., pp. 219–213.
Cook, R. 2006. Botanical collections as a resource for research. *The Public Garden*, 21(1), 18–21.
Cornell Plantations. 2014. http://cornellplantations.org.
Crews, S. 2006. Economic development grown in a Public Garden. *Public Garden*, 21(1), 12.
Daley, R. 2011. Earned income opportunities. In D. Rakow and S. Lee (eds.), *Public Garden Management*. Hoboken, NJ: John Wiley & Sons, Inc., pp. 205–218.
Dallas Arboretum. 2014. *Plan Your Adventure*. http://www.dallasarboretum.org/education/childrens-school-programs.
Dawson, B. 2010. The city and the garden feeding the movement. *Public Garden*, 25(1), 11–13.
DeBuhr, L. 2011 Continuing, professional, and higher education. In D. Rakow and S. Lee (eds.), *Public Garden Management*. Hoboken, NJ: John Wiley & Sons, Inc. pp. 205-218
Eden Project. 2014a. Eden's Great Day Out Programme. http://www.edenproject.com/whats-it-all-about/people-and-learning/social-inclusion/great-day-out.
Eden Project. 2014b. Healthy Walking Group. http://www.edenproject.com/whats-it-all-about/people-and-learning/social-inclusion/healthy-walking-club.
Fairchild Tropical Botanic Garden. 2014. Fairchild Challenge Annual Report 2013–2104. http://www.fairchildgarden.org/Portals/0/docs/Education/2013-2014%20%20Fairchild%20Challenge%20Annual%20Report-HI%20RES.pdf.
FeedMore. 2014. Lewis Ginter Botanical Garden Begins 5th Year of Garden Benefitting FeedMore. https://feedmore.org/lewis-ginter-botanical-garden-begins-5th-year-of-garden-benefitting-feedmore/.
Flanagan, C. 2011. The history and significance of public gardens. In D. Rakow and S. Lee (eds.), *Public Garden Management*. Hoboken, NJ: John Wiley & Sons, Inc., pp. 15–29.
Franklin Park Conservancy. 2014. Growing to Green. http://www.fpconservatory.org/The-Experience/Gardening-Programs/Growing-to-Green.
Gough, M. and J. Accordino. 2012. The role of public gardens in sustainable community development. Report to the Institute for Museum and Library Services. http://publicgardens.org/files/files/Sustainable%20Communities%202011.pdf.

Havens, K. 2011. Research at public gardens. In D. Rakow and S. Lee (eds.), *Public Garden Management*. Hoboken, NJ: John Wiley & Sons, Inc., pp. 272–283.

Heffernan, M. 2006. Uniquely wonderful places. *Public Garden*, 21(1), 9–10.

Hobhouse, P. 1997. *Gardening through the Ages: An Illustrated History of Plants and Their Influence on Garden Styles—From Ancient Egypt to the Present Day*. New York: Barnes and Noble.

Ithaca Children's Garden. 2014. Bulb Labyrinth Memorial Garden. http://ithacachildrensgarden.org/index.php/bulb-labyrinth-memorial-garden.

Kennedy, K. 2005. Holden's wellness program: Connecting plants and human well being. *Public Garden*, 20(2), 25–26.

Klemmer, C. and S.M. Skelly. 2006. The changing face of education. *The Public Garden*, 21(2), 8–12.

Kohlleppel, T., J. Bradley and S. Jacob. 2002. A walk through the garden: Can a visit to a Botanic Garden reduce stress? *HortTechnology*, 12(3), 489–492.

Lacerte, S. 2011. Public gardens and their communities: The value of outreach. In D. Rakow and S. Lee (eds.), *Public Garden Management*. Hoboken, NJ: John Wiley & Sons, Inc., pp. 175–189.

Landscape for Life. 2014. http://landscapeforlife.org.

Let's Move. 2014. Learn the Facts. http://www.letsmove.gov/learn-facts/epidemic-childhood-obesity, accessed December 2014.

Lewis, C. 2011. Case study: The fairchild challenge. In D. Rakow and S. Lee (eds.), *Public Garden Management*. Hoboken, NJ: John Wiley & Sons, Inc., pp. 202.

Lewis, J. 2006. Fascinating a child leads to a Steward of the environment. *Public Garden*, Issue 1, 13.

Matheson, M.P. 2011. The process of organizing a new garden. In D. Rakow and S. Lee (eds.), *Public Garden Management*. Hoboken, NJ: John Wiley & Sons, Inc., pp. 41–53.

Michener, D. 2011. Collections management. In D. Rakow and S. Lee (eds.), *Public Garden Management*. Hoboken, NJ: John Wiley & Sons, Inc., pp. 253–271.

Millennium Ecosystem Assessment. 2005. *Ecosystems and Human Well-Being: Synthesis*. Washington, DC: Island Press.

Monticello. 2014. *House and Gardens*. http://www.monticello.org/site/house-and-gardens, accessed December 2014.

Morris Arboretum. 2014. Internship Program at Morris Arboretum. http://www.business-services.upenn.edu/arboretum/ed_internships_core.shtml, accessed November 2014.

National Association for Interpretation. 2014. Mission, Vision, and Core Values. http://www.interpnet.com/NAI/interp/About/What_We_Believe/nai/_About/Mission_Vision_and_Core_Values.aspx?hkey=ef5896dc-53e4-4dbb-929e-96d45bdb1cc1, accessed November 2014.

New York Botanical Garden. 2014a. Bronx Green-Up. http://www.nybg.org/green_up/ accessed December 2014.

New York Botanical Garden. 2014b. NYBG Certificate Programs. http://adulted.nybg.org:8080/cart65/jsp/static.jsp?p=nybgcertified, accessed December 2014.

Niehaus, J. and L. Hassler. 2005. Programming for the community: Horticultural therapy at Tucson Botanical Gardens. *Public Garden*, 20(2), 20–23.

Owen, P. 1994. The influence of a botanical garden experience on human health. MS thesis. Kansas State University, Manhattan, NY.

Phipps Conservatory. 2014. Center for Sustainable Landscapes. http://phipps.conservatory.org/project-green-heart/green-heart-at-phipps/center-for-sustainable-landscapes.aspx, accessed December 2014.

Queens Botanical Garden. 2014. Queens in Story. http://www.queensbotanical.org/Education/56902/57016, accessed November 2014.

Rakow, D. 2011. *What is a Public Garden?* In D. Rakow and S. Lee (eds.), *Public Garden Management.* Hoboken, New Jersey: John Wiley & Sons, Inc., pp. 3–14.
Reichard, S. 2011. Conservation practices at public gardens. In D. Rakow and S. Lee (eds.), *Public Garden Management.* Hoboken, NJ: John Wiley & Sons, Inc., pp. 284–295.
Stanley Smith Horticultural Trust. 2014. About the Trust. http://www.adminitrustllc.com/stanley-smith-horticultural-trust/, accessed November 2014.
Sustainable Sites Initiative. 2014. http://www.sustainablesites.org/ accessed December 2014.
Tomlinson, K. 2003. Grief, loss and the healing powers of public gardens, Presenter. *American Association of Botanical Gardens and Arboreta Annual Conference*, Boston, MA. June 2003.
United States Green Building Council. 2014. What is LEED? http://leed.usgbc.org/leed.html?gclid = COTJ7IC7q8MCFZABaQodpyAA7g, accessed December 2014.
Universal Design 2014. What is Universal Design? http://www.universaldesign.com/about-universal-design.html, accessed December 2014.
Waylen, K. 2006. Botanic Gardens: Using Biodiversity to Improve Human Well Being. Botanic Gardens Conservation International, Richmond, UK.
Wells, N. and G. Evans. 2003. Nearby nature: A buffer of life stress among rural children. *Environment and Behavior*, 35(3), 311–330.
WHO. 2003. Medicinal Plants. Factsheet no 134. Geneva, Switzerland: World Health Organization.
Wilkening S. and J. Chung. 2011. Who goes and doesn't go to public gardens and why. *Public Garden*, 26 (Fall), 9–10.
Wyse Jackson, P. 1999. Experimentation on a large scale—An analysis of the holdings and resources of Botanic Gardens. *BGC News*, 3(3), 27–30.

FURTHER READING

Rakow, D. and S. Lee. 2011. *Public Garden Management.* Hoboken, NJ: John Wiley & Sons, Inc.
Raven, P. 1999. Plants in peril: A call to action. *Public Garden*, 14(4), 28–31.
Raven, P. 2006. Research in botanical gardens. *The Public Garden*, 21(1), 16–17.

5 Horticultural Displays at Zoos and Amusement Parks

Jennifer Campbell Bradley

CONTENTS

OBJECTIVES

Upon completion of this chapter, the reader should be able to

- Describe the history and importance of horticultural displays at zoos and amusement parks.
- Discuss landscape design as it relates to zoos and amusement parks.
- Investigate the uses of plant materials and landscape plantings in park spaces.
- Compare and contrast the benefits of horticultural displays for animals, people, and the environment.
- Explore the types and levels of educational programs involving horticulture at zoos and amusement parks.
- Analyze the importance of plant species conservation programs.
- Evaluate the steps involved in initiating, designing, constructing, and managing horticultural displays at zoos and amusement parks.

KEY TERMS

- Amusement park
- Animal enrichment
- Bonsai
- Conservation
- Desire line
- Desire path
- Ecology
- English garden design
- English hill and fence technique
- Environmental enrichment
- Habitat
- Ha-ha wall
- Interpretation
- Labyrinth
- Menagerie
- Moat
- Pergolas
- Theme park
- Topiaries
- Zoology

HISTORY OF HORTICULTURE AT ZOOS AND AMUSEMENT PARKS

I don't like formal gardens. I like wild nature. It's just the wilderness instinct in me, I guess—Walt Disney (Peterson, 1940, p. 56)

FIGURE 5.1 Horticulture in an elephant enclosure. (Photo by Jennifer C. Bradley.)

Plants surround people at home, indoors and outside, in towns, cities… everywhere. Where people are, there too are plants that have been cultivated, designed, planted, and maintained. In places providing entertainment, it should not be any surprise that plants will be found. In visiting an amusement park, visitors will notice rolling lawns, broad splashes of color, interesting plant materials, and grand displays that are truly impressive. Landscape plantings can beautify a park like nothing else, effectively setting the stage and fostering the appearance of natural settings within and around parks (Figure 5.1). Green spaces, plantings, and gardens provide a psychological escape from urban areas as well as a place to rest and enjoy the lovely views (Abkar et al. 2010). Beautiful gardens and landscape plantings are synonymous with these institutions. However, the look of modern zoos and amusement parks is quite different from when these parks got their start.

Animal Collections and Menageries

Zoos have been around through the centuries, stemming from people's fascination with animals and nature. Before the age of widespread travel, the Internet, or television, humans were limited in learning about and seeing unusual animals and places. People have long held a desire to see animals that are not common to their local geographical area. Early societies, including Chinese, Greek, and Egyptian have kept wild creatures as a way of satisfying human curiosity for the exotic. The excavations and images from ancient Egyptian cultures indicate that wild animals were kept in confined enclosures. The remains of hippopotamuses, elephants, and wildcats have been uncovered from historic digs as far back as 3500 BC, signifying that the collection of nonnative species did occur.

In biblical accounts, well-known collectors of exotic animals include King Solomon, Judah, and King Nebuchadnezzar of Babylonia. Alexander the Great is reported to have captured and sent animals back to Greece from his military expeditions. In medieval England, records indicate that the Tower of London once housed wild animals. There, the public could gain admittance to see these creatures for three half-pence, or by barter of food for the kept animals.

The era of exploration further brought about the capture and display of foreign animals. From the very smallest reptile to the largest of mammals, any creature was fair game. The first zoos were called *menageries* (diverse collection of wild animals on exhibition) and were the result of wealthy individuals gathering and keeping animals they found interesting. Little thought was given to the humane handling of animals because more often than not, the capture and ultimate captivity of these animals was simply for the sport, for curiosity, or for extravagant purposes. Early functions of menageries were to symbolize wealth, royalty, and power. Consider King Louis XIV who kept a menagerie at the Palace of Versailles. As history evolved, so too did the idea of collecting and confining animals for the purpose of public display. Paris had the oldest public animal collection, founded during the French Revolution.

Nineteenth century Europe was the location for the first modern zoos. These early zoos began focusing on public exhibits that were educational, and where guests could be entertained and gain inspiration. Horticultural displays and plantings were not much of a consideration. Animal confinements were small, simple enclosures that securely contained the animals, while providing a view to the public.

The world's oldest scientific zoo is the London Zoo, founded in 1828. At that time, due in part to the urbanization and tremendous growth in London, as well as a growing fascination with *zoology* (a branch of biology that deals with the scientific study of animals) and natural history, a demand for greater options for public entertainment evolved. The need for scholarly research into animal biology and behavior, along with the public interest in such facilities, led to the creation of this first modern zoo, which became an integral, cultural part of the city.

The first zoological park in the United States was the Philadelphia Zoo which opened its doors in 1874. The first public zoos in America were built on sections of large public parks (Hanson 2002) and these zoos took inspiration most directly from zoos in Europe. As public perception and environmental concerns began shifting, so too did zoos in the way they displayed these animals. Since the late nineteenth century, zoo directors have acted on the belief that the best way to display wildlife was in a natural setting (Hanson 2002). Horticultural displays and simple flower bed areas were the beginning of establishing the feel of an animal's actual, or perceived, *habitat* (the natural environment where a species or group of species lives).

Fairs and Expositions

An *amusement park* is a group of elaborate attractions for large numbers of visitors and may include live entertainment, theatrical presentations, and rides. Evolving from European fairs that were developed for people's recreation, amusement parks got their start. World fairs and expositions also influenced the development of the amusement park industry. *Theme parks* are a specific type of amusement park that is much more detailed in theme. The first amusement park that was designed with the intention of promoting a theme was Santa Claus Land, which opened in Indiana in 1946. Disneyland Park, originally Disneyland, was unveiled in California in 1955 by Walt Disney. The park was conceived after Disney visited several amusement parks in the 1930s and 1940s. Upon hiring a consultant to aid in site needs, a 160-acre area of land was purchased near Anaheim, California, with Disney himself being involved throughout the entire process. The idea for his park blended several themed areas into a single, large park. In 1971, Disney World (Walt Disney World Resort) was added near Orlando, Florida, including four theme parks and two water parks. Since Disney World's conception, the model for the amusement park industry as it is known today was firmly established.

Evolution of Horticulture in Zoos and Amusement Parks

In the 1920s, Richard Addison, an early curator of the San Diego Zoo, believed that visitors would appreciate exhibits that were naturalistic. "The Animal's home and natural surroundings are as interesting to most people as the animal itself, and to display it in an old style cage is realizing on only half its value" (Addison 1924, p. 129). Horticulture in zoos and amusement parks has continually evolved since the earliest days of the amusement park industry. Today, most modern zoos embrace *conservation* (the protection, preservation, and management of wildlife and plant species), and are focusing on research, dedicated breeding programs, and public education.

Landscape plantings have become an integral part of zoos and amusement parks. Carefully planned gardens and displays successfully screen between animal enclosures or themed spaces and also serve as a seamless transition between park areas. In amusement parks, harsh lines of rides and hardscape rails are softened with the use of plant materials. Plantings effectively conceal park infrastructure including power lines, fences, security barriers, and utilities throughout. Beautifying the park as a whole, horticultural plantings have become a valuable and expected part of these parks.

Many displays within these parks are designed to entertain through theming, such as desert, jungle, savanna, alpine, woodland, arctic, and ocean themes. The range of possibilities is endless and only help in creating a thought-provoking, comprehensive area within a park. Plants provide the visual background in supporting the impression of a certain place. Horticulturists at zoos and amusement parks are now beginning to understand the economic value of horticulture and the subsequent importance of design in creating additions that complement the themed attraction. Many amusement parks have successfully created gardens that have become attractions worthy of merit themselves, with visitors photographing interesting horticultural displays as frequently as the attractions.

As society is embracing the value of nature and the environment, so too are zoos. The desire for institutions to distance themselves from the stereotypical zoo ideology that predominantly exhibits animals behind bars has brought about relatively new terms for zoos including "conservation" and "park" zoo names. In 1993, the New

York Zoo changed its governing name from the New York Zoological Society to the Wildlife Conservation Society. With conservation, environmental awareness, and natural enclosures a forefront for zoos, there is no doubt that horticulture is playing a more significant role in changing public perception of zoos. Creating beauty and fostering an appreciation for the value of habitats and the environment is an idea that zoo managers have embraced.

DESIGN AND HOW PLANTS ARE USED

Design principles used by landscape architects help guide the overall design of parks. These strategies are invaluable in creating any beautiful and comfortable outdoor area and helping determine how a space will be perceived. Line, form, texture, repetition, balance, focal point, color, proportion, scale—these are design principles used extensively in creating beautiful, naturalistic, and extraordinary zoos and amusement parks. Landscape designers and architects use these design principles, just as all artists do, including sculptors, painters, illustrators, and architects. Design elements and principles are often overlooked as a part of horticulture, but the creative use of these tools is what landscape designers incorporate to make a park esthetically appealing.

In any landscape, plants can be used to define areas, hide a view, create privacy, provide a natural look, form habitat, control noise, and direct the speed and flow of traffic. Critical to successful designs in zoos and amusement parks is careful consideration for the large numbers of visitors the park will accommodate. Consequently, these parks are unique with small and grand spaces, where great quantities of trees, shrubs, grasses, vines, and flowers each are selected for the characteristics that they contribute. From a designer's applied-use perspective, plants are functional tools in influencing the overall vibe of such parks. There are four categories in how plants can effectively be utilized in a zoo or amusement park: (1) establishing pedestrian traffic and movement, (2) creating natural settings, (3) defining spaces and concealing infrastructure, and (4) providing interesting and fun zones.

Establishing Pedestrian Traffic and Movement

For any public space, the flow of people within that space deserves thorough and thoughtful consideration. Starting with a blank canvas, designers can draft and establish major and minor walkways and surrounding areas that flow into each space. At zoos and amusement parks, the sheer number of people trying to navigate through those spaces makes proper design paramount. It is also necessary to ensure that there are safe and enjoyable transitions between attractions. Careful understanding must be given to the overall ease of access for visitors, and effectively dispersing pedestrian traffic with minimal bottlenecks. Providing handicapped accessibility and ensuring that everyone can move safely throughout the park is critical. Proper planning for pedestrian flow around each area ensures that visitors have a chance to see the entire park, and that any given space is not overwhelmed with a mass of people. Designing for pedestrian direction and movement is an important function that is established early on by the designers and horticulturists who thoughtfully choose plants and design landscape beds that guide, direct, and slow people as they walk or funnel into attractions and various areas within the park.

For landscape designers, considering layout and the potential traffic patterns upfront results in a finished park that has less of a tendency for people to double back to visit unseen or unexplored parts of the park. For visitor enjoyment, successful transitioning is important such that each space blends into the next, appearing logical in sequence. Design techniques such as the use of color, line, form, texture, and focal point can each be utilized to spark interest and draw attention, effectively enticing visitors to move around and through the park.

The width of sidewalks, entrances and exits to exhibit areas, and overall walkway arrangements contribute to pedestrian speed of flow. Creating successful walkway transitions is necessary for safely moving large numbers of visitors through park spaces. The Federal Highway Administration (2014) has guidelines for establishing widths of walkways within public spaces. A minimum of 5 feet is the width to accommodate two people walking side by side. For some parks, the minimum is set at 8 feet. However, due to the large numbers of visitors navigating walkways, this minimum may not be adequate. Consequently, sidewalk requirements may be much larger, with widths often being 10, 20, and 40 or more feet wide. Ensuring that the walkway surfaces are safe and proportionate to the number of visitors is essential.

In a landscape, the path that people walk in the most direct route is known as a *desire line*, also called a *desire path*. This is a path created as a consequence of continued foot traffic over an area. This path may or may not be on an intended walkway. Similar to a cow path through a grassy field, these trodden trails are the shortest and most easily navigated path from one area to another. It does not matter how beautiful a landscape is designed, if the park does not have a convenient way to get from point A to point B, people will make their own trail, carving a pathway through grassy areas and over flower beds. The result continues as many more visitors follow. The consequential paths are unsightly, promote erosion, and because they are uneven surfaces, are potentially unsafe. Shrubs and plantings can prevent or reduce people from leaving walkways by reinforcing pathways, essentially serving as living borders. The use of hedges and plant material is in great contrast to the use of walls or metal fences and can be effective tools for keeping visitors on established walkways. Landscape plantings alongside sidewalks aid to reinforce the overall park design, provide beauty, and help make visiting the park more enjoyable.

Good design can also help in preventing visitors from straying into hazardous areas. Large shrubs can be effective tools both visually and physically in blocking entrance into staff-only or behind-the-scene areas. The three-dimensional line that a hedge row has essentially funnels people through spaces and makes the process of physically moving through the park one that is relaxed and pleasant (Figure 5.2).

Vegetation can be useful for providing interest when attraction wait lines are lengthy and slow-moving, making the necessary congregating areas more pleasurable. Plants and green spaces have been found to make human surroundings more

FIGURE 5.2 Park walkway with natural grasses. (Photo by Jennifer C. Bradley.)

pleasant, make people feel calmer, improve mood, and reduce stress (Stigsdotter and Grahn 2004; Abkar et al. 2010) (Figure 5.2).

Landscape plantings can be used to restrict the views of masses of people awaiting entrance into a park attraction. Plants can obscure views, giving visitors standing in line the illusion of a shorter wait. Dealing with long lines at parks is inevitable;

FIGURE 5.3 Designing to set the habitat scene. (Photo by Jennifer C. Bradley.)

however, waiting in a meandering line that incorporates landscape plantings can make the experience more bearable. For summertime crowds, large shrubs and trees strategically placed will provide shade, helping to cool areas naturally. Visitor frustration and discontentment is a concern for management at all parks. Finding solutions for making the wait less burdensome is a priority.

Creating Natural Settings

Undeniably, animals and amusement park rides are the star attraction of most parks. However, horticultural displays vividly set the scene. There can be little doubt about the value of proper design for creating a natural setting within an urban environment. For many decades, ornamental and cultivated plants have been introduced to animal displays to reduce the harsh appearance of the enclosures and to establish a natural looking atmosphere (Figure 5.3).

Landscape immersion became a prominent concept and an industry standard in the 1980s. This is a technique for enveloping the zoo visitor into the animals' environment. The goal is for visitors to feel as if they themselves are in another place (Hanson 2002). Prior to this, a visitor would stand next to an animal enclosure, looking into a setting that portrayed scenery and the included animal. Landscape immersion recreates places such that all around a display, visitors feel like they are transported to another continent and are themselves a part of the experience. Landscape immersion techniques are designed to provide a memorable, esthetically enjoyable experience.

Horticulture at zoos as well as amusement parks has evolved impressively. To create a stunning retreat, modern parks are designed by licensed landscape architects. Plants are laid out and chosen to mimic the natural environments that would be seen in a particular area or even a distant geographical region. The plants selected must, however, be safe for both the human visitor and the animal resident. Of concern is toxicity, but additional plant characteristics may cause injury or harm to guests. Designers should avoid plants with any poisonous parts, prickly or spiny vegetation, as well as nuisance plants with dropping fruits.

The use of plants for esthetic purposes is well known. However, zoos are now focusing more on incorporating plant conservation into their programs. For decades, zoos have made animal conservation a priority. More recently, zoos are initiating conservation programs focusing on rare and endangered plant species. Zoos acquire most of these plants from suppliers and other conservation groups that often include native plants and wild-origin species. To reduce pressure on wild plant populations, zoos are using growers and suppliers who have domesticated and cultivated these plant species using seeds and plant materials originally collected from the wild. This practice is a natural component of sustainability programs embraced by environmentally conscious institutions.

In designing and creating natural habitat settings, understanding the *ecology* (the science of relationships between animals and their environment) is helpful in successfully duplicating the look of specific native flora. It requires a thorough understanding of horticulture, animal and plant biology, as well as animal interactions. Of particular importance in zoos is safety for the animal within the enclosure.

Specialists related to animal nutrition, veterinarians, and animal behaviorists are consulted to ensure plant suitability, and as necessary, eliminating any potential hazards to the animals. Expert horticulturists are needed to research and run trials to determine what plants will grow and thrive. In some instances, plants that are native to environments being replicated just do not grow in a park's geographical location. It is up to specialized horticulturists to cultivate these wild plants or to find suitable replacements resembling plants from the animal's native habitat, effectively mimicking the desired plants' characteristics.

In traditionally designed residential or commercial landscapes, turf grass and precisely pruned shrubs are the norm. In contrast, zoos and amusement parks often make use of plants with more natural forms. Hardy ornamental and native grasses foster a naturalistic look and are extensively seen in savanna and prairie themes. Plant materials with rapid growth and vertical height, such as palms, offer an authentic tropical appearance and are especially suitable for concealing nearby attractions and providing background in naturally enclosing a space. Proven trees and shrubs with a fast growth rate, specific attributes and appearances, and plants which provide color and interest are also essential in zoos and theme parks.

In the design and construction phase, zoos may opt for installing large-size plant material. Although this is a financially substantial option, the results are immediate. By choosing almost fully grown specimens, the landscape appears already mature soon after planting. An additional technique for making the park appear lush is to plant vegetation much denser than necessary. Rather than waiting years for a park to have an established look, selecting mature plants or planting species tightly to eliminate gaps in bed areas are useful techniques that allow parks to have a natural, fully grown look, with interesting features beginning year one.

Plant displays are only part of the equation in recreating a wild space. Sights and sounds work together. Landscape designers understand the value in a total sensory experience. Colorful, unusual, or aromatic blooms, as well as exotic foliage certainly provide intrigue. Vibrant flowers such as sunflowers (*Helianthus annuus*) and eye-catching pot marigolds (*Calendula officinalis*) are a visual delight. Displays that encourage touching or smelling of plants can be educational and fun. Plants such as lambs' ear (*Stachys byzantina*), for its furry leaves, and lavender (*Lavandula angustifolia*), for its wonderful aroma, add to the overall sensory experience. Edible plants such as rosemary (*Rosmarinus officinalis*) and spearmint (*Mentha spicata*) are also good choices for their highly fragrant leaves.

Natural sounds assist with blocking out external noises. The mass of the plants themselves are effective tools for noise reduction. Dense plantings can serve to buffer noise from nearby attractions or necessary utilities such as air handling unit motors. Through noise reduction, a visitor is more fully immersed in the surroundings and the natural feel of the space. Often "ambient" noise is piped throughout the park and can include music, the sounds of rain, wind, and animal vocals. Water is another feature that allows for natural noise buffering. Splashing and physical movement involved in water displays contributes to the overall scenery experience. Fountains, along with natural appearing ponds and streams, provide an additional horticulture opportunity with water gardens. These unique plant displays along with the lure of a water feature make such attractions a favorite for visitors.

For overall layout and design of parks, *English garden design* is a technique that promotes a naturalistic look. English garden design is a style that originated in Europe in the nineteenth century and had a major influence on the forming of public parks and gardens. Early on, zoos began seeking methods for promoting a more natural effect by reducing bars and fences as barriers.

A prominent landscape architect, Frederic Law Olmstead, popularized this naturalistic design approach. Zoos and parks throughout the United States began using this natural style of design. English design style promotes naturalistic techniques that zoos use in abundance today. English garden design is best expressed with a natural, open layout that eliminates rows of pruned hedges and perfectly spaced trees. Olmstead's ideals focused on creating perspectives within a setting such that the sense of space appeared more expansive (National Association for Olmstead Parks 2014). Presenting an idealized view of nature with rolling lawns and extensive views out, English garden design can make a small landscape seem even larger and can be adapted to spaces within a park. English garden design style creates the perception of distance, even within the restricted confines of a zoo or amusement park setting.

English garden design influenced two practices that zoos still use today in their park enclosures: *moats* (deep wide ditch that is usually filled with water) and *ha-ha walls* (below eye-level design that creates a barrier vertically while preserving views). Moats are used in zoos because they give the impression of a naturalistic barrier. A moat is often a suitable option, and because of the elevation change or the water itself, animals remain within one area. Animals in these displays appear relatively free, and there is a clear sight line between the animal and the visitor. This allows for both an appreciation for the animal's beauty and the delight of nearness to a wild animal. As a landscape technique, ha-ha walls rely on subtle vertical dividers. Ha-ha walls (also known as the *English hill and fence technique*) incorporate a hill or slope in addition to a nonpassable ditch or other physical barrier. The resulting visual appearance is one of an unblocked view to the animals. Techniques for enclosing animals are regulated by the Association of Zoos and Aquariums to ensure the safety of people and the animals. Visitors may not readily notice the physical barrier and the below eye-level wall. The effect is a feeling of openness and being close to the animals and their habitat. Often with these techniques, visitors may actually be compelled to wonder just how the animals are separated from each other or from the visitors (Figures 5.4 and 5.5).

Defining Spaces and Concealing Infrastructures

At the simplest of uses, plants can serve as a beautiful backdrop to animal enclosures or themed areas. However, plants offer uses in so many other ways. Horticultural displays provide the backdrop, forefront, and stage to the individual zones, whether an animal enclosure or themed attraction.

Due to the sheer number of visitors to a park or zoo, defining and establishing separate spaces within a park is a challenge. Properly designed, landscape plantings create scale and balance naturally by fostering the feel of intimate spaces, such that a guest may be unaware of what lies mere feet away. Blocking extraneous views, whether beyond the park boundaries or within the park and its different themed

FIGURE 5.4 Walled enclosure limits picturesque views. (Art by Eli Bradley.)

areas, deserves careful consideration. Much like visiting a museum, visitors move through the facility, stopping to experience displays and meandering in an orderly fashion throughout. Visitors should be engrossed in the display at hand and remain oblivious to activities in other adjacent areas.

While outdoors, visitors benefit from the comfortable spaces that are made using trees, shrubs, and other plants of various sizes. Vegetation can be positioned to establish outdoor rooms, plants at all height levels lend themselves to visually mimicking structural components such as ceilings, walls, and floors. Trees or *pergolas* (vertical posts or pillars that support cross-beams with woody vines) can provide the feeling of a ceiling within a space. Shrubs and small trees are placed in the landscape to provide the visual walls to enclose a given area. Ornamental grasses, low annuals and perennials, groundcovers, and turf grass all provide a cool, natural floor. These plants provide a visually soothing green distraction, serving to soften the necessary concrete expanses.

To construct an unspoiled natural experience, plants are effective tools in hiding and beautifying necessary park infrastructure. Blocking extraneous views, including power, water, and sewer utilities, backstage and staff areas, garbage zones, roadways, and any other potential eyesores are often addressed with plant materials and large-scale plantings. Similarly, plantings of water plants as well as ferns alongside a water feature will soften necessary structures, effectively incorporating and naturalizing the working features of a constructed pond or stream. For low-level issues, thick groundcover conceals various power supplies, drainage, and sewer areas. Dense shrubs cover necessary fences, parking areas, and utilities. Large trees help block outside park views. The aim is for visitors to notice what is highlighted and should be seen. Plantings can artfully conceal the park necessities; essentially making them

FIGURE 5.5 Ha-ha wall promotes picturesque views. (Art by Eli Bradley.)

hidden in plain sight. Similarly, the experience for those people above ground level those on amusement rides, is important. The view for visitors high up who are looking down should be just as beautiful. Consequently, a thorough analysis of all parts of parks from top to bottom is paramount.

Creating comfortable human scale is especially critical at amusement parks where rides can go up several stories. Often these rides are predominantly anchored in a hardscape, with expansive metal components and support structures. The mass of metal machinery consists of harsh lines and mechanical workings that are better softened with natural materials or hidden altogether. Consider the relationship between the human who is standing next to a massive ride. Incorporating trees and shrubs at varying levels provides for transitions in height from the top of a towering structure down to the visitor's eye level, while softening the overall aspect of the imposing structure, making the space feel more comfortable.

Tall materials including massive grasses and towering trees are selected by landscape designers for their size, growth rate, and vigor. With careful design consideration, giant looming rails and rides can have an overall pleasing appearance from ground level when softened by plants that are incorporated into the same attraction area. Creating an intimate scale is important for ground spaces as well as the views as experienced at ride level. Plantings alongside rides and rails also need to be designed from the rider's view, carefully minimizing views of infrastructures so as to enhance the experience of the ride. Visitors can catch a glimpse of the top rails of a roller coaster among tall soothing green vegetation. This certainly sparks interest and mystery, as the ride appears to collide with a wall of trees. The effect of plant material alongside a ride is in stark contrast to bare rails with no surprises. While plant materials can be used to achieve a comfortable scale within parks, plants can also be used to create a perception of great height, mass, and distance. Depending on the theme or desired effect, plant materials are useful tools for park designers.

Providing Fun and Interactive Spaces

A visit to a zoo or amusement park is sure to lend itself to finding entertainment that is imaginative and fun. Parks are realizing the draw and interest that the plants themselves have in attracting and entertaining visitors. Having areas that focus on fun and whimsy is found at parks worldwide. The resulting entertaining spaces draw the crowds and provide another value to the park guest. Visitors to all parks have a diverse range of interests. Children may race from animal to animal; teens may congregate near themed rides, and plant lovers may prefer to leisurely stroll through the park identifying and taking photographs of the different plants. Parks must consider the individuality of their guests and should provide alternative spaces of interest. Examples of such spaces that have proven popular include theme gardens, display plantings, and play spaces. The popularity of landscaped displays has prompted parks to start using labels to identify plants, thus combining educational opportunities with entertainment and esthetics.

Theme gardens follow a specific idea and vary widely at different parks. Theme gardens can take on the attraction they are near. A great example of the use of theme gardens is at Disney's EPCOT. Each of the countries at the World Showcase area feature gardens depicting that country's garden style. For example, Mexico has large, colorful flowers and tropical plantings. France has highly manicured Versailles-inspired beds. China has water lilies while Norway uses white flowering plants, giving the illusion of snow. Italy has a whimsical garden made up of container gardens. Throughout these themed gardens, the delight from children and adults is apparent. Plants are often used in supporting roles in such areas but certainly have value in establishing or complementing a park themed area.

For theming, many parks often integrate *topiaries* (living sculptures made up of plants that have been clipped to develop and maintain clearly defined shapes). These creations used at zoos and amusement parks worldwide are a visually impressive and eye-catching exhibition. Topiaries are an art form made by shaping plants into discernable three-dimensional shapes, often with the appearance of sculptures. Designing topiaries has been practiced for thousands of years, having first appeared in Roman gardens. Geometric shapes such as spheres, cones, and cubes are common, although the possibilities are endless. Similarly, perfect for themed displays, animal or character shapes bring about awe and intrigue by visitors of all ages. Mazes and *labyrinths* (maze with only a single, easy path to the center) are created out of shrubs and are also topiaries. Mazes have geometric-shaped walls that create a wonderful escape and adventure. While mazes gained popularity during the Renaissance, they have been a fascinating landscape technique since ancient times.

Traditional topiaries are created by shaping trees or woody shrubs such as boxwood, holly, and yew. Plants used for topiaries are evergreen and typically woody; they have small leaves or needles, and have dense foliage. Careful training, shaping, and pruning are required over a period of time to achieve live sculptures from ordinary shrubs. Topiaries can also be formed within a few months by using quick-growing vines, such as climbing fig (*Ficus pumila*), grown on a frame or wire cage filled with a moss/soil mixture. The latter yields a quicker topiary and allows for displays to be moved and changed as needed. Additional labor, care, and maintenance are required with all types of topiary displays. As characters, animals, or geometric shapes, the themes and styles of topiaries are endless, making them a popular option.

Other theme ideas incorporate and invite local and visiting animals into the horticulture display. Butterfly gardens, hummingbird gardens, and fish ponds are all themed gardens with strong visitor appeal. All these gardens have a strong "wow" factor with the plant displays and the visiting fauna as well. Themed gardens with showy butterflies, amazing hummingbirds, unusual plants, colorful fish, visiting reptiles, and the occasional duck or goose appeal to many visitors.

Seasonal plantings are always of great interest. From giant displays of tulips in the spring, to showy chrysanthemums in the fall, or colorful poinsettias during the holidays, the impact is impressive. Large-scale displays of the season's most colorful plants bring out visitors and can become the star attraction, even during the off season. Avid gardeners often visit parks around these specialized, annual displays. Having seasonal displays does involve careful planning. Growers must have the plants ready all at once and on schedule. Often to ensure the color lasts throughout the intended season, additional replantings must occur such that tired or spent plants are quickly whisked away and replaced with flowers at their peak. Unusually hot weather, cold weather, or excessive rain can all impact seasonal displays drastically. Parks promote their seasonal displays during festivals or flower shows and from the first day of the show until the end of the scheduled event, the displays must be in top form the whole time. No doubt that the paying public expects to see the displays in perfect splendor. There are a lot of behind-the-scenes replacements that must occur to ensure that guests are able to experience the spectacular seasonal displays they are anticipating.

Play gardens that incorporate colorful and unusual plants can be a treasured destination for the smallest of visitors. Children's gardens are a wonderful play space where children can explore and immerse themselves in the displays. Zoos and amusement parks are integrating these spaces where children can pretend, play, interact, and discover. These spaces give children and families a place to stop, rest, and enjoy the park. Along with playful youngsters, adults will be intrigued with such spaces, and the result is that play gardens are often popular with children of all ages. Safety is paramount. So everything from plant toxicity to structural hidden dangers must be well-thought-out.

BENEFITS OF HORTICULTURAL DISPLAYS

There should not be any doubt that horticultural displays provide beauty to the surroundings. One of the primary reasons people visit gardens is for the appeal of immersing themselves in such splendor. Certainly the vegetation provides a natural loveliness that cannot be replicated with other materials. Beyond esthetics, there are specific benefits of having landscape plantings for the zoo animals, the people visiting the parks, as well as the environment.

BENEFITS TO ANIMALS

Take a close look at the plants in zoos. The vegetation is a whole lot more than displays for visual appeal. For animal enclosures, plants often take on an important role beyond beauty (Hanson 2002). While visitors enjoy the green settings that plant displays create, these plants and natural materials may also serve as valuable food sources for animals within the enclosures. Plants can serve as natural shelters and also provide opportunities for *environmental enrichment* (providing stimulating objects in zoo enclosures to promote natural behaviors and promote animal well-being). All these factors blend together in promoting more environmentally appropriate enclosures.

In zoo enclosures, plant materials may be used for supplemental food or as a nutritional component for a specialized diet. Whether growing inside the enclosure or simply offered and placed inside the area, plants provide and allow for animals to exhibit expected foraging behavior. These plants supplement nutritional needs in a way that animals may not obtain otherwise. Incorporating plants typical of a native habitat and normally consumed in the wild provides nutritional value while at the same time educates visitors about native habitats. Consider *Eucalyptus* species trees and koalas. Australia is a unique country with varied environments. One defining aspect in many regions of Australia is the sight and smell of eucalyptus trees. Incorporating eucalyptus in zoo animal enclosures promotes visual appreciation for one distinctively Australian plant species. Using this preferred food source, eucalyptus, including branches and foliage placed in the enclosures, allows visitors to see natural behaviors and interactions of the koalas with this particular plant species.

Another example is the native grasses used in enclosures to help exhibit natural grasslands while also providing for the diet of North American bison. No doubt these native food sources serve as more natural nutrition than the animals might get otherwise and allows for normal grazing behaviors. Larger enclosures of bison or elk showcasing tall grasses help depict prairies and the unique ecosystem that once dominated the American landscape.

Plants in an animal display serve as valuable structures within the enclosure when used to replicate natural features much like those found in specific ecosystems. Shelter from the elements is an obvious benefit and includes providing shade from the sun, shelter from the snow and rain, and wind breaks for cold temperatures. Beyond this, plants and plant structures can provide enrichment for the animals. Environmental enrichment (also called *animal enrichment*) is one of the most important aspects in an animal enclosure and is notably beneficial to the animals' welfare. Providing stimulating environments for the animals promotes species-typical behaviors and allows the animals an opportunity to have choice within their enclosure while enhancing the animal's well-being. Large logs and trees promote activities such as scratching, climbing, nesting, rubbing, and even hiding from curious onlookers. For environmental enrichment, landscape plant materials serve as a valuable resource in zoos. Enclosures should include plant materials that are varied and diverse in height and complexity to promote animal interest and visitor interest, alike.

The geographical location of zoos often makes incorporating specific plants into zoo enclosures tricky. In zoos throughout the world, the addition of aviaries, biomes, and other environmentally controlled structures provides opportunities for plant diversity and yields numerous benefits to the animals. These spaces allow for more comprehensive and appropriate depictions of ecosystems, whether arid or tropical. With birds and insects being free to fly throughout and appropriate plant species thriving within such controlled environments, the relationship between plants and animals is easily identifiable.

Plants are vital to the survival of animal species in zoos, whether for food, shelter, habitat, or environmental enrichment. The ability of the visiting public to witness these interactions helps foster an appreciation for the plight of endangered animals and their fragile ecosystems.

Benefits to People

The impression of freedom created with unobstructed animal viewing within an open enclosure helps the visitor to gain the point of view much like a field naturalist observing animals in the wild. For visitors, landscape plantings certainly yield a glimpse into the wild (Figure 5.6).

Natural beauty is soothing for anyone visiting a zoo or an amusement park. Numerous studies have been conducted illustrating the effects of horticulture and green spaces on people (Kaplan and Kaplan 1989; Heerwagen and Orians 1993; Abkar et al. 2010). The Association of Zoos and Aquariums reports that visitors

FIGURE 5.6 A glimpse into a natural gorilla scene. (Photo by Jennifer C. Bradley.)

experience a stronger connection to nature after a park visit (Falk et al. 2007). Simply visiting a green space can serve as natural refuge in an urban environment. Although visitors to a zoo or amusement park come to see the park's main attractions—animals, rides, and shows—the impact that the landscape plantings surrounding these areas have on visitors is also important. These spaces often provide diversion, or a sense of escape from urban noise and distractions.

Benefits to the Environment

Just like threatened or endangered animals face challenges for survival, so too are many plant species. As wild spaces are impacted by human activities, it is understood that the animals living in these environments are threatened. But when an animal's habitat is reduced or eliminated, the plants that grow in that habitat are certainly impacted as well. Habitat loss is a leading cause of species becoming threatened or endangered (US House Committee on Natural Resources 2014). Many wild species are directly impacted and the only hope of saving some plant populations is through conservation efforts. Plant conservation and cultivation programs help globally as well as locally. It is important to help safeguard plant species, protecting and ensuring biodiversity in wild plants. Zoos throughout the world are partnering with growers and horticulturists to grow, propagate, and reestablish plant species. In turn, these parks are designing and including such plant species in their animal collections and displays.

The North Carolina Zoological Park is just one such institution that has strong efforts in plant species conservation through its Rare Plants and Conservation

Program (North Carolina Zoological Park 2012). The program works threefold, through its on-site, regional, and international initiatives. For example, at the zoo they are working to grow and display the formally endangered native plant, Smooth Purple Coneflower (*Echinacea laevigata*). Regionally, the zoo works with private and public partners throughout the state and helps host the annual NC Rare Plant Discussion Meeting, fostering discussion on statewide rare plant conservation issues. Internationally, the zoo is working in Uganda, East Africa, with the Tooro Botanic Gardens to help preserve and restore rare and endangered plant populations.

Historically in the United States, entertainment led the list of goals for zoos. However, today's mission is firmly based on conservation, education, science, and recreation. Exhibits in zoos today successfully depict threatened habitats and those animals and plants facing possible extinction. It is this notion that since such habitats are threatened, public accessibility for viewing these species at zoos is even more valuable. The American Zoological Society is in the forefront of the industry, and has adopted conservation of the world's wildlife and environments in its highest priority (Hanson 2002).

EDUCATIONAL PROGRAMS

Botanical gardens and arboretums often incorporate permanent classroom space, ongoing lectures, community outreach, and complex scholarly research. Historically, amusement parks and zoos did not dedicate valuable real estate to educational programming. Yet, in recent years, many zoos and amusement parks have been exploring the opportunities available in offering educational programs. Widespread efforts to promote horticulture as its own attraction have advanced very slowly. While many zoos and amusement parks are strictly in the business to provide thrilling animal encounters, rides, and entertainment, some are recognizing the benefits and are embracing educational programs and the added benefit that such programs have on park perception, conservation efforts, and the park's overall profitability.

In zoos and amusement parks, education occurs in many ways. While horticulture information can be shared easily through plant identification tags and signage, many other educationally diverse programs exist. The types of educational programs include *interpretation* (an educational technique of communicating and providing factual information), guided tours, how-to classes, and horticultural internships. Educational programs vary from basic gardening to more complex horticultural lectures and curricula. The audience that these programs attract includes garden clubs, school children, horticultural professionals, Master Gardeners, garden societies, and plant enthusiasts.

At zoos, education is certainly a focus for promoting an understanding of the animals. Interpretation most often includes written information and signage that communicates about specific animals including scientific names, origin, conservation efforts, interesting facts, etc. However, plants can also serve in an education role. Written information, interesting signage, interactive displays, and the like give people an opportunity to also learn about the plants. Consider one such space, the butterfly display. The plants in these areas mimic a lush garden and are often labeled

FIGURE 5.7 Educational interpretive habitat sign. (Photo by Jennifer C. Bradley.)

so that information is given as to what species of plant the butterflies prefer at different stages of their lives. Many zoos have written information situated throughout the park that informs people about all types of plants and habitats displayed with the animals. Labeling plants is an easy and informative way to continue educational awareness for the animals and the plants that encompass their natural habitats (Figure 5.7).

One successful example of interpretation that links animals with the plants that surround their natural habitat is pandas and bamboo. Mostly all children and adults alike understand about pandas and the intimate relationship that exists with a specific plant, bamboo. Pandas are one of the most recognized but also one of the most endangered animals on the planet, with less than 2000 remaining in the mountain bamboo forests of central China (Smithsonian National Zoological Park 2014). Information about pandas' special diet and this unique plant typically go hand in hand. Undeniably, the panda is closely associated with bamboo, representing an easily recognizable relationship that exists with animals and specific plant species.

Guided tours for zoos and amusement parks are an integral part of park offerings. Clearly, the public is interested in learning more about the animals as well as park infrastructure and the management of such parks. For garden enthusiasts, the interest is on horticultural practices and gardening techniques. Parks are catering to garden enthusiasts and home gardeners who have a strong desire to learn just how such massive spaces are maintained so expertly. Behind-the-scenes and themed tours provide an additional option for parks wanting to provide a unique experience to those visitors truly interested in horticulture. Tours provide a fun opportunity for guests to learn about plant collections in the park, to see the greenhouse and nursery facilities, to understand about gardening practices and scheduling, and to talk with the horticulture staff.

Educational classes at zoos and amusement parks are taught by staff as well as invited speakers and professionals. These classes and programs allow horticulture staff an opportunity to share with their guests gardening techniques about particular subjects. Such topics vary widely and might include rose gardening, growing *Bonsai* (artistic technique of pruning and growing dwarfed trees and shrubs in shallow containers), container gardening, annuals and perennials, or turf grass management. Whether offered as a weekly programming option or as a special seasonal offering, educational classes at parks have a special niche. For public relations, offering horticultural classes allows the public to gain an educational understanding while the park promotes itself, gives back to the community, and brings additional visitors into the park.

Zoo horticulture is a growing career option within traditional horticulture programs. The field combines animals, horticulture, and people and requires a thorough understanding of the interdependence of each. Students specializing in this field of study are better prepared for jobs in zoos, aquariums, state and national parks, and nature preserves.

College interns offer a unique educational opportunity for parks. For institutions that have successful intern programs, students can certainly gain critical on-the-job experience. Many of these programs are highly sought after and are quite competitive to obtain. Key aspects for hiring interns can be beneficial for the park and include the following:

- Interns being knowledgeable in academic horticultural methods and techniques.
- Interns offering youthful interest and energy.
- Interns providing seasonal work during busy summer months.
- Interns becoming a pool for highly qualified potential full-time employees.

Providing internships is a clever method for bringing in knowledgeable staff. Interns serve as a valuable resource and are well-suited for aiding zoos and amusement parks in many potential ways.

CONSERVATION

A less-recognized impact of horticultural displays in zoos and amusement parks is that these plantings can foster in the visiting public an appreciation for conservation. Certainly, plants help create the look and feel of a particular ecosystem. This effect is often a drastic change from the local and often urban environment surrounding the park. Properly designed and maintained, an animal enclosure can take on the look of another region or another country. Incorporating landscapes within each enclosure and nearby park spaces, these animal displays have the ability to mimic another environment in such a manner that the visitor has a clearer appreciation for the animal's native habitat (Figure 5.8).

A forerunner for the idea for conservation in zoos was the Arizona-Sonora Desert Museum. In the 1950s, this zoo set a goal of educating the public about plant life, wildlife, and the scenic value of the Sonora desert. The displays were designed to establish a connection between animals and their environments. Although a small

FIGURE 5.8 Plants complement the area for a desert theme. (Photo by Jennifer C. Bradley.)

zoo, directors from around the country began looking at this program as a model for conservation (Hanson 2002).

Zoos and aquariums are taking a much more active role in conservation and in promoting conservation-based learning for their visitors (Ballantyne et al. 2007). These institutions are making possible an opportunity for many urban residents to see the marvels of natural wildlife and gain awareness through education about ecosystems and conservation. It can be difficult to be concerned about an animal one has never seen. In a zoo, visitors can intimately see many species of animals and observe the animals' personalities, see young offspring, and watch the interactions of those creatures with each other and with their surrounding habitat. These observations become a personal link in developing environmental awareness. It is the hope that visitors will be inspired to help in environmental conservation efforts, contributing financially or acting on a local or global level.

Like no other experience (including mass-media), a zoo has the ability to emotionally and mentally transport the visitor to another place in the world. Even a two-dimensional video representation of an animal does not adequately yield the same effect as seeing one up close with all the senses, sights, sounds, and smells that encompass a visit. With the animal and the plants in place, a visitor can get a real appreciation for what that animal's environment is like in its native region and country. Properly designed and presented, these parks can recreate a wild habitat using threatened native plants alongside the threatened animals. Such exhibits help teach about the interdependence of animals, people, and plants, and the impacts of habitat destruction.

Some zoos recognize the opportunity that exists for incorporating and promoting environmental awareness through signage, brochures, plant lists, and self-guided tour maps. Education is fundamental in teaching environmental awareness, responsibility, and the value of being environmentally conscious.

For the majority of zoos, conservation efforts for animal species under their care are a priority. Breeding programs and international efforts have been in place to ensure diversity of species. Animals that would not be viewed in the wild by the majority of people are presented in zoos, and conservation is achieved by means of housing threatened and endangered species. Meanwhile, programs continue behind the scenes to ensure that species thrive in zoos as well as in the wild.

Habitats that are vulnerable for animal species are also vulnerable for the plant species occupying that same environment. Plants that are threatened or endangered can be cultivated and grown for display as well as reintroduction into the wild. For plant species, partnerships among conservation groups, zoos, and botanical gardens are critical in promoting biological diversity. Cultivation and plant breeding programs on an international level are also in place to ensure the conservation of those impacted plants.

Conservation programs within zoos are beneficial for the zoo animals as well as for the plant species that live alongside them. Such programs ultimately give visitors a glimpse into a specialized habitat that would be foreign to most people.

INITIATING AND MANAGING HORTICULTURAL DISPLAYS

Landscapes at zoos and amusement parks require extensive forethought, collaboration, and business management on a grand scale. From the initial idea of building the park, many intricate steps must be recognized: park planning and design, maintenance, and overall management. Each of these steps is a fundamental component to ensuring the success of horticulture displays within the park.

Park Planning and Design

The idea of a park is born. Now what? Even before the first conceptual drawings, careful planning must occur. It all begins with site visits to existing zoos, amusement parks, and a variety of other public and private gardens. Viewing these parks and visiting with park professionals are fundamental in effectively starting the process. To appreciate the scope, scale, and undertaking of a new park, visiting existing parks, from small to large, both near and far yields potential insight. This understanding can help refine ideas as the process unfolds.

Beginning with the initial idea, the scope and design for these places require specialized landscape architects and expert design firms. With such detailed, large-scale designing, utilizing quality partnerships is essential. So much must be considered in the design stage because the results have long-term implications. A successful park layout with all the intricacies of each space is dependent on an all-encompassing design. Designers should consider the park as a whole in addition to the small and large spaces within, as well as considering the management of the park and expectations of large numbers of visitors. Zoos and amusement parks are unique in that beyond designing for esthetic beauty, they must consider if the animal enclosures are environmentally adequate. Architects and landscape designers must have an expert understanding of wild plant species and be skilled in utilizing plants native to an animal's specific geographic region. Finally, the design must ensure that varied and appropriate design techniques are used throughout the park to achieve the overall beauty that will engage the visitors, with the ultimate goal of enticing the public to want to come back.

A critical component for landscape architects to consider is the maintenance requirements that will directly result from the design, which will continue throughout the life of the park. The maintenance of these places is extensive, enduring daily wear and tear. A proper design can reduce a lot of landscape costs and labor expenses later on. Understanding the ongoing maintenance that results from each design idea is critical and must be thoughtfully anticipated in the design phase.

Managing Park Horticulture

Even with a solid, well-thought-out design, maintenance of the landscapes at zoos and amusement parks is a substantial budgetary concern. Horticulturists must continually provide maintenance and park upkeep annually, seasonally, and daily to keep the park looking sharp for visitor enjoyment. Maintenance needs vary from season to season with expected as well as unexpected factors. All parks and gardens must contend with environmental factors. These include temperatures, extreme weather, seasonal climate variations, rain or lack thereof, insects, disease, weeds, and natural plant death and decline. Overall maintenance continues throughout the seasons and includes rotational plantings of annual flower beds, pruning, mowing miles and miles of turf grass, edging, removing spent blooms, weeding, mulching, and general plant upkeep. Mowing is a substantial maintenance requirement that continues throughout the growing season, requiring scheduling once or sometimes twice a week to maintain the visitor-ready look. When the parks open each day, all trees, shrubs, groundcover, and turf areas must be maintained in top form.

Part of what makes a visit to such parks thrilling is the initial action of walking into the park, physically entering the gates and taking in the sights, sounds, and smells. Often a park will commit significant resources in setting the stage to making that initial impression one that inspires awe. To achieve this goal, the flowers, plants, and all display areas must appear flawless. The goal is to continue that initial sentiment throughout the park. One of the greatest tools for achieving this appearance is through impressive plant displays. However, having a flawless landscape directly correlates with park maintenance requirements. Routine mowing, edging, and trimming are done at night. Special accommodations such as lighting are critical to maintain and allow for proper attention to detail as well as for worker safety. Scheduling becomes crucial in ensuring that the park is ready when the first visitor enters the park.

Zoos and amusement parks must have dedicated space for greenhouses, nursery facilities, trash collection, recycling bins, compost piles, and maintenance equipment. Horticulture programs continue behind the scenes at these facilities as gardeners ready plants to go into the park. Workers must continually prepare everything that will eventually be on display, including planting container gardens, training topiaries, and growing seedlings. Additionally, these facilities serve as office space for horticulture programs where all scheduling, landscape designing, ordering, business management, and even equipment maintenance occur.

In addition to regular maintenance within the park, horticulturists are needed to identify and find solutions for often head-scratching challenges. These professionals have the knowledge for addressing pests and environmental problems. Spraying an environmentally safe chemical on weeds or choosing drought-tolerant summer flowering annuals are practices that assist horticulturists with landscape maintenance. Composting and organic gardening programs also aid in maintenance and sustainability while positively impacting visitor perception and education.

Financially, keeping a park looking extraordinary is costly. The expense involved with labor and materials for maintaining the dazzling landscape plantings is immense, and requires specific budgetary foresight and management. Thousands, if not millions, of annual flowers are planted and replanted each year. So, while it may be necessary to replant an entire display after a record rainfall or hailstorm, other factors beyond the environment may too impact the beauty of plants in the park. Environmental influences on a landscape can wreak havoc, but zoos and amusement parks have additional factors that can impact the success of horticultural plantings; these include animal effects and visitor effects.

Landscape maintenance is unique in zoos where the animals on display greatly impact the plants within the animal's enclosure. Each animal species offers different challenges for horticulturists trying to keep the space looking attractive. Consider primates climbing trees, bears stripping bark, giraffes foraging tree leaves, burrowing groundhogs, and profuse urine and excrement throughout the displays. These are just some examples of the management challenges that animals impart on a landscape. Creative solutions help, but often horticulturists simply must continually replace plant material.

People impact the landscape as well, especially with the sheer numbers of visitors moving through a park daily. Foot trails exist as people cut through landscape beds, trampling and damaging plants in the process. Understanding visitor effects on horticulture displays is essential. The lure of colorful displays often brings visitors too close to fragile bed areas. Touching, picking flowers, standing on delicate vegetation for the perfect photo opportunity are daily occurrences and just some examples of what horticulturists must anticipate. Such continual factors do impact plant growth and overall beauty.

Horticultural programs at zoos and amusement parks vary and can range from mediocre to phenomenal, depending upon park management and the resources available. The success of impressive displays requires business skills and understanding beyond basic plant biology. Managing park horticulture includes working with employees in various areas such as full time groundskeepers, growers, designers, educators, interns, volunteers, and countless additional seasonal helpers. Successful horticulture programs at zoos and amusement parks have dedicated business professionals to coordinate the financial aspects, personnel management, and logistics oversight. While horticulturists are skilled and knowledgeable in all areas of making the park look attractive, proven horticulture programs are best managed alongside business specialists. The overall success of horticulture programs in zoos and amusement parks hinges on strong management and the vigorous daily upkeep of the parks.

CASE STUDY 1 Disneyland, Los Angeles, California

Disneyland in California opened in 1955, and since its beginning, landscaping has been integral for all Disney parks. Plants are as much a part of park theming as is its historic rides including Jungle Cruise or Splash Mountain. Disney is arguably the forerunner of amusement parks promoting and fully embracing horticulture. The landscaping sets the scene in every area of Walt Disney World's theme parks. As an extraordinary standard for park horticulture, Disney's landscape innovation and overall gardening programs are paramount.

CASE STUDY 2 Epcot, Orlando, Florida

Epcot is one of four theme parks at Walt Disney World in Orlando. The World Showcase is a major component of the park where seven countries are showcased along with beautiful gardens effectively representing the cultures. The displays are stunning and horticulturally accurate. Epcot also promotes its annual Flower and Garden Festival. This event brings horticulture professionals and garden enthusiasts together for educational programs and an opportunity to experience Disney horticulture at its finest. The festival takes place in the spring when the growing conditions are optimal but also when visitor numbers are traditionally low.

CASE STUDY 3 Disney's Animal Kingdom, Orlando, Florida

Disney's Animal Kingdom in Orlando is a combination of a nature-based theme park and a zoo. Around 100,000 trees and over 2.5 million grass plants were planted when Animal Kingdom was built. Incorporating native species wherever possible, the park embraces a naturalistic plant look throughout all themed areas. Shortly before the park opened in 1998, large trees that had died were purposely left in the park among the living species. Much like a true forest or wild growth-area, plants in various stages of growth, both alive and dead, are found side by side. Done properly, this technique only adds to the illusion that the space is truly "wild." The decaying plants provide habitat for some animal species as well. In one Animal Kingdom area, DinoLand USA, the park uses plants to complement a Jurassic theme. There, some of the oldest genera of plants known grow alongside rollercoasters and children's play spaces. Ferns, cycads, and early angiosperms help set the scene and educate visitors on plant evolution.

CASE STUDY 4 Busch Gardens, Williamsburg, Virginia and Tampa, Florida

This amusement park has two parks in Virginia and Florida. Both are unique in effectively combining amusement park attractions, zoo displays, and public gardens within one park. Horticulture is a focus of Busch Gardens, where gardens are showcased throughout with lovely landscape plantings, detailed displays, and beautiful overall theming. For most visitors to this park, the visual stimulation provided by the park's attractions and performances is intense. Consequently, the horticulture must be equally impressive to be noticed (Brown 2014). Sustaining such extraordinary displays is daunting. In the spring, for example, tulip displays include over 50,000 bulbs, only to be replaced after the last frost with annual plants. Begonias, geraniums, salvia, lantana, petunias, and other annual bedding plants provide peak color throughout the summer with some 60,000 blooming plants. In the fall, these plants will once again be removed, with chrysanthemums added to take their place. Certainly, colorful displays are paramount at Busch Gardens, involving massive efforts to keep color throughout the seasons. The spectacular scenery complements the larger-than-life characters and imagination of this place. For people interested in careers or for those wanting to learn more about zoo horticulture, Busch Gardens in Virginia offers a blog which accurately details daily horticulture in the park: http://www.buschgardensvablog.com/I

CASE STUDY 5 North Carolina Zoological Park, Asheboro, North Carolina

The North Carolina Zoological Park is an example of a zoo that is involved in active plant conservation. Its Rare Plant Conservation Program along with the North Carolina Zoological Society is collaborating worldwide with growers and zoos working toward the protection and cultivation of rare and endangered plant species.

Plant Conservation in Africa is a project that continues through the North Carolina Zoological Park involving the safeguarding of plant diversity of the Albertine Rift. Located in parts of Rwanda, Democratic Republic of Congo, Burundi, and Tanzania, the Albertine Rift is of important environmental value. The Rift is unique in that more than 5,800 plant species (approximately 14% of Africa's plant species) grow in that area.

To promote conservation of these plant species, the North Carolina Zoological Society began partnering with Tooro Botanical Garden (Fort Portal, Uganda) in 2006. Through this partnership, the Zoo has provided funding and training to support the Botanical Garden's staff as well as its conservation programs. One resulting program is a medicinal plant demonstration garden, which helps to educate visitors about the healing properties and value of these local plants. The Botanic Garden promotes sustainable practices for the harvest of these plants from the wild. Since many of the local people have limited access to healthcare, understanding the medicinal properties of the local plants is valuable. The Botanic Garden is expanding its exhibits, adding an arboretum that showcases tree and shrub species only found in this area of Africa. More information can be found at http://www.nczoo.com/ConservationAnimals/AfricanPlantConservation.aspx.

FUNDING IDEAS

Funding for horticultural displays at zoos and amusement parks generally is included within a park budget. As most of these institutions are for profit, the money generated from ticket sales is divided among various departments within the park. Entrance fees are the primary source of revenue for public institutions. The largest of theme parks may also have stock traded publicly through the Stock Market. Regardless of park size, horticulture is a large portion of any park budget. From the initial design to the ongoing yearly maintenance, financial resources are immense.

These for-profit parks typically generate the funds necessary for horticulture programs, staff, maintenance, equipment, and plant materials. From the parks inception, funds are typically appropriated for horticulture and landscape maintenance with an annual horticulture budget.

PRIVATE DONATIONS

Philanthropic investors provide funding for large and small projects. Additionally, some parks have distinct fund raising campaigns just for the gardens. Private donors can offer special funding for specific landscape beds, seasonal displays, or themed areas. Zoos and amusement parks may also receive sizable funding from patrons, businesses, or nonprofit groups within the community. Philanthropic community investors have greater interest in supporting local projects. For a substantial donation, naming rights or recognition near the landscaped area is an incentive.

CONSERVATION GRANTS

Conservation and environmental preservation is an area of interest throughout the United States. There are grants that specifically target programs helping endangered species' habitat and preserve affected plants. Zoos and amusement parks that either work with such plants or collaborate with other programs, nationally and internationally, are good candidates for such funding. National and international groups, local and regional plant societies, and environmental associations are potential sources for funding. The following are examples of funding organizations or resources for further information:

- America in Bloom—grants (http://www.americainbloom.org/resources/Grant-Opportunities.aspx)
- Native Plant Society of New Mexico (regional) (http://www.npsnm.org/conservation/grants/)
- Scott's Miracle Grow—grants (http://scottsmiraclegro.com/corporate-responsibility/gro1000/)
- Stanley Smith Horticulture Trust (http://www.adminitrustllc.com/stanley-smith-horticultural-trust/)
- Terra Viva Grants Directory (http://www.terravivagrants.org/)
- The Hardy Plant Society of Oregon (regional) (http://www.hardyplantsociety.org/grants)

- The Walt Disney Company—Conservation Funding (http://thewaltdisneycompany.com/content/conservation-funding)

SUMMARY

Zoos and amusement parks have changed tremendously over the past century. Beginning as menageries and traveling fairs, these parks have grown into huge attractions visited by millions of people annually. As the exhibits and attractions have evolved over the years from simple animal cages, bare bones displays and amusements, so too have horticulture displays. Today's park patrons expect not only lush landscapes and manicured lawns along the walkways, but also horticultural displays that match the exhibits found within the park.

Modern parks are a horticulturally impressive experience for all the senses. Plants are front and center from the entrance displays throughout the entire park area, with landscaping that serves to sooth the visitor, often providing a calming space within an urban environment. Landscapes are designed to incorporate plants to facilitate the illusion that the guest is actually visiting another place.

Plants are the tools that designers of parks can successfully use to transform a landscape. For animals on exhibit, plants can serve as a food source, habitat, or provide environmental enrichment. Plants offer shelter and shade for people as well as the zoo animals. Plantings that are strategically placed help block views as well as create intimate spaces. Foliage provides for good backdrops and help to lessen mechanical noise from machinery, providing lush lawns with soft textures pleasant to the touch and bright, fragrant blooming plants to please the eye. All senses are engaged.

Landscape plantings at zoos and amusement parks have obvious esthetic benefits. A closer examination shows that the plants are much more than greenery dressing up the park. The plantings influence perception of the environment and the animals on exhibit. Conservation of plants, animals, and habitats is fostered with a focus on sustainability, not only at the park site, but linked to the origin of the plants and animals on exhibit. Horticulturists working alongside zoologists promote a fuller understanding of the importance of saving ecosystems.

Educational programs promote understanding of a variety of environmental and horticultural topics and gardening techniques. While parks attract visitors due to their interesting animals and entertaining rides and shows, plants help set the scene, giving a visual and total-sensory experience. Horticulture is a prominent part of the overall experience that accurately complements the animal habitat, themed area, and park as a whole.

REVIEW QUESTIONS

1. Discuss how horticulture was used in early zoos and amusement parks.
2. Describe the evolution of horticulture in zoos and amusement parks.
3. Outline five ways horticulture is used in zoos and amusement parks.
4. What are three benefits that horticulture provides for zoo animals?

5. Provide examples for how horticulture displays in zoos and amusement parks benefit people.
6. Describe four educational techniques used in zoos and amusement parks.
7. Discuss conservation efforts at zoos and amusement parks and their implications for plant species.
8. Describe the challenges of landscape maintenance at zoos and amusement parks.

ENRICHMENT ACTIVITIES

1. Take a field trip to a local zoo or amusement park. Take photos of examples for how plants are used throughout, including esthetic as well as functional uses.
2. Research a local zoo or amusement parks and examine its history. If possible, analyze old photos and recent ones to observe any differences in the use of landscape plantings.
3. Locate a copy of a general landscape design. Familiarize yourself with the symbols and layout of the plan. Carefully analyze the lines, walkways, and various areas, as well as how the plants are arranged within the space.
4. Choose any zoo animal and research its habitat and diet, and make a list of plants that could safely be placed inside the enclosure for food, shelter, and enrichment activities.
5. Through an online source, evaluate and compare horticulture programs offered at two different zoos or amusement parks.
6. Interview a horticulturist at a zoo or amusement park to learn about career opportunities. Find out what training or education is required for the job.

REFERENCES

Abkar, M.K., M. Mariapan, S. Maulan, M. Sheybani, and S. Beheshti. 2010. The role of urban green spaces in mood change. *Australian Journal of Basic and Applied Sciences*, 4(10), 5352–5361.

Addison, R.A. 1924. Showmanship and the zoo business. *Parks and Recreation*, 8(2), 129.

Ballantyne, R., J. Packer, K. Hughes, and L. Dierking. 2007. Conservation learning in wildlife tourism settings: Lessons from research in zoos and aquariums. *Environmental Education Research*, 13(3), 367–383.

Brown, P.S.S. 2014. *Herculean Horticulture*. Retrieved from http://www.virginialiving.com/home-garden/Outside/herculean-horticulture/.

Falk, J.H., E.M. Reinhard, C. Vernon, K. Bronnenkant, J.E. Heimlich, and N.L. Deans. 2007. *Why Zoos and Aquariums Matter: Assessing the Impact of a Visit to a Zoo or Aquarium*. Silver Spring, MD: Association of Zoos and Aquariums.

Hanson, E. 2002. *Animal Attractions, Nature on Display in American Zoos*. Princeton, NJ: Princeton University Press.

Heerwagen, J.H. and G.H. Orians. 1993. Humans, habitats, and aesthetics. In S.R. Kellert and E.O. Wilson (eds.), *The Biophilia Hypothesis*. Washington DC: Island Press/Shearwater Books, pp. 138–172.

Kaplan, R. and S. Kaplan. 1989. *The Experience of Nature: A Psychological Perspective*. New York: Cambridge University Press.

National Association for Olmstead Parks. 2014. *Design Principles.* Retrieved from http://www.olmsted.org/the-olmsted-legacy/olmsted-theory-and-design-principles/olmsted-his-essential-theory.

North Carolina Zoological Park. 2012. *Rare Plant and Conservation Program, 2012 Annual Report.* Retrieved from http://www.nczoo.org/conservation/2012Rare_Plant_Newsletter.pdf.

Smithsonian National Zoological Park. 2014. *Giant Pandas.* Retrieved from http://nationalzoo.si.edu/Animals/GiantPandas/PandaFacts/.

Stigsdotter, U. A. and P. Grahn. 2004. A garden at your workplace may reduce stress. In A. Dilani (ed.), *Design & Health III—Health Promotion through Environmental Design.* Stockholm: Research Center for Design and Health, pp. 147–157.

The Federal Highway Administration. 2014. *Designing Sidewalks for Trails and Access.* Retrieved from safty.fhwa.dot.gov/ped.../swless13.pdf.

U.S. House. *The Committee on Natural Resources.* 2014. Retrieved from http://democrats.naturalresources.house.gov/issue/endangered-species.

Walt Disney. 2015. The Biography.com website. Retrieved 10:18, February 11, 2015, from http://www.biography.com/people/walt-disney-9275533.

FURTHER READING

Hanson, N. 2014. *The Last Generation of Kids that Played Outside* [Blog post]. Retrieved October 15, 2014 from http://www.huffingtonpost.com.

Health Council of the Netherlands. 2004. *Nature and Health: The Influence of Nature on Social, Psychological and Physical Well-Being.* The Hague: Health Council of the Netherlands and RMNO.

Heerwagen, J.H. and Orians G.H. 2002. The ecological world of children. In P.H. Kahn and S.R. Kellert (eds.), *Children and Nature.* Cambridge: MIT Press, pp. 29–64.

Kahn, P.H. and S.R. Kellert (eds.). 2002. *Children and Nature: Psychological, Sociocultural and Evolutionary Investigations.* Cambridge, MA: MIT Press.

Kaplan, R. 1993. The role of nature in the context of the workplace. *Landscape and Urban Planning*, 26(1–4), 193–201.

Kaplan, S. 1995. The restorative benefits of nature: Toward an integrative framework. *Journal of Environmental Psychology*, 15(3), 169–182.

Kirkby, M. 1989. Nature as refuge in children's environments. *Children's Environments Quarterly*, 6, 7.

Lohr, V.I. 2011. Greening the human environment: The untold benefits. *Acta Horticulturae (ISHS)*, 916, 159–170. http://www.actahort.org/books/916/916_16.htm.

Lohr, V.I. and P.D. Relf. 2014. Horticultural science's role in meeting the need of urban populations. *Horticulture: Plants for People and Places*, Vol. 3. Netherlands: Springer, pp. 1047–1086.

Maller, C.J., C. Henderson-Wilson, and M. Townsend. 2009. Rediscovering nature in everyday settings: Or how to create healthy environments and healthy people. *Ecohealth*, 6(4), 553–56.

Miller, B., W. Conway, R.P. Reading, C. Wemmer, D. Wildt, D. Kleiman, S. Monfort, A. Rabinowitz, B. Armstrong, and M. Hutchins. 2004. Evaluating the conservation mission of zoos, aquariums, botanical gardens, and natural history museums. *Conservation Biology*, 18, 86–93.

Peterson, Elmer T. 1940, January. At Home with Walt Disney. *Better Homes and Gardens.* (n.v.), pp. 13–15, 56.

Plant Conservation in Africa. 2011. NC Zoo Society website. Retrieved February 9, 2015 from http://www.nczoo.com/ConservationAnimals/AfricanPlantConservation.aspx.

Scheu, D.L. 1996. *The Role of Horticulture in Theme Parks.* MS thesis, University of Delaware. Newark, ED: University Delaware.

Industry Associations

American Horticultural Society (AHS) (www.ahs.org).
American Public Garden Association (APGA) (www.publicgardens.org).
Association of Zoos and Aquariums (AZA) also offers a Zoo Horticulturist Certification Program (www.aza.org).
Association of Zoological Horticulture (AZH) (www.azh.org).

6 Prison Horticulture

Deborah Rutt

CONTENTS

OBJECTIVES

Upon completion of this chapter, the reader should be able to

- Describe activities, benefits, and objectives of prison horticulture programs.
- Understand the history of horticulture and gardens within correctional facilities.
- Identify historical incarceration trends in the United States.
- Describe characteristics of inmate populations.
- Identify key partnerships and logistical considerations for developing a prison horticulture program.

KEY TERMS

- Adult basic and secondary education
- Gate clearance
- Horticultural therapy
- Life skills training
- Material safety data sheet
- Prison horticulture
- Recidivism

- Security levels
- Vocational training

WHAT IS PRISON HORTICULTURE?

Prison horticulture refers to a variety of plant-based activities and programs that take place in any detention facility such as a prison, jail, or youth detention facility. Prison horticulture programs generally include a focus on the rehabilitation of incarcerated individuals, and have at their foundation a belief that plant-based activities can provide opportunities for reform of the individual as well as improved institutional conditions.

Prison horticulture activities and the scope of programs may vary widely. Examples of prison horticulture activities include edible and ornamental gardening, landscaping, horticultural therapy, family garden programs, greenhouse activities, composting, and native plant propagation for habitat restoration (Table 6.1). Prison horticulture may occur as small-scale, informal projects initiated by individual facility staff members or inmates, or as part of a more formal program involving a collaboration of inmates, facility staff, volunteers, academic institutions, and other community partners.

Prison horticulture offers a number of potential benefits for inmates, facility staff, institutions, and the larger community. Horticulture programs may provide inmates the opportunity to increase their education level, as well as learn specific skills in preparation for employment in the community, and develop improved teamwork and interpersonal communication skills. Horticulture activities may improve the physical and mental health of inmate participants, and prison gardens will generally improve the esthetics of facility grounds, benefitting facility staff and inmates alike. Prison horticulture may offer cost savings for correctional facilities and taxpayers by growing produce for inmate meals and through the composting of food waste and plant debris. Horticulture programs may provide assistance in meeting legislatively mandated requirements to employ inmates, and the potential for positive media coverage for institutions. Many prison horticulture programs reflect the larger societal focus on environmental stewardship, and may offer additional cost savings for institutions through resource conservation and more sustainable facility operations. Programs which focus on environmental stewardship may provide participants the opportunity

TABLE 6.1
Prison Horticulture Activities

Edible gardening
Ornamental gardening
Landscaping
Horticultural therapy
Family gardens
Greenhouse activities
Composting
Native plant propagation

to assist in the restoration of natural habitats outside the facility and may give incarcerated individuals a sense of connection and responsibility to the larger community. The ultimate benefit of any effective prison program will be a reduction in *recidivism*, which refers to the subsequent arrest and incarceration of an individual who has been released from custody. When previously incarcerated men, women, and youth are able to live a productive, crime-free lives outside of prison, the benefits extend beyond the individual or institutional level, and create safer, healthier families and communities.

HISTORY AND BACKGROUND

Prison horticulture has been an element of detention facilities throughout history, including prisoner-of-war and internment camps. Large-scale agriculture through forced inmate labor was historically used as a cheap source of food, while some early, reform-minded prison wardens developed horticulture programs with the goal of inmate rehabilitation. Prison gardens have also been initiated by incarcerated individuals as a source of healthier food, and as a way to beautify their surroundings and stay occupied physically and mentally. Gardening has served inmates as a means of creative expression, but also as an expression of individual and cultural identity, and as an act of resistance against incarceration and the rigid structures of the institution. In Nelson Mandela's autobiography "Long Walk to Freedom," he describes the personal significance of the garden he tended during his imprisonment, "A garden was one of the few things in prison that one could control. To plant a seed, watch it grow, to tend it and then harvest it, offered a simple but enduring satisfaction. The sense of being the custodian of this small patch of earth offered a small taste of freedom."

EARLY PRISONS

Incarceration as a form of punishment for criminal behavior is a relatively modern development. Before the eighteenth century enlightenment in France and England gave rise to new ideas about individual liberty and human rights, the response to criminal behavior often included corporal punishment, torture, slave labor, and social ostracism. This was the case in colonial America, and continued until the establishment of the first penitentiaries by the Pennsylvania Quakers in the early nineteenth century. The term "penitentiary" reflected the view of prisoners as religious penitents, serving time as punishment for their sins. While the creators of the first penitentiary system may have had moral aims, the conditions in these institutions were generally inhumane and unsanitary, and continued to include forced labor.

In the mid-nineteenth century, a reformatory movement arose in response to poor prison conditions and attempted to reframe the role of the prison as rehabilitative. With an emphasis on cleanliness, order, hard work, and education, this early reform movement aimed to transform criminals into law-abiding citizens. Again, while the goals of the reformers may have been well intentioned, prison conditions remained harsh and inhumane.

During the twentieth century, additional prison reform movements arose in an ongoing effort to create more humane conditions and provide rehabilitative programs for inmates. In the 1930s, prison construction was more likely to take place in rural settings which were seen as better environments for inmate reform. In the 1950s, the word "correctional" began to be used to describe institutions of incarceration, at the same time that social scientists were raising questions about the treatment of prisoners, and advocating for improved facilities and expanded services for prisoners such as counseling and education. In 1955, the United Nations Standard Minimum Rules for the Treatment of Prisoners was published, providing minimum standards of accommodation for discipline and use of restraints, housing conditions, food, hygiene, clothing, medical care, religious practices, working conditions, and a number of other issues. The rules were not legally binding and the minimum standards they describe are not met in many modern prisons.

EARLY PRISON HORTICULTURE

Throughout the twentieth century, reform-minded prison wardens attempted to use horticulture as a means of creating more humane conditions and as a form of inmate rehabilitation through meaningful work, improved diet, and "fresh air treatment." At the Sing Sing Penitentiary in Ossining, New York, the 1920s saw the establishment of extensive landscaping and gardens, led by Warden Lewis Lawes and inmate Charles Chapin who had been a newspaper editor in New York City and was serving a life sentence at Sing Sing for the murder of his wife. Chapin came to be known as the "Rose Man of Sing Sing" for the extensive rose gardens he developed on the prison grounds. Chapin was also a bird enthusiast and with the permission of Warden Lawes and the financial backing of his former business associates, Chapin built an aviary that included a space for inmates to sit and observe birds.

Gardens were also a key feature of the prison on Alcatraz Island in San Francisco Bay, California. After the Civil War when Alcatraz served as a fort and military prison, gardens were developed on the island to provide respite for officers and their families. In 1924, the California Spring Blossom and Wildflower Association initiated an island-wide beautification effort, employing military prisoners to plant trees and shrubs, while being trained in gardening and pruning techniques. In 1933, the military prison on Alcatraz closed and the facility was transferred to the jurisdiction of the Federal Bureau of Prisons. A year later, a federal penitentiary was opened on Alcatraz, and the island's gardens became the responsibility of Fred Reichel, the prison warden's secretary. Reichel supervised the care of the island's gardens and received gardening advice and plants from the California Horticultural Society. In 1941, Alcatraz Inmate Elliot Michener began a nine year stint as a gardener on the island, during which time he built a greenhouse and continued to develop the island's extensive landscapes.

US Attorney General Robert F. Kennedy ordered the closing of the penitentiary at Alcatraz in 1963, and 10 years later, the island was opened to the public for the first time when it became part of the Golden Gate National Recreation Area under the jurisdiction of the National Park Service. In 1992, the initial efforts to restore the gardens of Alcatraz began as joint effort of the National Park Service and the Golden Gate National Parks Conservancy to inventory the island's plants and begin the propagation of plants for garden restoration. The Alcatraz Historic Gardens Project formally began in 2003, and the restored gardens continue to be preserved and maintained by a team of volunteers led by the staff of the Garden Conservancy.

US PRISONS 1980: TODAY

The social reformers and reform-minded prison wardens of the early and mid-twentieth century could not have foreseen the wave of mass incarceration that would begin to overtake US prisons in the 1980s. In 1982, the federal government declared the "War on Drugs" in which new policies were developed in an effort to reduce illegal drug use and distribution, resulting in a sharp increase in the US prison population, with a larger number of inmates serving long-term or life sentences. This dramatic growth in the prison population was a consequence of an increase in the number of drug prosecutions, along with the implementation of increased sentence lengths, mandatory minimum sentences and "three strikes" laws. In 1986, the average sentence length for a federal drug offense was 22 months. By 2004, sentences had nearly tripled, averaging 62 months in prison. Although the incidence of violent crime has declined, the population of inmates serving life sentences has more than quadrupled since 1984. Communities of color were, and continue to be, disproportionally impacted by the new drug laws and enforcement policies, despite the fact that illegal drug use is approximately equal across race groups. In the United States, African-Americans are over six times as likely to be incarcerated as whites, and Latinos are more than twice as likely.

The United States leads the world when it comes to the number of incarcerated people. In 2013, there were 2.2 million people in prison or jail, representing a 500% increase over 40 years (Figure 6.1). The US rate of incarceration, at 716 per 100,000 people, is also the highest in the world. The US imprisons its citizens at a rate that is 5–8 times that of Canada and western European nations (Figure 6.2).

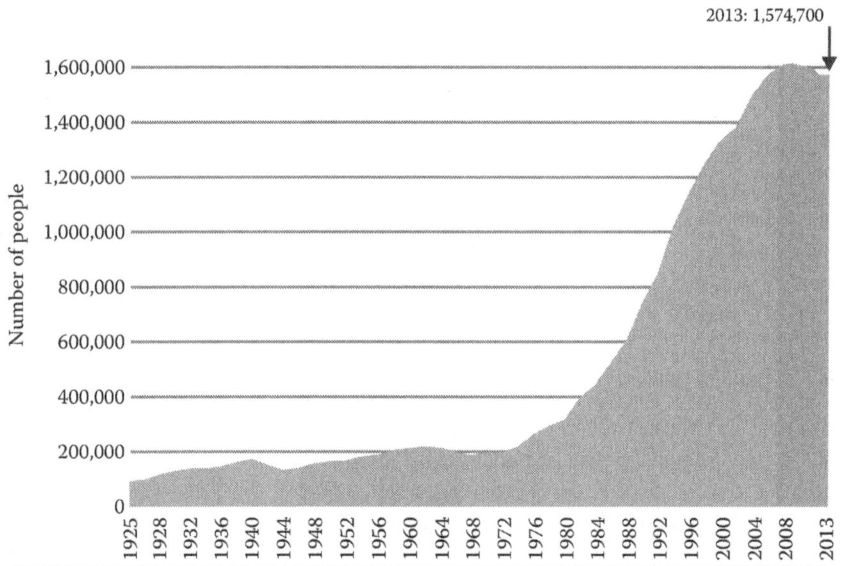

FIGURE 6.1 U.S. State and federal prison population, 1925–2013. (Adapted from Bureau of Justice Statistics, Prisoners series.)

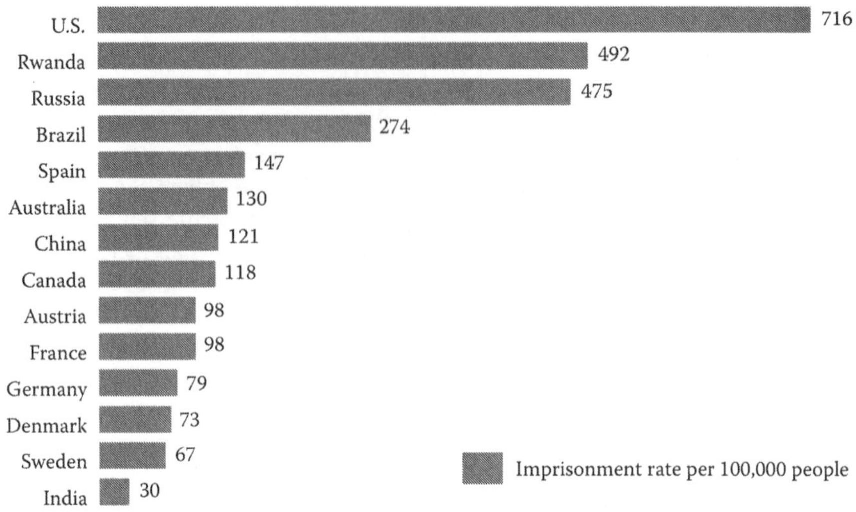

FIGURE 6.2 International rates of incarceration, 2012/2013. (Adapted from Bureau of Justice Statistics, Prisoners series.)

According to US Bureau of Justice Statistics, 93% of people in prison are male, although the rate of incarceration for women has been increasing at a higher rate than for men as a result of drug policy changes, with the number of women in prisons growing by over 800% since 1977. African-Americans represent approximately 38% of people in prison, while 35% are white and 21% are Hispanic. Approximately half of all inmates are serving time for nonviolent drug offenses, property or public order crimes, while two thirds of women are in prison for nonviolent crimes. Mental illness is prevalent among the incarcerated, with 55% of male prisoners and 73% of women prisoners diagnosed with a mental health disorder. Educational attainment for prisoners is significantly lower than that of the general population, with 41% of prison inmates having no high school diploma or GED, compared to 18% for the general population.

EDUCATION AND TRAINING IN PRISONS

The value of providing education to inmates has been recognized since the first penitentiaries were opened in the United States. Initially referred to as "Sabbath schools," these early prison education programs focused on religious training. As prison reform movements came and went, and as the focus on prisons shifted between punishment and rehabilitation, investment in prison education has fluctuated correspondingly. The 1970s are considered the golden age of correctional education when the expansion of programs reflected a belief in cognitive and vocational skill development as the most important tool of inmate rehabilitation.

In the 1980s, as prison populations began to increase sharply, funding for inmate education and training programs was declining. Congress passed the Violent Crime Control and Law Enforcement Act of 1994, resulting in the elimination of Pell Grant funding for prisoners, all but eliminating higher education opportunities in correctional facilities. A survey of prisoners carried out by the Bureau of Justice showed that the percentage of prisoners participating in education and vocational training programs fell between 1997 and 2004, despite evidence of the effectiveness of educational and vocational programs in prisons. On average, inmates who participated in prison education programs showed a 43% reduction in recidivism. Research showed that each dollar spent on prison education potentially reduced corrections costs by more than four dollars during the three-year postrelease period when individuals were most likely to reoffend.

Education and training programs that do exist in correctional facilities encompass a variety of activities aimed at meeting the significant needs of the incarcerated population. *Adult basic education* programs offer training in math, reading, writing and English as a second language and may include special education programs for those with developmental disabilities or learning differences. *Adult secondary education* focuses on instruction for the GED test or other certificates of high school equivalency. *Vocational training* aims at preparing individuals for employment by providing general workplace competencies as well as skills for specific positions or industries. Additional programs offer *life skills training*, such as communication and interpersonal skills, parenting skills, and stress and anger management.

PRISON HORTICULTURE PROGRAM OBJECTIVES

Prison horticulture program objectives often parallel those of other educational and training programs taking place in correctional facilities. The focus of prison horticulture can vary widely, but generally have an overarching goal of providing inmates education, vocational and personal skills, all with the purpose of improving the chances of successful integration back into the community upon release.

Prison horticulture programs may have the objective of providing basic or secondary education curriculum, using plant-based activities to teach science, math, and other subjects. Horticulture programs may focus on vocational training, providing general workplace skills such as communication and teamwork, or training inmates in the specific skill set needed for employment in the horticultural industry. Prison horticulture programs may emphasize the value of environmental sustainability, developing competencies needed for employment in green industries. Programs may align with community partners working toward habitat restoration, growing native plants within the institution that can be planted out in restoration areas. Horticulture programs in prisons may focus on the development of life skills, such as interpersonal communication or parenting skills. In family gardening programs, participants and their children may take part in horticulture activities together, improving family bonds, while learning healthy, shared activities. *Horticultural therapy* programs in prison will focus on improving inmates' emotional and psychological well-being through guided, plant-based activities.

TABLE 6.2
Prison Horticulture Program Objectives

Basic or secondary education
Vocational training
Environmental sustainability
Habitat restoration
Life skills training
Horticultural therapy
Public health

Gardening in prison, as elsewhere, has been shown to provide significant therapeutic outcomes, including healing from trauma, addiction recovery, stress reduction, and anger management (see Chapter 7 on Horticultural Therapy). Prison horticulture may have a public health focus and aims to improve participants' physical health through exercise and outdoor activity, by growing produce to improve the nutritional content of inmate meals (Table 6.2). Prison horticulture programs may have a number of objectives, offering inmate participants a range of opportunities to work toward a healthier and more sustainable future for themselves, their communities and the environment.

CASE STUDY 1 Rikers Island GreenHouse and Green Team Programs, New York

Rikers Island, which serves as New York City's jail, is the largest jail complex in the world and is located in Flushing Bay adjacent to the runways of La Guardia Airport. The 413 acre island is home to a daily average inmate population of 11,300 and includes 10 jail facilities, seven of which are for male detainees awaiting sentencing or transfer to longer-term facilities, one for male inmates who have been sentenced to one year or less, one youth detention facility for male inmates between the ages of 16 and 18, and one facility for female inmates, which includes a 25-bed baby nursery. Rikers Island is in many ways a self-contained city with its own power plant, healthcare facilities, and bus system.

In 1996, The Horticultural Society of New York (HSNY) began a horticulture program for the men and women incarcerated at Rikers Island. HSNY's first project on the island, a horticultural program for adolescent males which ran for nearly seven years, lost its funding from a city youth grant in 1993. For four years, the Rikers Island greenhouse sat vacant until Anthony Smith, HSNY's newly appointed president and former assistant deputy commissioner of General Services under Mayor David Dinkins, began development of the new horticulture program with funding from private donors and charitable organizations. Smith had been a longtime supporter of rehabilitative programs for inmates, and was motivated to redevelop the HSNY program at Rikers because of his belief that "horticulture offers all the elements—education, job skills, self-esteem, creativity, confidence—of a successful program." He drew a distinction between the HSNY's horticulture program and the existing farm and landscape labor details that have been operated on Rikers since 1981, by emphasizing the potential for postrelease employment.

Once Smith had funding and the support of the Department of Corrections, the first order of business was to renovate the existing greenhouse which sat vacant for four years following the termination of the youth horticulture program, and was in a state of disrepair. While the electrical, plumbing, and glass repair work was being done on the greenhouse, there were meetings with the Department of Corrections over the course of several months to secure two officers to provide security for the program, and a bus to transport male inmates to the greenhouse, which sits adjacent to the women's housing unit.

After nearly a year of planning, securing funding, and completing structural repairs, the HSNY's GreenHouse Program at Rikers Island began with 15 women inmates who attended the program in the morning and 10 men who participated in the afternoon. The inmate participants volunteered for the program and earned 20 cents an hour as part of the work program. For four to six hours a day, they received horticultural training and worked to build flower and vegetable gardens. The inmates in the program provided labor for the development of gardens in the two and a half acres surrounding the greenhouse and adjacent classroom, but they also were central to the creative design process. Guided by HSNY program staff, more than 350 program participants built a complex of gardens over the course of six years, including a herb, butterfly and rock garden, a native forest and fruit orchard, a memorial garden to honor an instructor who passed away, as well as a gazebo and a pagoda.

The gardens at Rikers Island provide a setting for horticultural instruction and hands-on training to participants in the GreenHouse program with curriculum that covers botany, soil science, ecology, as well as garden design, construction, and maintenance. In addition to their work building and maintaining the gardens at Rikers, inmates in the program have the opportunity to grow plants and build planters for use in city schools and parks, as well as constructing bird feeders, and bird and bat houses used to restore natural landscapes in the city's communities. A classroom structure adjacent to the greenhouse serves as a teaching space, office, gardening book library, workspace, and kitchen. When the inmate gardeners break for lunch, they are able to prepare food, integrating harvests from their own garden vegetable and herbs into staples provided by the jail. Participants also learn how to create products from herbs including lotions, lip balm, and soaps.

During the winter months when gardening work is limited to caring for plants in the greenhouse, the program offers workshops in math, writing, drawing, and computer literacy. These skills are put to use in carpentry and landscape design projects, and in the production of a newsletter which records students' experiences in the program. Newsletter articles may cover gardening, nutrition, or information about finding jobs in the community, and serve to document the students' work for Rikers inmates, facility staff, and potential funders.

Reflecting Anthony Smith's initial focus on developing participants' employment skills, the program at Rikers has partnered with several community nonprofit organizations to provide postrelease jobs and services. As most of the students are engaged in the GreenHouse program inside Rikers for only six to twelve months, Smith recognized the need for additional training and support in the community in order for the program participants to develop professional horticulture skills. HSNY's postrelease Green Team program is aimed at fulfilling that need, offering ex-offenders from the Rikers GreenHouse program the opportunity to participate in an internship program. Green Team participants are provided vocational training in horticulture, along with job search skills. Green Team interns are employed in the construction and maintenance of gardens and landscapes. The work teams are typically made up of six interns and one coordinator who is responsible for acquiring and maintaining client contracts, designing projects, and overseeing the work of the interns. The Green Team interns' responsibilities to the larger team and program go beyond their horticultural work. Interns also agree to be willing to learn and take direction, and to demonstrate professional work manners and a commitment to staying clean and sober. Interns are given time to attend GED classes and drug programs, and deal with their social services, housing and healthcare needs.

HSNY's Green Team has worked with a variety of agencies including New York City Parks and the Housing Authority, but much of their initial work took place at public libraries. The Green Team designed garden sites at 15 libraries in New York, and received a $40,000 contract from the library system to maintain the gardens. These funds were used to pay intern stipends and the salary of the Green Team coordinator, who transported the interns to worksites in a van donated by a HSNY board member. As the Green Team program developed, it became clear that the public agencies and nonprofits for whom they were working had shrinking budgets and would not be able to entirely support the Green Team's need to keep anywhere from four to 14 interns employed. In order to secure work from a broader sector, the Green Team had to submit bids in competition with other firms to earn contracts for maintenance of landscapes at apartment buildings, private residences, and parks. The Green Team essentially transformed itself into a professional landscape company, and within two years had secured $100,000 worth of contracts. In addition to these maintenance contracts, they continued to develop projects with community groups. At housing projects in Harlem and the Lower East Side, the Green Team worked with children in a community center to build vegetable gardens. At Frederick Douglass High School in Harlem, Green Team interns helped to prepare and teach a greenhouse curriculum with special education students.

While the Green Team coordinator has the challenge of bringing in a steady flow of work for the program, the interns face other, more personal challenges, such as securing housing, caring for family, and addressing child custody and other legal issues. Interns must also face the risk of relapse into addiction, and must avoid the hazards that come with returning to the communities where their drug use and crimes took place. Planning for reintegration into the community begins well before program participants are released from Rikers. The GreenHouse program staff work with participants to determine each person's needs for postrelease support, and help connect them to agencies that can assist with education, housing, addiction recovery, and other transition services. Program participants are also tracked on a database from the time they enter the GreenHouse program at Rikers. The program's database keeps profiles on all participants, including the date they enter the garden program and their release date from Rikers, as well as their education and employment history, skills, certifications, and interests. The tracking system follows the interns postrelease at 3 months, 6 months, 1 year, and 3 years, allowing the program to track their success in reducing recidivism. In the first four years of the Green Team program, interns had a recidivism rate of less than 10%, compared to a national average of 68%. The program interns' recidivism was also significantly better than the 28% national average for all ex-offenders who were engaged in employment, education or training shortly after release.

In 2006, James Jiler, the former director of the GreenHouse program, published *Doing Time in the Garden* a book detailing his work with the program. Jiler describes in detail the partnerships, planning and program elements that have been essential to the success of the programs, but in the final lines of his book Jiler focuses on the power of nature and its impact on those with whom he has worked. Jiler acknowledges that the process of personal change is slow and difficult, "But for those who work long enough in a garden to witness nature's renewal as it occurs on a seasonal basis, the capacity for self-renewal grows. At the least, transforming a bit of city landscape begins the process of helping ex-offenders transform themselves."

CASE STUDY 2 Coffee Creek Healthy Food Access Project, Wilsonville, Oregon

Coffee Creek Correctional Facility is Oregon's state prison for women and prisoner intake facility, located in Wilsonville, approximately ten miles south of Portland. The 108-acre facility opened in 2001 and houses an average of 1275 women, with about half of the population in a minimum security facility and half in an adjacent medium security facility. Coffee Creek also serves as the intake facility for all men entering the state prison system. All inmates remain in intake for about a month while going through a series of assessments to identify medical, mental health, substance abuse, educational, and cognitive needs. Once the intake process is complete, male inmates are relocated to one of the 13 men's prisons located throughout the state, while women are assigned to either the medium or minimum security facility at Coffee Creek. Women are generally assigned to minimum security if they have four years or less remaining on their sentence, and have demonstrated behavior appropriate for the lower security level.

Reflecting national trends, the incarcerated women at Coffee Creek have significant differences compared to male inmates. Women at Coffee Creek are more likely to be serving time for a nonviolent drug offense compared to their male counterparts, and are more likely to have been the primary care giver for a child prior to being incarcerated. More than 75% of women have a substance abuse history, and approximately 80% of women at Coffee Creek have been diagnosed with a mental health disorder.

The women at Coffee Creek frequently identify poor diet and weight gain during their incarceration as major concerns, and this caught the attention of state public health agencies. While there was little data on the issue of obesity at Coffee Creek, it was clear that the nutritional environment at the prison had room for improvement. The prison was serving 3000 calories per day to women, and inmates had access to additional high-calorie snacks through the prison canteen, where the average number of calories purchased per inmate per day is 1100, equal to the amount of calories in five servings of French fries from a fast food chain. State officials were also focused on the issues of inmate health and healthcare costs, and in their Workgroup on Corrections Health Care, they identified chronic obesity-related disease as a serious concern, noting that between 2006 and 2012 the inmate population had shown a 48% increase in diabetes and a 159% increase in cardiovascular disease.

The Healthy Food Access Project at Coffee Creek began in 2011 as a partnership between the Oregon Public Health Division and Department of Corrections to improve the nutritional environment at Coffee Creek. The Public Health Division received a 3-year, $180,000 grant from Kaiser Permanente to support the Healthy Food Access Project, with the stated goals of documenting the average amount of weight gain, evaluating different strategies for prevention, expanding the existing prison garden, integrating fresh produce from the garden into prison meals, and providing nutrition related education.

At the start of the project, an evaluation team from the Oregon Public Health Division carried out a survey which confirmed the concerns about obesity expressed by women incarcerated at the facility. The survey found that 89% of women incarcerated at Coffee Creek for 6–24 months were overweight (40%) or obese (49%), significantly higher rates than for nonimprisoned adult women in Oregon. The Healthy Food Access Project partners took steps early in the project to reduce the daily calorie count. The corrections department dietician and food services staff worked with a dietician from public health to reduce the number of calories served to women in the dining hall at Coffee Creek from 3000 to 2200. They implemented menu labeling in the dining hall and on the commissary order forms to help women make informed choices about what they eat. In order to inform inmates and earn their trust and support for the project, Healthy Food Access partners began publishing articles in the inmate newsletter about project activities, as well as information to promote healthy food choices and physical activity.

The centerpiece of Healthy Food Access Project is the organic garden inside Coffee Creek's minimum facility, which received funding from the Healthy Food Access grant to expand from the original 10,000 square foot garden space

FIGURE 6.3 The organic garden in the main yard of the Coffee Creek Correctional Facility. (Photo by John Valls.)

to a 23,000 square foot vegetable, herb, and flower garden which produces 5000 pounds of organic produce for inmate meals each year. The garden expansion took place in the main prison yard of the minimum facility (Figure 6.3) where over 600 women who live in that part of the prison, are able to observe and enjoy the year-round display of both edible and ornamental garden plants, as well as the abundant number of birds, frogs, butterflies, and other pollinating insects that the garden has invited in. The grant also provided for a half-time garden coordinator to manage the garden and aid in the development of additional community partnerships in support of the garden program.

Depending on the season, the organic garden at Coffee Creek employs a full-time crew of 6–12 women who are incarcerated in the minimum facility. The garden positions are coveted, and the hiring decisions are made by corrections staff based on the inmate's disciplinary record and work history within the institution. The hiring process includes an interview with the facility manager and garden coordinator. The women who are selected for the position are given a two-week trial period on the garden crew, after which they are asked to sign an agreement to commit to the garden position for at least nine months, and must agree to professional behavior and a willingness to learn and accept feedback on their work habits and team communication. The garden crew members work with graduate student interns from Lewis & Clark College's Ecopsychology counseling program to improve skills for interpersonal communication and conflict resolution. Inmate gardeners earn approximately $50 a month for working and training in the prison garden 30 hours per week.

The garden crew members develop crop plans and planting schedules each year, and are responsible for planting, maintaining, and harvesting from the

FIGURE 6.4 Gardening activities are part of the Healthy Food Access Project at Coffee Creek Correctional Facility. (Photo by John Valls.)

garden throughout the season (Figure 6.4). Garden crops focus on popular vegetables and herbs such as tomatoes, peppers, cucumbers, snap peas, lettuce, spinach, kale, cilantro, basil, and dill. Varieties are chosen which require minimal preparation in the prison kitchen, such as cherry tomatoes, lemon cucumbers, and loose-leaf lettuce. The garden crew also maintains the garden's composting system and a large worm bin built from broken down freezer chest. The worm bins are located in a garage space inside the facility which also provides space for grow lights, and storage for seeds, fertilizer, wheel barrows, hose reels, and other equipment. The garden's greenhouse allows for the propagation of plants for the garden, and doubles as a meeting space for the garden crew, and a hands-on garden classroom during Oregon's lengthy rainy season (Figure 6.5).

The garden program at Coffee Creek provides education to more than 100 incarcerated women each year in organic gardening classes covering topics such as composting, gardening with herbs and beneficial insects. A key education partner is the Oregon Food Bank which provides the curriculum for a five-week, beginning organic gardening course called Seed to Supper. The course is taught by the inmate garden crew with guidance and support from community volunteers. Each year the course provides 40 inmate students a hands-on experience in all aspects of organic gardening and offers written information on low-cost, home gardening in a textbook appropriate for the average inmate's seventh grade reading level. Another important education partner is the National College of Naturopathic Medicine (NCNM) in Portland which offers a hands-on nutritious cooking class called Food as Medicine Everyday, or FAME. FAME is taught in the prison kitchen by a naturopathic physician and integrates produce from the prison garden into simple recipes and lessons focusing on the value of whole grains, fruits, and vegetables.

FIGURE 6.5 Greenhouse activities at Coffee Creek Correctional Facility, Oregon's state prison for women. (Photo by John Valls.)

The garden program at Coffee Creek also offers incarcerated women the opportunity to take part in environmental action to fight extinction for threatened butterflies in partnership with the Oregon Zoo and the Institute for Applied Ecology. The inmate participants grow thousands of native plants for the zoo's Oregon Silverspot Butterfly propagation program, and for habitat restoration in Oregon's coastal prairies. The conservation project has been supported by funds from Toyota Together Green, a partnership with Audubon aimed at providing opportunities for conservation action to underserved populations.

As with the Rikers Island program, horticulture at Coffee Creek provides a mix of benefits. While the garden program was developed with the objective of improving access to healthy food, it also provides training in horticulture, teamwork skills and interpersonal communication, as well as the opportunity to work toward a healthier environment inside and outside the prison.

CASE STUDY 3 The Sustainability in Prisons Project, Olympia, Washington

The Sustainability in Prisons Project (SPP) is a partnership between the Washington Department of Corrections and The Evergreen State College in Olympia, Washington. SPP has its roots in the rain forests of Washington State's Olympic peninsula. In 2004, Dr. Nalini Nadkarni, a forest ecologist and Evergreen College professor, was becoming increasingly concerned about the environmental impact of harvesting mosses from old-growth forests for use in the floral industry, and was looking for partners in developing protocol for propagating mosses ex situ for commercial use. Nadkarni reached out to

Dan Pacholke, the superintendent at the Cedar Creek Correctional Center in Littlerock, Washington, with the idea that inmates could assist in research on growing mosses. Nadkarni was inspired to engage inmates because she knew that they had time on their hands, but she also had the sense that inmates could benefit from science education and a connection with the natural world.

By 2005, Nadkarni had engaged a team of inmates, graduate students, and community volunteers in observing moss growth and devising methods for moss propagation. As an outgrowth of the moss project, Nadkarni began a lecture series for inmates and corrections staff (Figure 6.6) which she called "Sustainable Futures, Sustainable Lives." She invited scientists from environmental agencies and universities to lecture on ecology and sustainability related topics. Nadkarni also began to connect corrections staff with experts in sustainable operations to help the department work toward their goal of reducing the environmental impact of their facilities. From these beginnings, SPP has developed into a program involving inmates and corrections staff in all 12 of the Washington Department of Correction's facilities, with the mission of bringing science and nature into prisons. The program focus has been on developing partnerships with corrections staff and scientists from educational institutions and environmental organizations, in order to engage incarcerated men and women in a variety of nature and sustainability related activities.

The SPP supports vegetable and ornamental gardens in all of the correctional facilities in Washington. Some of the vegetable gardens produce food for donation to local food banks, while others provide vegetables for inmate meals. The scale of the prison gardens vary widely, from small garden plots to large-scale agriculture. At the Washington State Penitentiary in Walla Walla, inmates and corrections staff have collaborated with a local grain cooperative to develop

FIGURE 6.6 Science lecture at a men's prison in Washington state. (Photo by Benj Drummond.)

FIGURE 6.7 Garden area for inmates serving life sentences in a Washington state prison. (Photo by Jose Morales.)

a no-till wheat farm on 250 acres surrounding the prison. The farm produces nearly 20,000 bushels of wheat each year, which is sent to a mill in Spokane to be ground into flour. The flour is used at the nearby Airway Heights Corrections Center to produce baked goods which are sent to prisons throughout Washington. Ornamental gardens serve a variety of purposes within Washington prisons. At the Mission Creek Corrections Center for Women, they have developed a "Mommy and Me" garden to provide shared activities for incarcerated mothers and their children. Some facilities have also created gardens in high security areas, including a herb garden in a mental health unit, as well as flower gardens engaging inmates serving life sentences (Figure 6.7).

Four prisons in Washington have more formal horticulture training programs. These nine-month programs work with local community college staff to offer inmates lectures, labs, and hands-on training in horticulture sciences, integrated pest management, and landscape maintenance and design. Inmate participants attend classes five days a week, and can earn 46–55 college credits upon completion. While the classes are college level, some of the programs have incorporated a General Educational Development (GED) Program component into the horticulture training for those inmate participants needing to earn a GED. The college credits earned in the horticulture training program can be transferred to a community college for inmates choosing to continue their education upon release. These credits may go toward an Associate's degree or a number of certification programs for horticulturalists, arborists, and other landscape professionals.

The SPP includes extensive production of native plants for habitat conservation (Figure 6.8). Since 2009, SPP has partnered with corrections staff and land management agencies to develop conservation nurseries at three prisons in

FIGURE 6.8 Growing native plants for SPP conservation projects in a Washington state prison. (Photo by Jose Morales.)

Washington. Inmate technicians work alongside graduate research assistants and scientists to grow over 50 species of native plants that are generally not available from commercial nurseries. Inmates are provided training on topics including botany, ecology, as well as soil science and cultivation techniques. In the first four years of the conservation nursery program, more than 70 inmates have grown over 950,000 plants for habitat restoration, including 16 species used to restore prairies in the South Puget Sound region of Washington. In addition to the plant-based conservation activities, SPP has trained inmates to rear endangered species inside prisons for release into the wild, including the Oregon Spotted Frog and the Taylor's checkerspot butterfly. SPP's habitat conservation programs have received support and funding from a number of partners including The Nature Conservancy, The Center for Natural Lands Management, The Woodland Park Zoo, The Washington Department of Natural Resources, and both the federal and state Departments of Fish and Wildlife.

After 10 years of growth, SPP began the development of a national network connecting correctional facilities and other organizations working on nature and sustainability related projects in correctional settings. With support from a National Science Foundation grant, they hosted conferences in 2012 and 2013, and published. The SPP Handbook, which offers guidance in the development of nature-based programs in prisons, covering topics such as program management, evaluation strategies and interacting with the media. The SPP Handbook shares specific SPP protocols and evaluation tools developed for a variety of prison programs including gardening, composting, beekeeping, rain-water catchment, conservation projects and a wide range of sustainable operations topics. These efforts have resulted in the development of SPP teams in a number of other states including Oregon, California, Maryland, Utah, and Ohio.

The SPP staff has identified what they see as essential components of successful nature-based programs in a correctional setting. SPP focuses on collaborations with corrections staff and inmates, as well as academic institutions and conservation organizations. They stress the need for project activities to provide mutual benefits for all partners, and to not place excessive burden on any one partner. SPP emphasizes the value of bringing nature into prisons, and the importance of providing training and education to inmate participants. In addition to connecting inmates to the natural world, SPP activities aim to connect participants to the outside community through contributions to scientific research and conservation efforts, and by growing produce for local food banks. From the beginning of the project, SPP efforts have included elements of evaluation to track the success of their programs, and to share results thereby encouraging and supporting other organizations working toward similar programs in prisons.

DEVELOPING A PRISON HORTICULTURE PROGRAM

Because of the wide range of potential goals and activities for prison horticulture, as well as differences in climate and facilities, each program will have a unique set of requirements and logistical hurdles to address in starting a horticulture program. The first step in initiating successful programs is to develop the essential partnerships that are necessary for working in a prison setting. Prison horticulture programs require collaboration between corrections staff, inmates and community partners. In order for these collaborations to be successful, there must be a focus on common goals and working together to develop a program that ensures ongoing support and initiative among all partners.

Because these programs will be located at a correctional facility, the program goals must align with the objectives of the institution and have support from facility staff. The first objective of any correctional institution, and the primary focus of the facility staff, will be safety and security. All program activities must therefore be developed in strict alignment with the facility's protocol for safety and security. The facility manager will ultimately determine the feasibility of a prison horticulture program, and will want to see a proposal detailing program elements and activities. Proposals should reflect an understanding of the specific needs of the inmate population at the facility, and clearly link program goals to those of the correctional institution, such as providing training, employment, educational, and therapeutic activities for inmates. Proposals should highlight the potential for institutional cost savings through food production for inmate meals, and improved sustainability in facility operations, as well as other benefits for institutions such as improved esthetics and the potential for positive media coverage. Proposals should also specify potential sources of funding, as the facility's budget is unlikely to have significant resource to direct to horticulture programs. Proposals for initiating programs should generally be modest in scope, as corrections staff will want to start small and build trust and confidence in community partners before expanding activities.

Prison horticulture programs should be developed in close collaboration with corrections partners who can assist in determining which activities will be feasible in terms of safety and security, and which will best serve the needs of the specific institution and inmate population. In addition to the facility manager, other useful contacts within the institution may include security staff, physical plant workers, kitchen staff, education program employees, and volunteer coordinators. It may also be helpful to identify staff with an interest in gardening and sustainability, perhaps members of a facility's green team, who may be willing to advocate for a prison horticulture program, assist in its development, and help facilitate the logistics of program activities (Table 6.3). Partnering with prison staff is essential for understanding how to maintain safety and security, how to work effectively and appropriately with inmates, and how to carry out horticultural activities within the prison environment.

Most prison horticultural programs are initiated with the aim of providing learning and developmental opportunities for inmates within the institution, therefore it is essential to work in collaboration with inmate participants, and develop a program with their specific needs in mind. For example, a program which aims to meet the vocational needs of inmates, must focus on activities that will engage participants and provide skills that will be of practical use in securing employment in the community for those with a felony record. Program curriculum should offer up-to-date content and should be accessible to those with lower reading levels. Training activities should be hands-on and develop skills that are in demand in the current job market. Programs should provide recognition of inmate achievement or completion of training modules with certificates, ideally trade certifications recognized outside of the prison. Programs with an educational or therapeutic focus should likewise develop activities that address the needs of the specific population they are serving. All programs should consult inmate participants in evaluating the effectiveness of the program and provide opportunities for inmates to share ideas for program development.

Community partners are essential in providing resources for developing a prison horticulture program and maintaining long-term viability. Community partners will have a range of roles and extent of involvement. Partners may provide financial resources, educational material or technical consultation, or they may be the host organization that initiates and develops the program, seeks and administers funding, and coordinates instructors and facilitators to work on-site with inmates and facility staff. Community partners may be individuals, institutions, or nonprofit groups with an interest in horticulture, sustainability, or addressing inmate needs. Examples of

TABLE 6.3
Corrections Staff Partners for Prison Horticulture Programs

Facility managers
Security staff
Physical plant staff
Kitchen staff
Education program staff
Volunteer coordinators
Green team members

potential community partners include horticulture and community gardening organizations, food banks, public health or environmental organizations, and local faith communities. Organizations with technical expertise are also potential community partners, such as local agricultural extension agencies, horticulture or ecology departments of nearby academic institutions, as well as research experts who can assist in developing evaluation tools for measuring a program's outcomes and effectiveness (Table 6.4). Examples of community partners who may provide funding for prison horticulture programs include horticultural and community gardening organizations, public health foundations, environmental groups, and land use agencies, to name a few. Any community member who will be regularly entering the facility and working with inmates will be required to receive security clearance, training, and a volunteer or contractor badge that corrections agencies provide to those entering their facilities.

Once key partnerships have been formed and the focus of the program is determined, the logistical considerations of starting a program can be addressed. As previously mentioned, security demands will impact every aspect of a program, and will vary depending on the security level of the facility and the location within a facility. *Security levels* in correctional facilities are generally described as minimum, medium or maximum security, and will determine the extent of restrictions and how much movement is allowed by inmates. Inmates are housed at specific security levels based on a number of factors including their disciplinary record while incarcerated and the length of time remaining on their sentence. A correctional facility may have only one security level, or may be divided into separate facilities with different security levels. Within each security level, there will be areas where only inmates authorized to participate in certain jobs or program activities are allowed. Horticulture programs located in higher security level facilities or within areas of the prison where all inmates are allowed to move will have much stricter limitations on activities, materials and tools, as well as the height and density of plantings.

At most correctional facilities, there are crews of lower security level inmates who have received *gate clearance* authorizing them to work or take part in program activities outside the prison fence. Working in these outer areas can allow for horticulture activities that require more space and fewer security restrictions, such as large-scale composting or agriculture. Some safety and security considerations will apply in all correctional settings, such as excluding any plants which are toxic

TABLE 6.4
Community Partners for Prison Horticulture Programs

Horticulture organizations
Community gardening organizations
Food banks
Public health agencies
Environmental groups
Local faith communities
Agricultural extension agencies
Academic institutions—Horticulture or ecology departments

or have hallucinogenic properties. It is critical that community partners have strong communication with corrections staff to ensure that all program activities and materials are appropriate for the prison environment and the facility's specific security level.

Depending on the scope of proposed activities, there may be an available garden space at the facility, or significant site development may need to occur prior to getting started or expanding an existing program. Prison horticulture programs may begin with just a few raised beds or small area of landscaping, or may start with a larger growing area, a greenhouse, grow lights, classroom space, and composting area. Greenhouses and other structures that protect from the weather will allow for an expanded growing season, as well as year-round gardening activities and hands-on education. Greenhouses and other structures can be expensive to build and maintain. Facility staff will want assurance that all structures meet prison security requirements and local building codes, and that a source of funding is available for long-term maintenance. Construction activities on facility grounds will need to be carried out by physical plant staff and inmate crews, so any plans for building new structures or for the modification of existing structures will require significant advanced planning and close collaboration with facility staff.

Correctional facilities will have their own landscaping equipment, but community partners will generally be required to provide tools and materials for horticulture program activities, and will need to work with facility staff to arrange for secured on-site storage for all items. Approval for all tools and materials must be obtained in advance, and it is essential to consult with corrections staff to determine which items present a security risk as it is not always obvious to the nonexpert. For example, smaller tools are often of greater concern as they can be hidden in clothing. Irrigation equipment such as hoses and drip irrigation line may be restricted or limited in length because they can be used as aids to climbing over a security fence. Items such as landscaping fabric, plastic sheeting and garden twine may also be restricted as they can be made into ropes.

There may be limited storage for materials inside the secured area of the facility, and bulk items such as compost and wood chips may need to be stored elsewhere on the grounds and brought in by facility staff as needed. Other soil amendments and fertilizers may be purchased in bags or in bulk, depending on the size of the program, storage availability, and the preference of facility staff. Again, approval and storage logistics should be carefully arranged with corrections staff in advance, and an inventory record should be maintained for all program materials along with *Material Safety Data Sheets*, which provide essential information from manufacturers on their product's ingredients and hazards, as well as recommendations for safe handling and dealing with accidental ingestion or exposure.

While there are numerous challenges and considerations in creating horticulture programs within correctional facilities, there is a strong history and potential for developing programs that serve the significant needs of inmate populations. Prison horticulture programs should start small, work in close consultation with corrections staff and inmate participants, and secure support from community partners in order to ensure success and maintain long-term viability.

SUMMARY

Prison horticulture refers to a range of plant-based activities and programs occurring in detention facilities such as prisons, jails, or youth detention facilities. Examples include gardening, farming, landscaping, horticultural therapy, family garden programs, greenhouse activities, composting, and native plant propagation.

Benefits of prison horticulture may include education, improved vocational, teamwork and communication skills, as well as physical and mental health improvement. Correctional institutions may realize cost savings and increased sustainability in facility operations, along with other benefits.

Incarceration as a form of punishment for criminal behavior is a relatively modern development. Prior to the eighteenth century Enlightenment, response to criminal behavior generally involved corporal punishment, torture, slave labor, and social ostracism. The first penitentiaries were developed by the Pennsylvania Quakers in the early nineteenth century. The nineteenth and twentieth centuries saw a number of prison reformatory movements in response to poor prison conditions, and as an attempt to reframe the role of the prison as rehabilitative.

Prison horticulture has been an element of detention facilities throughout history, including large-scale agriculture through forced inmate labor. Reform-minded prison wardens attempted to use horticulture as a means of creating more humane conditions. Incarcerated individuals have used horticulture as a way to meet their own needs, and as an expression of resistance to incarceration. Early examples of prison horticulture programs were seen at the Sing Sing Penitentiary in Ossining, New York, and on Alcatraz Island in San Francisco Bay, California.

In 1982, the federal government declared a "War on Drugs" resulting in a sharp increase in the US prison population and increased sentence lengths. Although the incidence of violent crime has declined, the population of inmates serving life sentences has more than quadrupled since 1984. Communities of color have been disproportionally impacted by the new drug laws and enforcement policies, despite the fact that illegal drug use is approximately equal across race groups. The US rate of incarceration is the highest in the world. Mental illness is prevalent among the incarcerated, especially female inmates, while educational attainment for prisoners is significantly lower than that of the general population.

Education programs in prisons have existed since the first penitentiaries were opened in the United States, with early prisoner education programs focusing on religious training. Modern prisons generally include some degree of inmate education programs including adult basic and secondary education, along with vocational and life skills training.

Prison horticulture programs may use plant-based activities as part of basic or secondary education curriculum, as vocational training, for development of life skills, or as horticultural therapy. Prison horticulture may also be used as a public health effort to improve inmates' physical health and to improve the nutritional content of inmate meals.

The first step in initiating prison horticulture programs is the development of collaborative partnerships between corrections staff, inmates and community organizations. Program goals must align with the objectives of the correctional institution, with a focus on safety and security.

Proposals for prison horticulture programs should specify potential sources of funding, and demonstrate how program activities will address the specific needs of the inmate population. Proposals should highlight the potential for cost savings and other institutional benefits, and should be modest in scope, as correction staff generally prefers programs to start small, with the potential to expand once success has been demonstrated.

Community partners play a number of roles in prison horticulture programs, including providing financial resources, educational material or technical consultation. Community partners may also serve to initiate, develop, and administer programs. Examples of community partners include food banks, community gardening organizations, public health or environmental organizations, churches, agricultural extension agencies, and academic institutions. Examples of potential funding sources include horticultural and community gardening organizations, public health foundations, environmental groups, and land use agencies.

Programs in higher security level facilities will have much stricter limitations on activities. Some prison horticulture programs will include activities outside the prison fence, involving inmates with gate clearance. Prison horticulture programs may begin with a small garden or a may include a larger growing area, a greenhouse, grow lights, classroom space, and composting area. Community partners will usually need to provide tools and materials for horticulture program activities, and will need to collaborate with facility staff to arrange secured storage for all items. Partnering with prison staff is essential for maintaining safety and security, and in the development of horticultural programs within the prison environment.

REVIEW QUESTIONS

1. Name three activities that may occur as part of prison horticulture programs and provide examples of each from the case studies included in this chapter.
2. What are the key events in the historical development of prisons?
3. How has horticulture been used historically in prisons?
4. What is the "War on Drugs" and how has it impacted incarceration trends in the United States?
5. Describe characteristics of US inmate populations and differences between male and female inmates.
6. What are typical education and training programs that exist in correctional facilities?
7. Name three objectives for prison horticulture programs and how they meet the needs of inmate populations.
8. Describe essential partnerships needed for the development of prison horticulture programs and provide examples of each from the case studies included in this chapter.
9. What are the different security levels found in correctional facilities and how do they impact horticulture activities?

10. Describe the logistical considerations for the development of prison horticulture programs.

ENRICHMENT ACTIVITIES

1. Create a map of the correctional facilities in your state or region, including the security levels of each facility, the inmate population incarcerated there, and information on the education and training programs available at each institution.
2. Investigate recent media coverage of correctional facilities and prison reform advocacy groups in your region. Summarize current efforts to address the needs of the incarcerated population in your area.
3. Interview a correctional employee or a member of a prison reform advocacy group. Ask them to identify the major needs of the inmate population with whom they have worked, and how prison horticulture might meet those needs.
4. Draw a prison garden design map, including edible and ornamental landscapes, and a greenhouse. Describe year-round activities for inmate participants.
5. Write a mock letter to a correctional facility manager proposing a prison horticulture program. Describe how the program aligns with the goals of the institution and how it will serve the needs of the inmates.

FURTHER READING

Bureau of Justice Statistics. 2013. *Prisoners in 2012: Trends in Admissions and Releases, 1991–2012*. http://www.bjs.gov/index.cfm?ty=pbdetail&iid=4842, accessed July 24, 2014.

Bushway, S. 2003. *Reentry and Prison Work Programs*. Urban Institute Reentry Roundtable, New York University Law School. http://www.urban.org/UploadedPDF/410853_bushway.pdf, accessed August 17, 2014.

Davis, L., R. Bozick, J.L. Steele, J. Saunders, and J. Miles. 2013. *Evaluating the Effectiveness of Correctional Education: A Meta-Analysis of Programs That Provide Education to Incarcerated Adults*. https://www.bja.gov/Publications/RAND_Correctional-Education-Meta-Analysis.pdf, accessed July 24, 2014.

Garden Conservancy. n.d. *The Gardens of Alcatraz: History of Gardening on the Island*. http://www.alcatrazgardens.or/history.php, accessed August 17, 2014.

Garden Project. 2013. http://www.gardenproject.org/, accessed August 10, 2014.

Helphand, K.I. 2006. *Defiant Gardens: Making Gardens in Wartime*. San Antonio, Texas: Trinity University Press.

Jiler, J. 2006. *Doing Time in the Garden: Life Lessons through Prison Horticulture*. Oakland, California: New Village Press.

Lamberton, K. 2000. *Wilderness and Razor Wire: A Naturalist's Observations from Prison*. San Francisco: Mercury House.

LeRoy, C.J., K. Bush, J. Trivett, B. Gallagher. 2012. *The Sustainability in Prisons Project: An Overview (2004-12)*. Olympia, Washington: Gorham Publishing.

LeRoy, C.J., K. Bush, J. Trivett, B. Gallagher. 2013. *The Sustainability in Prisons Project Handbook: Protocols for the SPP Network*, 1st edition. Olympia, Washington: Gorham Publishing.

Lindemuth, A. 2011. *Can Prison Landscapes be Secure, Restorative, and Ecologically Sustainable?* [Blog post]. January 6. http://www.healinglandscapes.or/blog/2011/01/can-prison-landscapes-be-secure-restorative-and-ecologically-sustainable-guest-post-by-amy-lindemuth/, accessed July 22, 2014.

Mandela, N. 1994. *The Long Walk to Freedom: The Autobiography of Nelson Mandela.* Boston: Little, Brown.

Morris, C. 2013. *Growing New Lives: A Gardening Program on the Largest Penal Colony in the world. Edible Manhattan*, July. http://www.ediblemanhattan.com/departments/ideas/growing-new-lives/, accessed July 29, 2014.

Nadkarni, N.M. 2006. The moss-in-prisons project: Disseminating science beyond academia. *Frontiers in Ecology and the Environment*, 4, 442–443.

National Association of State Budget Officers (NASBO). 2013. *State Spending for Corrections: Long-Term Trends and Recent Criminal Justice Reform Policies.* http://www.nasbo.org/sites/default/files/pdf/State%20Spending%20for%20Corrections.pdf, accessed July 24, 2014.

Oregon Department of Corrections. 2014. *Workgroup on Corrections Health Care Costs Report.* http://olis.leg.state.or.us/liz.2014R1/Downloads/CommitteeMeetingDocument/33378, accessed July 29, 2014.

Sentencing Project. 2014. *Trends in Corrections Fact Sheet.* http://sentencingproject.org/doc/publications/inc_Trends_in_Corrections_Fact_sheet.pdf, accessed July 24, 2014.

Therapeutic Landscapes Network. n.d. *Gardens in Prisons.* http://www.healinglandscapes.org/Gardens/prisons.html, accessed July 22, 2014.

Ulrich, C. and N.M. Nadkarni. 2009. Sustainability research and practices in enforced residential institutions: Collaborations of ecologists and prisoners. *Environment, Development and Sustainability*, 11(4), 815–832.

United Nations Congress on the Prevention of Crime and the Treatment of Offenders. 1955. *Standard Minimum Rules for the Treatment of Prisoners.* https://www.unodc.org/pdf/criminal_justice/UN_Standard_Minimum_Rules_for_the_Treatment_of_Prisoners.pdf.

7 Horticultural Therapy with Special Populations

Leigh Anne Starling

CONTENTS

OBJECTIVES

Upon completion of this chapter, the student should be able to

- Define horticultural therapy (HT).
- List and describe the benefits of horticultural therapy participation.
- Compare and contrast the different types of HT programs.
- Understand the application of HT activities.
- Describe and define the American Horticultural Therapy Association Therapeutic Garden Characteristics.
- Understand how to adapt the horticultural environment for accessibility.

KEY TERMS

- Horticultural therapy
- Therapeutic horticulture
- Vocational horticulture
- Horticultural therapist
- Treatment/Rehabilitation
- Participant's plan
- Accessible and adaptable
- Americans with Disabilities Act of 1990
- Principles of Universal Design
- American Horticultural Therapy Association

HORTICULTURAL THERAPY

Horticultural therapy brings together two professional disciplines: horticulture and therapy. Horticulture is defined as the cultivation of flowers, fruits, vegetables, or ornamental plants. Therapy is defined as the treatment of disease or disorders, as by some remedial, rehabilitating, or curative process. Combining these two definitions leads to an understanding of the professional discipline of HT. Horticultural therapy involves the use of horticulture, horticultural practices, and the horticulture environment to assist in the healing, rehabilitation, or amelioration of persons requiring treatment.

Horticultural therapy has been defined as a process through which plants, gardening activities, and the innate closeness people feel toward nature are used as vehicles in professionally conducted programs of therapy and rehabilitation (Davis 1998, 3).

A more comprehensive definition describes Horticultural therapy as a professionally conducted client-centered treatment modality that utilizes horticultural activities to meet specific therapeutic or rehabilitative goals of its participants. The focus is to maximize social, cognitive, physical, and psychological functioning and to enhance general health and wellness (Haller 2006, 5).* The American Horticultural Therapy Association defines horticultural therapy as the engagement of a client in horticultural activities, facilitated by a trained therapist, to achieve specific and documented treatment goals (American Horticultural Therapy Association 2007, 3). The first definition acknowledges the people–plant relationship, that is, the basis of HT, the innate closeness people feel toward nature. The second definition directs attention to the purpose and the participant. The third definition identifies the trained therapist and eliminates the aspect of rehabilitation; however, the definition of therapy includes rehabilitation. This chapter will discuss horticultural therapy in the context of the second definition.

The definition of HT includes separate components and connected parts. For a program to be defined as HT, four components must be present: the client, documented treatment goals, the horticultural activity, and the trained therapist (Relf and Dorn 1995, 98–101; Haller 2006, 5). The client represents an individual or group of individuals of all ages, backgrounds, and abilities who are receiving services for an identified disability, illness, or life circumstance requiring treatment or rehabilitation. Treatment or rehabilitation goals are developed cooperatively with the client to help achieve desired outcomes. The horticultural activities are primarily based on the cultivation of plants (Relf and Dorn 1995, 98–101; Haller 2006, 5). The therapist is trained in the use of horticulture as a medium to assist the client in achieving their treatment goals. These four program elements are that which differentiate a HT program from other types of programs using horticultural activities.

HISTORY OF THE HT PROFESSION

In the early history of the United States, people with disabilities were placed in asylums, poorhouses, almshouses, and sanitariums. Farming was a means of self-sufficiency and patients were often involved in the growing and harvesting of crops. As early as 1812, Dr. Benjamin Rush, considered the Father of American Psychiatry, acknowledged improvement in male patients who worked in the garden (AHTA 2007, 3). Dr. Rush is credited with being one of the first to document the therapeutic benefits of gardening in a hospital setting. During the 1800s, two hospitals notable in the history of HT began to use horticulture for therapeutic purposes. Established in 1813, The Asylum for Persons Deprived of the Use of Their Reason (today Friends Hospital, Pennsylvania) involved patients in gardening and is thought to have established the first known greenhouse for use in patient activities in 1879 (Davis 1998, 5; AHTA 2007, 3). Michigan's Pontiac State Hospital, established in 1878 as the Eastern Michigan Asylum for the Insane, used portions of its farmland for patient's treatment, and gardening was part of the patient's work program from the beginning (NCTRH 1976).

The 1900s saw the use of gardening shift from a form of work to a form of therapy (Figure 7.1). The Menninger Clinic, a notable hospital in the history of HT, was opened in 1919. Dr. C.F. Menninger and his son, Karl, established the Menninger Clinic in Topeka, Kansas, and patients were involved in horticultural activities as part of their daily treatment (Lewis 1996, 78; Davis 1998, 6). The Menninger Clinic was instrumental in the advancement of HT in psychiatric hospitals and played a future role in the establishment of academic training in horticultural therapy. During the 1920s and into the 1940s, horticultural therapy was largely thought of as part of occupational therapy. Horticultural activities were incorporated into training and textbooks and in 1942, what has come to be known as the first course on using horticulture for therapy, was taught in an occupational therapy program at Milwaukee Downer College in Wisconsin (NCTRH 1976; Davis 1998, 7).

A turning point in the profession came in the late 1940s and early 1950s with the return of World War II veterans and an increase in the number of Veteran's hospitals. Occupational therapists, volunteers, and garden club members brought gardening and plants into the hospitals to provide activities for returning veterans. Garden club members recognized the important role of plants in rehabilitation, and by 1951, the National Council for State Garden Clubs had identified HT as one of their objectives (Davis 1998, 7). Still an objective today, the National Council promotes therapeutic gardening and horticultural therapy in local communities through offering program assistance and an awards program.

Two individuals are noted during this time and given credit for much of the advancement of the HT profession in the 1950s, Rhea McCandliss and Alice Burlingame. Working under Dr. Karl Menninger from 1946 to 1953 at the Menninger Clinic, and again from 1959 until her retirement in 1972, Rhea McCandliss became one of the first professional horticultural therapists (Lewis 1996, 78–79). Alice Burlingame, a psychiatric social worker and occupational therapist, started the first HT program at Michigan's Pontiac State Hospital in 1951. She pursued her interest in working with people and plants as a student in landscape architect and horticulture at Michigan

Perkins School for the Blind, located in Watertown, Massachusetts, is the first school in the United States for people who are visually impaired. Horticulture was part of student life at Perkins from 1912 through 1951 as part of vocational training. Students were involved in tending the orchards and gardens as a means toward independence. In the 1970s, Perkins began providing horticultural therapy services and in 1979, Perkins received grant funding to construct its first greenhouse and teaching center. The horticultural therapy program became nationally recognized and eventually outgrew its original greenhouse. In 2003, funded through a grant and donations, the current 5000-square-foot state of the art Pappas Horticultural Center was built. The fully accessible Center includes a large geothermally heated greenhouse, water fountains, an accessible garden, and a garden shop. The current program provides students with leisure activities, sensory training, prevocational training, and job opportunities. Horticulture therapy benefits not only students attending Perkins programs but also those involved in the outreach services for elders and mainstreamed students too. Today, Perkins' horticulture program is a national model in education for people who are blind.

FIGURE 7.1 History in the present. (Adapted from Research Library, Perkins School for the Blind, in email correspondence to author, August 29, 2014.)

State University where she joined with Dr. Donald Watson, Horticulture Department Faculty, to present the first HT workshops held at Michigan State University in 1952 (Lewis 1996, 79; Davis 1998, 8). Rhea McCandliss was one of the speakers for these workshops. Soon afterward, Michigan State University began offering coursework in HT in conjunction with clinical training through Pontiac State Hospital's Occupational Therapy program and in 1955, awarded the first Master of Science degree in horticultural therapy (NCTRH 1976). By this time, public gardens and arboretums had started offering programs, the most notable of which were Arnold Arboretum, in Boston, and Holden Arboretum, in Ohio. A few years following, in 1959, the Institute for Rehabilitation Medicine at New York University Medical Center, in New York City, opened a HT greenhouse. Howard Rusk, the institute director, fully integrated the HT program into the treatment programs for individuals who had physical disabilities (NCTRH 1976; Davis 1998, 8).

Through the 1960s, the profession experienced growth through published knowledge and the collective efforts of dedicated individuals. The momentum created from the 1950s HT workshops continued and in 1960, Alice Burlingame and Dr. Watson published the first textbook in HT, *Therapy through Horticulture*, followed by the National Council for Garden Clubs' *Handbook for Horticultural Therapy*, published in 1964 (Davis 1998, 8; AHTA 2007, 4). Rhea McCandliss' active involvement in HT at the Menninger Foundation led her to conduct a survey in 1968 that is noted as one of the first attempts to establish the prevalence of, and interest in, HT programs in the United States. This survey indicated the need for trained horticultural therapists and highlighted the need for communication among programs (Lewis 1996, 8). By the end of the 1960s, HT had become recognized as a beneficial activity to meet the specific needs of patients and began to be thought of independently from occupational therapy.

The 1970s saw the formal establishment of the profession. Two leading figures in academia emerged during this time, Dr. Richard Mattson and Dr. Diane Relf. Dr. Mattson, professor at Kansas State University, Karl Menninger and Rhea McCandliss worked with Menninger Clinic staff and KSU horticulture faculty to develop the first horticultural therapy bachelor's degree in 1971 (Davis 1998, 9–10). Dr. Mattson directed the K-State program for the next 45 years. Following this notable event in the academic community, other university programs began to follow. During the early 1970s, South Carolina's Clemson University, Michigan State University, University of Maryland, and Texas Tech University all offered either horticultural therapy options or degrees. A pivotal event occurred in 1973 when a group of 20 people led by Earl Copus, the director of Melwood Horticultural Training Center in Maryland, Paula Diane Relf, a University of Maryland student working on her PhD in horticultural therapy, Rhea McCandliss, and others came together and formed the National Council for Therapy and Rehabilitation through Horticulture (NCTRH). The NCTRH formalized much of the information about HT in the early years of the profession and established the first professional registration standards for horticultural therapist in 1975. In 1976, the University of Maryland awarded the first PhD in HT to Paula Diane Relf (AHTA 2007, 4). Over the next three decades, while a professor at Virginia Polytechnic Institute and State University, Dr. Relf researched, published, and promoted the therapeutic benefits of horticulture all the while assisting NCTRH with its professional development and publications.

With the establishment of the NCTRH, the profession now had an identity. The 1980s saw programs develop and jobs increase. During this time, NCTRH was focused on bringing public awareness to the profession through grant-funded projects designated to develop jobs for persons with disabilities in the horticulture industry. Beginning in 1982, and over the next 9 years, NCTRH administered five federally funded employment programs through the U.S. Department of Education and U.S. Department of Health and Human Services that resulted in the hiring of 2269 individuals with disabilities in the horticultural industry (Davis and De Riso 1992, 185–186). In 1987, the NCTRH changed its name to the American Horticultural Therapy Association (AHTA).

The 1990s was a time of change for the AHTA and the profession. Government funds were no longer available and professional horticultural therapists and botanic gardens developed training programs in horticultural therapy to meet the demand for specialized training. It was evident that there was a need for research and consolidated HT information.

In 1990, Dr. Relf was instrumental in the establishment of the People–Plant Council (PPC) whose mission was to facilitate and promote research and communication around the discipline of people–plant relationships. Today PPC continues to promote research in HT as well as in all dimensions of plants in human well-being. AHTA focused on providing leadership and education to its members and two important books presenting the current practices in the profession were published during this time. Through the Friends of Horticultural Therapy, an AHTA support organization, *Horticultural Therapy and the Older Adult Population*, was published in 1997. Recognized leaders in the profession contributed their knowledge and experience to write *Horticulture as Therapy: Principles and Practice*, published in 1998, which AHTA promoted as the textbook representing the profession of HT.

By the turn of the century, HT, as a profession, and AHTA, as the representative organization, were moving toward increased professional standards, achievement of professional recognition, and collaboration with allied professions such as landscape architecture, and environmental psychology. Today, HT is practiced throughout the world.

BENEFITS OF WORKING WITH PLANTS

The process of horticultural therapy occurs through the relationship that develops between the person and the plant, the specific plant-related activities a person does or completes when taking care of plants, and the interactions that occur when people are engaging with plants in a horticultural environment. Ultimately, the therapeutic process is experienced through direct interaction and nurturing of plants and the therapeutic element is the benefit a participant experiences as a result of this interaction (Relf and Dorn 1995, 101). Participants in HT may experience emotional, intellectual, social, and physical benefits.

Emotional: By taking care of plants, a person may

- Increase self-esteem, self-awareness, and personal responsibility
- Experience decreased stress and anxiety

- Recognize life's cycle of change and transition and apply this awareness to one's self
- Become motivated through recognition of the parallel nature of human and plant growth
- Experience increased self-confidence
- Release aggression and anger

Intellectual: By completing plant-related tasks, a person may

- Learn a new skill or relearn an old one
- Develop problem solving skills
- Increase memory and attention to detail
- Practice structuring and sequencing to allow for impulse control
- Develop coping mechanisms to manage stress and anxiety
- Practice and improve initiation of tasks

Physical: By working with plants, a person may

- Improve coordination
- Exercise and build strength
- Increase stamina and endurance
- Develop and improve fine motor skills involving hands, wrists, fingers, or feet
- Develop and improve gross motor skills involving arms, legs, feet, or entire body
- Reconnect with one's physical body after experiencing accident, illness, or trauma

Social: By engaging in plant activities, a person may

- Develop relationships with other participants
- Engage in cooperative experiences
- Develop appropriate boundaries with others
- Learn healthy work attitudes and behaviors
- Decrease feelings of isolation
- Practice communication skills

All benefits overlap. For example, a horticultural therapist working with someone who has a physical or intellectual disability can develop an activity to address all four benefits. A participant can work on self-esteem through successfully growing plants and, at the same time, receive physical benefits through planting, intellectual benefits through measuring planting depth, and social benefits through the group activity. These identified benefits also help to understand types of HT programs and settings.

HT PROGRAM TYPES AND SETTINGS

The first component in the definition of horticultural therapy involves the client. Horticultural therapy programs provide services to individuals who have a broad range of abilities and different life circumstances. A person may participate in an

HT program for reasons such as a disability, a medical condition, or a traumatic life event resulting in an emotional, physical, social, or cognitive impairment.

Individuals might have a physical condition such as recovering from stroke or have complications from chronic illness or disease. For example, horticultural therapy has been found to be useful in reducing stress and improving mood of patients in a cardiac rehabilitation program (Wichrowski et al. 2005, 273). Individuals might have a developmental disability such as an intellectual disability or autism. Others might have neurocognitive disorders such as Alzheimer disease or cognitive impairment resulting from brain injury. For example, by participating in horticultural therapy, positive effect on mood and increased cognitive functioning in long-term care residents with Alzheimer disease was achieved (D'Andrea et al. 2007–2008, 16). Participants might have a mental health disorder such as depression, anxiety, or are in treatment for substance use. In a study with patients diagnosed with psychiatric disorders receiving HT services at a state hospital, patients reported increased self-esteem and improved social interactions, and demonstrated using horticulture as a leisure skill (Sellers 2001, 19).

Program Types

To aide in distinguishing HT from other programs that use horticultural activities, AHTA has defined the following four types of programs:*

Horticultural therapy: In a horticultural therapy program, clients are directly involved in horticultural activities to meet identified goals. Activities are facilitated by a trained professional. AHTA believes that horticultural therapy is an active process which occurs in the context of an established treatment plan where the process itself is considered the therapeutic activity rather than the end product.

Therapeutic horticulture: In a therapeutic horticulture program, goals are not clinically defined and documented but the leader will have training in the use of horticulture as a medium for human well-being. Therapeutic horticulture is a process that uses plants and plant-related activities through which participants strive to improve their well-being through active or passive involvement.

Vocational horticulture: A vocational horticulture program, which is often a major component of a horticultural therapy program, focuses on providing training that enables individuals to work in the horticulture industry professionally, either independently or semi-independently. These individuals may or may not have a disability.

Social horticulture: Social horticulture, sometimes referred to as community horticulture, is a leisure or recreational activity related to plants and gardening. No treatment goals are defined, no therapist is present, and the focus is on social interaction and horticulture activities. A typical community garden or garden club is a good example of a social horticulture setting.

* Copyright 2007. American Horticultural Therapy Association (AHTA). AHTA Definitions and Positions.

Three of these programs are led by a professional horticultural therapist and one is not. This chapter will focus on those programs that meet the requirements to be defined as a HT program: horticultural therapy, vocational horticulture, and therapeutic horticulture.

Program Settings

Horticultural therapy programs provide services to diverse groups of people in different settings. Clients are referred to an HT program based on what services they require. Referrals may come from a community provider like an outpatient mental health agency, from within the facility by the treatment team, from a school providing transition services to students with developmental disabilities, or even a family member wanting their elderly parent to live in a facility with gardens.

The settings in which HT programs take place define the overall program goal and thus guide the individual's treatment or rehabilitation goals. The HT program goals are aligned with the mission and purpose of the organization such as vocational training, rehabilitation or restoration, and maintaining health and wellness. The horticultural therapist supports the client within this context.

Mental Health

Programs in the mental health setting are considered therapy programs with the goal of improving mental health and personal wellness. Mental health programs can take place in, for example, an inpatient hospital setting or a community mental health day program. Both settings serve different groups of people. In an inpatient setting, patients have acute psychiatric disorders and generally limited hospitalization stays, on average as short as 1 week. Common patient goals are developing coping mechanisms by using horticulture, self-expression through using plants as a metaphor for self, or exploring personal responsibility through plant care. A community day treatment program generally serves individuals who have chronic mental illness. Participation is based on individual needs and can range from weeks to months. Common patient goals in this type of program are learning social skills through working in groups, learning problem-solving skills by completing horticultural tasks, maintaining healthy behaviors by participating in the program, and increasing quality of life through horticultural activities.

Long-Term Care

In the long-term care setting, the program may provide either horticultural therapy or therapeutic horticulture services, and in some instances both. Programs may be found in rehabilitation hospitals, assisted living facilities, and residential facilities, with the program goal of assisting persons with their medical needs or daily activities. HT programs in this setting may serve individuals with cognitive disorders like Alzheimer disease or persons with chronic illness. Common HT goals include socialization through group activities, memory stimulation through touching, smelling, and tasting plants, and physical goals such as maintaining strength and building stamina by gardening. Programs may also serve individuals in the advanced stages of illness or disease where passive interaction with plants is considered therapeutic. Studies with nursing home residents with dementia have demonstrated that allowing access to gardens has been shown to decrease agitation and inappropriate behaviors in nursing home residents who previously did not have outdoor access (Detweiler et al. 2012, 104–105). In the same study, having access to gardens is believed to have provided residents with increased autonomy by being able to choose to go outside.

Rehabilitation

Programs in this setting may include therapy, therapeutic, or vocational services. Depending on the type of rehabilitation service provided, programs take place in a medical setting such as a hospital or skilled nursing facility and assist patients who are recovering from accidents, cardiac complications, brain injuries, and other physically related traumas. Physical restoration is the program goal. Common patient goals include building strength through gardening, problem solving and learning new skills through completing horticultural activities, and stress reduction by working with plants.

For example, a rehabilitation clinic serving patients with chronic musculoskeletal pain investigated the use of horticultural therapy to improve physical functioning, mental health, and ability to cope with chronic pain. This study found that the patients in the control group showed significant improvement in the areas of physical functioning and bodily pain and were able to use horticultural activity to control pain (Verra et al. 2012, 47–48). Another type of rehabilitation program serves individuals with substance use disorders; this type of rehabilitation program is sometimes referred to as a recovery or treatment program. The program goal is the physical, medical, and mental health restoration of the person by overcoming substance use. Client goals in this setting include using horticultural activities as coping skills, learning new job skills such as landscaping skills, using plant metaphors to increase self-awareness, and developing anger management skills through physical activity.

Corrections

Correctional facilities include state and federal prisons or local city and county jails. Persons in correctional facilities may include individuals diagnosed with mental illness, substance use, chronic physical problems, or other trauma-related disorders. Programs taking place in a correctional setting are usually vocational horticulture programs but not all are vocational HT programs. The difference between the two is more than just learning about horticulture and acquiring vocational skills; it is the personal growth in "values, beliefs, and attitudes" that can be applied to other areas of the participant's life (Rice et al. 1998, 263–264). The program goal is rehabilitation through vocational training, and in this setting, the program goals apply to the group as a whole and individualized plans are usually not developed. Examples of program goals applicable to all participants are to develop job skills, work behaviors, social skills, and problem-solving skills. Other goals may include developing anger management skills and impulse control.

Vocational

Vocational programs serve individuals of all abilities. The diversity of vocational programs is such that they may be found in schools, nonprofit organizations, residential facilities, healthcare facilities, or on-site at the place of employment. A person recovering from an injury may need to learn a new skill and might participate in a vocational program, or people with developmental disabilities may choose vocational training as an option in school. All vocational programs include plant production and all provide training in work-related skills and behaviors; some programs offer paid employment while others assist participants in obtaining paid employment in the community. For example, a facility serving adult men in a low security mental health unit developed a worker's cooperative to teach vocational skills through growing food for cooking groups and flowers and plants for sale. It was observed that the men demonstrated cooperative skills and motivation increased among patients (Page 2008, 28–30). In a study with juvenile offenders participating in a vocational training and rehabilitation program, the participants who chose horticulture experienced increased self-esteem and had lower recidivism rates compared to those in the non-horticulture group (Hale et al. 2005, 46).

Community

Depending on the organization, community-based programs provide therapy, therapeutic, and vocational services to other community organizations. Programs may be provided through botanic gardens, arboretums, park and recreation departments, or community colleges. These programs are usually fee based and may be offered on a one-time or ongoing basis. Program goals are designed to meet the specific needs of the client and either offer services on-site or bring services to an organization. These types of programs are an important asset in a community often providing services that would otherwise be unavailable to an organization with limited resources or no access to outdoor space.

GOALS AND OBJECTIVES

The second component in the definition of horticultural therapy involves treatment or rehabilitation goals. Goals exist within the participant's plan. Depending on the individual served and the program setting, a participant's plan may be referred to as either a treatment plan, an individualized rehabilitation plan, and in some settings, an individual service plan. The participant's plan is the guiding plan-of-action. The plan defines which services are provided, who is responsible for providing the service, and the progression of services provided. Common to all types of participant plans are the goals and objectives.

Goals are strength-based and reflect the needs and desires of the participant. Goals should have personal meaning to the participant. They are defined in terms of the desired outcome, what is expected to be accomplished. Depending on the length of time a person receives HT services, goals can be short term or long term; they should be reasonable and attainable during the time frame specified. Objectives are defined as the steps taken to meet the goal(s) and are used to determine whether progress is being made toward the goal. Objectives are written for each goal and specify in concrete terms what is to be done and the measurable changes to be observed; objectives include the projected date or time of achievement and are generally written in sequential order. The goals and objectives formalized into a treatment or rehabilitation plan gives purpose to the therapeutic relationship.

In a therapeutic horticulture program where the goals are not clinically defined, often the HT program goal is the goal of the organization. For example, in a residential care facility, all residents might participate in a therapeutic horticulture program as a means to experiencing a positive quality of life while living in the care facility. The goal is defined in terms of the client participating in horticultural activities designed to promote and provide opportunities for socialization. A vocational goal is a type of program goal and can be individualized for the participant but is also not clinically defined; rather it is described in terms of skill acquisition and work behaviors learned. The key word here is "clinical." Clinical refers to medical treatment and goals are described within that context. In HT programs, clinical goals, social goals, or vocational goals are part of the participant's plan and documented in the participant's plan.

The following case studies are examples of three different types of HT programs: Legacy Health is an example of a therapy program in a rehabilitation hospital; Melwood is an example of a vocational program in a nonprofit organization; Missouri Botanic Garden is an example of a community-based program offering both therapy and therapeutic services.

CASE STUDY 1 Legacy Therapeutic Gardens and Horticultural Therapy Program

Location: Portland, Oregon.

The Legacy Therapeutic Gardens and Horticultural Therapy program is part of the Legacy Rehabilitation Services, Legacy Health, a nonprofit healthcare system. The HT program began in 1991 and expanded to include the Rehabilitation Institute of Oregon in 1996.

HORTICULTURAL THERAPY PROGRAM

The HT program provides services to patients in acute care behavioral health, acute care rehabilitation, and outpatient rehabilitation utilizing eleven therapeutic gardens throughout the six Legacy hospitals.

The inpatient HT services include treatment sessions for groups and individuals of all ages. Patients include individuals who have a physical or cognitive disability resulting from spinal cord injury, brain injury, stroke, multiple trauma, burn injuries, multiple sclerosis, and amputation. Behavioral health patients also include individuals who have a mental or emotional disorders resulting from depression, anxiety, adjustment disorder, bi-polar disorder, and posttraumatic stress. Patient length of stay varies with adult rehab about ilitation of 10–14 days and behavioral health 5–10 days depending on diagnosis and rehabilitation services provided. Annually, an estimated 800 patients are served by the HT program.

Patients are identified and referred for HT services through an interdisciplinary team conference approach. Patients receive HT services once or twice a week; sessions are conducted year-round both indoors and outdoors, depending on patient goals and weather conditions. Patients work with the HTR on the goals identified in their individual treatment plan. Patient goals for adult rehabilitation may include the following:

- Increasing strength and endurance
- Improving balance
- Focusing of attention
- Improving memory
- Sequencing of tasks
- Increasing problem solving skills
- Developing communication skills
- Exploring leisure adaptations
- Reintegrating into community

THERAPEUTIC GARDENS

The therapeutic gardens are designed to meet the specific needs of the patients served in each of the six medical facilities. The gardens also facilitate use by treatment team members (physical, occupational, recreation, child life, and art therapies and speech and language pathologists) and to promote independent leisure activity. Most gardens are public gardens and also open to patients and their families and staff members and offer special programs and events year-round. Three of the gardens are secured gardens and offer special programs and events during specified open garden times during the summer. Garden plantings include annuals, perennials, trees, and shrubs selected for their use in HT activities, sensory qualities, and wildlife benefits. Planting structures and pathway dimensions, grading, and materials are specific to accessibility requirements of patients who use wheelchairs and other mobility devices.

Details about the Legacy Therapeutic Gardens and Horticultural Therapy Program can be found on their website (Figures 7.2 and 7.3).

Resources

Teresia Hazen, in email message to author, July 22, 2014.

Legacy Health. Horticultural therapy. Available at: http://www.legacyhealth.org/health-services-and-information/health-services/for-adults-a-z/horticultural-therapy/professional-information/what-is-ht.aspx. Last accessed on July 20, 2014.

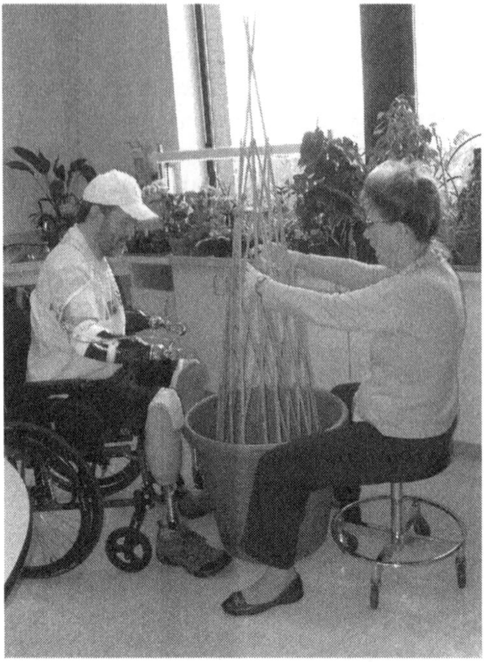

FIGURE 7.2 Horticultural therapist Teresia Hazen teaches a patient to build a trellis to assist with increasing coordination and to build strength. (Courtesy of Legacy Health.)

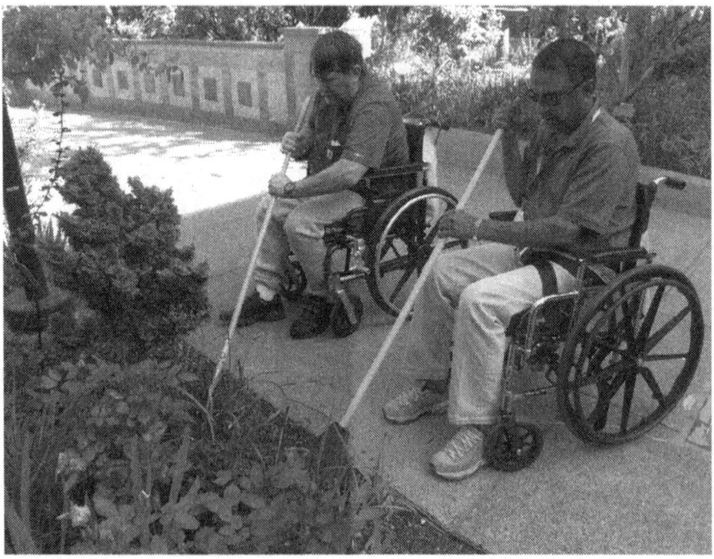

FIGURE 7.3 Patients engage in raking to build upper body strength. (Courtesy of Legacy Health.)

CASE STUDY 2 Melwood Horticultural Training Center

Location: Upper Marlboro, Maryland.

Melwood, a nonprofit organization, began in 1963 as the Melwood Agricultural Training Center. The name was changed in 1968 to Melwood Horticultural Training Center and, as services have expanded beyond horticultural training, the organization is now known as Melwood. The HT program is located within the larger Community Services division.

HT PROGRAM

The HT program provides services to individuals with a broad range of abilities. Clients include individuals who might have neurological, intellectual, sensory, or physical disabilities which require work, social, and environmental adaptations to meet their specific needs.

People make a request to participate in the HT program and go through an intake process with Melwood to ensure adequate funding is available. The person works with a case manager from Melwood and his/her individual team to develop an individual plan. Within the plan are the person's goals. Examples of goals include

- Interacting with the community
- Learning greenhouse skills
- Gaining work skills for full time employment
- Participating in activities in the garden center

A person may choose to work in the HT program with the HTR in support of achieving their goals. Persons working in the HT program are employees and Melwood is the employer. Employees are paid minimum wage and work a maximum of 4 h a day, up to 5 days a week. The Horticultural Center employs between 36 and 40 individuals with the length of employment ranging from months to several years depending on the person's goals.

The Center employees are involved in all aspects of plant production and plant sales. Community plant sales are held throughout the Washington, D.C. area in spring and fall and the garden center is open to the public year-round. On average, the center sells about 8000 plants annually. The center also grows plants for use in Melwood's Landscaping and Horticultural Services contract sites such as the U.S. Department of Agriculture.

HORTICULTURAL FACILITY

The Melwood Horticultural Training Center has two distinct areas: the garden center and the conservatory. The garden center facilities include four large production greenhouses and 3200 square feet of outdoor production area. Plant production includes annuals, perennials, herbs, and vegetables. Seasonal production includes mums, poinsettias, pansies, hanging baskets, geraniums, and coleus. The Horticultural Center also includes a conservatory with plant collections

FIGURE 7.4 Melwood conservatory, Melwood, Upper Marlboro, Maryland. (Photo by Leigh Anne Starling.)

representing varieties of interior plants, a pond, and a group activity space. The conservatory is used by Melwood participants and staff for Melwood events. The Garden Center includes a classroom area, areas for activities, and a library for the employees of the Garden Center.

Details about the Melwood Horticultural Training Center can be found on their website (Figures 7.4 and 7.5).

Resources

Personal interview: Sheila Gallagher, HTR, interviewed by Leigh Anne Starling, July 18, 2014.

Melwood. Available at: http://www.melwood.org. Last accessed on July 21, 2014.

CASE STUDY 3 Missouri Botanical Garden Therapeutic Horticulture Program and Horticulture Therapy Contract Services

Location: St. Louis, Missouri.

Horticulture Therapy Contract Services is part of the Therapeutic Horticulture Program offered at Missouri Botanical Garden. The Therapeutic Horticulture Program began in 2004–2005 and is part of the education programs.

HORTICULTURE THERAPY PROGRAM

Horticulture Therapy Contract Services provides programs to community-based organizations located in the local St. Louis area. Community programs served include, but are not restricted to, senior living centers, healthcare facilities,

FIGURE 7.5 Horticultural therapist Sheila Gallagher conducts a horticultural therapy group with individuals to promote socialization. (Photo by Leigh Anne Starling.)

rehabilitation centers, and nonprofit organizations. Individuals receiving HT services include older adults with a wide range of physical and cognitive abilities, patients receiving medical treatment for cancer, children recovering from accident or illness, individuals with sensory or intellectual disabilities, adults with chronic disease, and adults and children recovering from physical and emotional trauma.

The Horticulture Therapy Contract Services is a fee-based program. Upon contacting the Therapeutic Horticulture Program, an organization may request HT services and activities designed to meet the specific needs of their clients. Annually, 10 to 12 community organizations receive ongoing HT services on a weekly or monthly basis. Programs may be on-site at Missouri Botanical Garden or off-site utilizing the organization's garden. Program length for both session time and frequency of services is dependent upon client goals and services requested. The HTR works with the organization to identify the client's goals and program services and works with the organization's staff to ensure documentation of the participant's goals and objectives. Depending on the organization and the clients served, goals might include the following:

- Reducing stress
- Developing coping skills
- Socializing
- Integrating into community
- Increasing physical strength
- Problem solving

- Stimulating memory
- Sensory stimulation
- Following instructions
- Developing social skills

While some of the organizations have gardens on-site, Horticulture Therapy Contract Services is responsible for supplying all plants and materials to conduct programs, transporting all supplies and materials, and for the care and maintenance of plant material utilized in the activities. The organization must be able to provide staff coverage to care for gardens on their premises.

HORTICULTURAL FACILITY

The horticultural facility includes a sensory garden and a small production greenhouse. The sensory garden plantings include annuals, perennials, and herbs selected for their use in HT activities and sensory qualities. The garden includes a fountain, pergola, and shade trees. Planting structures include raised beds, containers at varying heights, and ground beds. Planting structures and pathway dimensions meet accessibility requirements for clients who use wheelchairs and other mobility devices.

The greenhouse is used primarily for plant production for both on-site and off-site programs. Grow lights are used for starting seeds and plant propagation.

Details about Missouri Botanical Garden Therapeutic Horticulture Program and the Horticulture Therapy Contract Services can be found on their website (Figures 7.6 and 7.7).

Resources

Personal interview: Donald Frisch, HTR, interviewed by Leigh Anne Starling, July 22, 2014.

Missouri Botanical Garden. Horticulture Therapy Contract Services. Available at: http://www.missouribotanicalgarden.org/learn-discover/adults/therapeutic-horticulture.aspx. Last accessed on July 21, 2014.

HT ACTIVITIES

The third component in the horticultural therapy definition involves the horticultural activity. The basis of the activity is the plant and the cultivation of plants. A wide range of available plant-related activities like planting seeds, propagation, planting, or deadheading gives the horticultural therapist options for planning short- or long-term projects. The extension of activities from plant products such as dried flowers, herbs for cooking, natural dyes, or arts and crafts provides variety; however, these are not direct plant-related activities and somewhat removes the people–plant relationship from the activity.

The versatile nature of plants makes HT an accessible activity that can be adapted to meet the needs of the participant or group. Horticultural therapists must have a knowledge and understanding of the persons they serve. Knowledge of the specific disability, illness, or condition of the participant and an understanding of what that

FIGURE 7.6 Horticultural therapist Donald Frisch works with a participant to plant in order to develop problem-solving skills and learn a new skill. (Photo by Donald Frisch.)

means to the person is necessary to plan an effective activity. It is the horticultural therapist's responsibility to choose and provide activities in support of the goals and objectives defined for the group or individual (Table 7.1). Activities should have a purpose and be purposeful; if possible, clients should be aware of why they are doing what they are doing. In all activities, the focus is on the therapeutic process of the activity. Three things to consider when conducting activities include the following:

- Focusing too much on the end product will be at the expense of interacting with the client(s) and paying attention to the process.
- Horticultural therapists are facilitators. While education may be part of the activity or program, avoid expectations such as perfect performance and creating competition among participants.
- Recognize personal circumstances which might be counterproductive to the group. Activities are designed to meet the needs of the clients, not the therapist.

FIGURE 7.7 Zimmerman Sensory Garden, Missouri Botanical Garden, St. Louis, Missouri. (Photo by Donald Frisch.)

Activity Planning

Activity planning starts with the participant's goals and is planned in context of the overall horticultural component of the program. Activities have a short-term focus, to meet the goals of the participant, and may have a long-term focus of sustaining horticultural resources for program use.

Traditionally, HT is conducted as a session with a group or individual. Sessions are generally time limited, 1 hour to 1½ hours and accommodate short-term activities. Short-term activities are generally completed in one session, have a defined beginning and end, and allow a participant to observe results and experience a sense of completion. Activities are broken down into steps, not only to achieve the end result, but each step should be designed to achieve some specific benefit. If group members are regular attendees, thereby offering continuity between sessions, activities can be broken up into parts and completed over a series of sessions. When programs are not conducted in sessions, such as vocational programs where clients may work for several hours a day, activities can still be short-term, for example, completed in 1 day, and are considered components in the long-term production process. Vocational activities such as mowing a lawn can be considered a short-term activity in the overall process of providing landscaping services to a customer on a regular basis.

In that plants produce more plants and plants produce plant products, planning must be in place to accommodate this ongoing process. Planting calendars and growing schedules are important tools for this purpose. An example of a long-term plan is growing flowers for flower arranging. The short-term activity is planting the seeds in one session. This long-term plan of growing flowers for cutting can take several

TABLE 7.1
Examples of HT Activities

Task Activity: Propagating Geraniums from Cuttings

Therapy program: This activity can be used with patients to assist in developing eye–hand coordination and increasing strength. The goal is clinically defined and identified in the individualized treatment plan.

Therapeutic program: This activity can be used in a group setting with residents to promote socialization and enhance cognitive function through sensory stimulation. The goal supports the overall organizational goal of increasing quality of life.

Vocational program: This activity could be used with clients to assist in learning how to follow directions and complete a task. The goal is defined in vocational terms in support of learning job skills as part of a client's rehabilitation plan.

Metaphor Activity: Taking Care of "Sick" Plants

Taking care of "sick" plants is an opportunity for patients to take care of themselves. Removing brown leaves, pruning, repotting, and deadheading are tasks involved in plant maintenance. The relationship between plants and people is nurturing. Through helping plants, participants are able to fulfill the desire to nurture and care for living things. The extended benefit is felt when the plant puts out new leaves, sprouts, or branches. A person can see the direct results of caring for the plant and experience the positive benefits of helping.

Social Activity: Making a Dish Garden

Making a dish garden has many tasks that can be assigned individually to produce the end product. Using a large dish garden bowl, each group member chooses a small houseplant. Like a production line, one person fills the bowl with charcoal, another fills the bowl two-thirds with soil, everyone takes a turn placing their plant in the bowl, someone fills in soil around the plants, and to finish the project, someone waters. By each person having an assigned part, group members not only contribute to the whole process, but also practice taking turns. Once the dish garden is finished, it can be placed in a communal space and serve as a discussion piece.

months and includes activities like thinning seedlings and transplanting along the way. Activities like this take planning months in advance of the final activity.

Activity planning requires more than just planning for supplies and materials; there are client considerations to attend too as well. Considerations such as safety, setting, medications, outdoor exposure, work space, and duration of time should all be taken into consideration. Balancing the needs of two living entities, people and plants, requires planning, coordination, communication, creativity, and adaptability to facilitate effective and therapeutic horticultural activities.

Plant Considerations

Horticultural therapy plant material is selected to meet the needs of the participant. The horticultural therapist considers the purposeful use of the plant and its ability to engage, stimulate, and respond to participants. These interactive qualities of plants can be categorized as sensory, functional, and responsiveness (Haas and McCartney 1996, 61–68). Haas and McCartney (1996) describe the interactive qualities of plants as:

Sensory: Plants can stimulate all five of our senses. By using plants with vibrant or pastel colors, a person may be energized or calmed by what he/she see. Smelling plants with different fragrances can bring about memory recall. Or promote relaxation. Listening to leaves rustling in the wind, or wisteria pods bursting, enhances the experience of plants in a way that most participants are unfamiliar with, sound. Touching plants allows a person to experience and interpret a plant on a personal level. Tasting and eating plants is at the core of our relationship with plants.

Functional: Functional plants are purposeful plants with multiple uses. Plants with functional qualities are those that can be used in activities to help participants develop or improve physical or cognitive skills. Multiple uses, for example, a flower, include the sensory experience, the physical act of cutting the flower, the cognitive problem solving of cutting the flower, the eye–hand coordination in making a floral design, or simply allowing the flower to go to seed and saving the seed for next year.

Responsiveness: People and plants engage in interactive relationships. On a fundamental level, plants respond to the care they are given and a participant responds in turn. The intimate relationship that can form between the person responsible for the plant and the acknowledgement of the plant's dependency upon their care can transform into feelings of increased self-esteem and self-confidence, hope, and a sense of control over one's environment. Regardless of a participant's ability, plants are nonjudgmental and do not discriminate; they respond to care.

These functional qualities of plants should be taken into consideration in planning activities and in choosing plant resources. It is one thing to have a beautifully designed garden on which to look but another to have a garden designed for active participation by clients with plant material that has multiple uses. The same should be considered for a greenhouse. A greenhouse with fully mature, healthy plants is beautiful to observe but a greenhouse full of plant material that can be propagated, transplanted, or watered provides opportunities for activities or vocational training.

THE HORTICULTURAL THERAPIST

The fourth component in the definition of horticultural therapy involves the trained therapist. The professional horticultural therapist facilitates the interaction between the three previously identified program components. Horticultural therapists are specially educated and trained professionals who involve the client in any phase of gardening—from propagation to selling products—as a means of bringing about improvement in their life. As members of treatment or care teams, horticultural therapists determine individual goals and work plans to help clients improve skills and maximize abilities (AHTA 2007, 3).

A horticultural therapist has the dual role of a therapist and a horticulturalist. As the horticulturist, the HT is responsible for ensuring plants, the plant environments, and all related program components are maintained and cared for in support of the program. As the therapist, the HT is responsible for treatment planning, program planning, and all associated documentation. Personal traits such as patience, empathy, compassion, and the ability to be nonjudgmental are essential. Effective communication skills and good listening skills are necessary as horticultural therapists are members of professional teams including doctors, psychiatrists, physical and occupational therapists, nurses, job coaches, teachers, employment specialists, and others working to provide the client services. They also have numerous roles, including community relations, coordinator, educator, consultant, designer, and fundraiser (Hazen 2014, 252–253).

The American Horticultural Therapy Association

AHTA is a nonprofit membership-driven organization whose mission is to promote and advance the profession of HT as a therapeutic intervention and rehabilitative modality. AHTA administers the professional credentialing of horticultural therapists; Horticultural Therapist—Registered (HTR) is the professional credential granted by AHTA. An HTR credential indicates a person has the education and skills necessary to conduct HT program activities.

A horticultural therapist studies plant science, human science, and horticultural therapy. In addition, training in the application of HT through an approved internship is required. Upon completion of these requirements, a person may apply to AHTA for professional registration. Universities and botanic gardens offer degrees or coursework in HT. AHTA is a resource for available educational opportunities.

HT PROGRAM COMPONENTS

Horticultural resources for an HT program include gardens, greenhouses, and grow lights. Horticultural therapists often work in collaboration with a design team consisting of facility staff and landscape architects to develop a therapeutic garden. Horticultural therapists may be involved in the design of a greenhouse structure and the interior space and work with buildings facility staff, for example, to build benches and to install watering systems. Horticultural therapists might even be involved in organizing a group of community volunteers to build raised beds in a courtyard. In all situations, it is the horticultural therapist's responsibility to ensure the design and the results meet the needs of the population served.

Design Guidelines and Considerations

There are three important sets of guidelines to consider in the designing of therapeutic gardens, the greenhouse interior or determining indoor growing spaces: (1) the AHTA Therapeutic Garden Characteristics, (2) the ADA Standards for Accessible Design, and (3) the Principles of Universal Design.

AHTA Therapeutic Garden Characteristics

In response to a growing number of articles and publications about gardens in healthcare settings in the early 1990s, AHTA formed a task force to identify common characteristics of therapeutic gardens. In 1995, AHTA formally presented the American Horticultural Therapy Association Therapeutic Garden Characteristics. The identified characteristics incorporated the concepts of accessibility and universal design and components of HT programming. These seven characteristics are as follows:*

1. *Scheduled and programmed activities*: A HT program guiding and promoting a program of activities and experiences in the garden is ideal. In addition to regularly scheduled HT activities, events, classes, workshops, and engaging community volunteers bring recognition to the garden by staff, family, and the broader community.
2. *Features modified to improve accessibility*: Garden elements, features, and equipment are selected or modified to provide accessible places, activities, and experiences to the greatest extent possible. Modifications made to improve accessibility also increases interactions with plant materials.
3. *Well-defined parameters*: Edges of garden spaces and special zones of activities within the garden are intensified to redirect the attention and the energies of the visitor to the components and displays within the garden.
4. *A profusion of plants and people–plant interactions*: Therapeutic gardens are designed to introduce individuals to planned, intensive, outdoor environments in which people–plant interactions are facilitated through simple design.
5. *Benign and supportive conditions*: Therapeutic gardens provide safe, secure, and comfortable settings for people. Avoidance of potentially hazardous chemicals and toxic plant materials while providing shade and other protected space can create a feeling of safety.
6. *Universal design*: Therapeutic gardens are designed for the convenience and enjoyment for people with the widest possible range of disabling conditions, of all ages and abilities. The therapeutic garden provides opportunities for a full range of sensory experiences.
7. *Recognizable placemaking*: Therapeutic gardens are frequently simple, unified, and easily comprehended places. Placemaking is an important strategy in landscape design efforts and assists the visitor in recognition of plant material and creates independence by providing direction.

ADA Standards for Accessible Design

The Americans with Disabilities Act of 1990 (ADA) is a landmark law which gives civil rights protection to individuals with disabilities and ensures equal access to employment, education, and community living. The ADA established accessibility standards in 1991, has since revised the standards, and the current standards, the 2010 ADA Standards for Accessible Design (2010 Standards) define the technical requirements for all new construction and alterations and architectural changes (Department of Justice 2010). The 2010 Standards apply to any aspect or device that

* Copyright 2007. American Horticultural Therapy Association (AHTA). AHTA Definitions and Positions.

limits a participant with a disability full access and the ability to perform a task. This includes, but is not limited to, the dimensions of pathways, entry and exit ways, ramps and slope, curbs, and access to and from restrooms, work rooms, or activity room. The Standards apply to the height and placement of hand railings, benches, drinking fountains, water spigots, door knobs, and signage.

When AHTA first published the AHTA Therapeutic Garden Characteristics, the ADA was still relatively new. Society as a whole was learning how to incorporate the standards, how to design things different, and how to adapt to a new way of life that included individuals with disabilities as full members of society. Horticultural therapists, landscape architects, horticulturalists, and professionals working in health care or rehabilitation services had already been adapting gardens and horticulture environments to become accessible. Today there is a much clearer understanding of the ADA requirements and therapeutic gardens often reflect design elements above the ADA requirements.

Principles of Universal Design

The concept behind universal design is that with adaptations or alternations to design, all people regardless of age or ability can use a product or participate in an environment. The Principles of Universal Design are rooted in the barrier-free movement in the 1950s when environmental barriers limited access for people with mobility. As public awareness changed and persons with disabilities integrated into society, and federal legislation such as the ADA brought equal access to public awareness, universal design guidelines became necessary as society moved forward (The Center for Universal Design 2008).

The Center for Universal Design began in 1989 as it was funded by a grant from the National Institute on Disability and Rehabilitation Research, U.S. Department of Education. The Principles of Universal Design were published in 1997 by The Center for Universal Design at North Carolina State University. The seven principles are identified and described in Table 7.2.

Adaptable and Accessible

A brief explanation of the terms adaptable and accessible is helpful in understanding the horticultural environment. Facilities must be accessible and structures adapted to meet the needs of participants. Applying accessibility to the HT environment means the environment is approachable (widened walkways), able to be entered (widened doorways) and used, or reached easily (lower water spigots). Adaptable or adapted, applies to a structure (raised beds and greenhouse door), equipment (pruners and rake), or activity (color coding and assembly line procedures) that can be, or has been, changed or modified to meet requirements or conditions for the horticultural environment.

Therapeutic Gardens

As the benefits of gardens and being in nature have been promoted to the general public over the past several years, many terms are being applied to gardens to

TABLE 7.2
Principles of Universal Design

Principle One: Equitable Use
Definition: The design is useful and marketable to people with diverse abilities.

Principle Two: Flexibility in Use
Definition: The design accommodates a wide range of individual preferences and abilities.

Principle Three: Simple and Intuitive Use
Definition: Use of the design is easy to understand, regardless of the user's experience, knowledge, language skills, or current concentration level.

Principle Four: Perceptible Information
Definition: The design communicates necessary information effectively to the user, regardless of ambient conditions or the user's sensory abilities.

Principle Five: Tolerance for Error
Definition: The design minimizes hazards and the adverse consequences of accidental or unintended actions to correct mistakes without penalty.

Principle Six: Low Physical Effort
Definition: The design can be used efficiently and comfortably and with a minimum of fatigue.

Principle Seven: Size and Space for Approach and Use
Definition: Appropriate size and space is provided for approach, reach, manipulation, and use regardless of user's body size, posture, or mobility.

Note: The principles were compiled by advocates of universal design, in alphabetical order: Bettye Rose Connell, Mike Jones, Ron Mace, Jim Mueller, Abir Mullick, Elaine Ostroff, Jon Sanford, Ed Steinfeld, Molly Story, and Gregg Vanderheiden.

Source: Copyright 1997 NC State University, The Center for Universal Design.

describe these benefits such as "healing gardens," "restorative gardens," "therapeutic gardens," and "enabling gardens." Seldom is a garden called a horticultural therapy garden in literature or labeled as such in a public space. However, a therapy garden is different from these other gardens through its defined purpose. The following are general descriptions for these types of gardens in comparing them to a HT garden:*

Healing Garden: Healing gardens are plant-dominated environments including other aspects of nature like water and wildlife. They are generally found in hospitals and other healthcare settings. Healing gardens are accessible to all persons and benefits are experienced simply by being in the garden; the garden is enjoyed through passive use (AHTA 2007, 1; Cooper Marcus and Sachs 2014b, 3).

Restorative Garden: A restoration or meditation garden may be a public or private garden that is not necessarily associated with a healthcare setting. This type of garden employs the restorative value of nature to provide an

* Copyright 2007. American Horticultural Therapy Association (AHTA). AHTA Definitions and Positions.

environment conducive to mental repose, stress-reduction, emotional recovery, and the enhancement of mental and physical energy. The design of a restorative garden focuses on the psychological, physical, and social needs of the users (AHTA 2007, 1).

Therapeutic Garden: A garden can be described as being therapeutic in nature when it has been designed to meet the needs of a specific user or population. Like healing gardens, they can be found in healthcare settings, but they are also found in residential facilities and community organizations. The garden may provide for both horticultural and nonhorticultural activities, therefore, may be enjoyed through passive or active use. This type of garden may be part of a larger healing garden or may exist on its own (AHTA 2007, 1; Cooper Marcus and Sachs 2014b, 3).

Enabling Garden: An enabling garden eliminates all physical barriers and allows people of all abilities to participate in the garden. Enabling gardens are designed for gardening and benefits are derived from active participation. These gardens can be found in hospitals, rehabilitation facilities, and in public gardens. Generally, gardens in HT programs are also enabling gardens (Cooper Marcus and Sachs 2014b, 3).

The HT Garden

In a garden designed specifically for horticultural therapy, the horticultural therapist facilitates the achievement of the participant's goals through plant-related activities. The gardens are designed to support primarily horticultural activities on an ongoing basis and participants derive benefits through active use of the garden (AHTA 2007, 1). The garden is designed to be fully accessible not only through pathways and doorways, and access to water and work space, but also by utilizing adapted structures like raised beds and planting containers and adaptive tools.

Types of Planting Containers

Various types of accessible planters are used in HT gardens, these include raised beds, containers, table top planters, and vertical walls. Raised beds allow someone who uses a wheelchair, or someone who needs to stand, to garden alongside the bed. Raised beds can also be adapted for seating. Free standing raised beds should be at least 30″ in height and 2′ to 3′ in width, thereby allowing access to the middle. The raised bed can be of any length but generally 4′ long is recommended. Raised beds that are part of the garden design follow the same guidelines except length.

Containers can include hanging baskets, pots, window boxes, dish garden bowls, and barrels. Containers of different heights are important for bringing plant material within reach or up to eye level. Hanging pots can be hung at varying heights and hung with a pulley system that allows the client to lower or raise the pot for watering or for viewing.

Vertical walls are generally used with plants that naturally climb and are constructed as a trellis or fence with the planter box at the bottom. Vertical planters are different and are constructed in such a way that individual plant containers are combined to make a wall and held together in a frame. The advantage to gardening with a vertical wall or vertical planter is that it eliminates the need for bending. They are also good for settings with limited space and can be used as privacy screens.

Table top beds are beds that allow the participant to sit while gardening. Table top planters need a minimum of 27″ clearance underneath and generally 24″ width for one person. If two people will be working on opposite sides, the bed needs to be 48″ wide and at least 6′ long to comfortably accommodate two wheelchairs. Typically, table top beds allow for 6″ planting depth and are best for short-term plantings.

Widened Pathways and Accessible Routes

Pathways should be a minimum of 36″ width for one-way use to meet ADA requirements; however, in a therapy garden, wider pathways are desirable to accommodate two wheelchairs and clients and staff walking side-by-side. When designing the garden, the landscape architect works with the horticultural therapist and other professionals to determine the layout of the garden to meet the ADA requirements. Ideally, the garden is located within a short distance from the main building and restrooms.

Seating and Work Space

Sitting benches along the pathways or in alcoves in the garden are important as well. A shaded work area for group activities with tables and chairs and a private work space located within the garden is important to minimize the need to transport supplies and materials. Water spigots placed throughout the garden eliminates having to move hoses and if possible, a sink should be included in the work area. The size of the work area ought to be large enough to accommodate the group and participants who use mobility devices including canes, manual wheelchairs, and power chairs; the HT work area may need to be located away from pathways if the garden is open for public use during sessions times (Hazen 2014, 256).

Plant Materials

In the therapy garden, trees are essential for shade. Plants can be used to create screens and borders for pathways. Plant collections arranged in themes can provide a client with direction and a sense of control. Flower gardens, sensory gardens, vegetables gardens, and butterfly gardens are all examples of theme-based HT gardens. Strategic plant placement to maximize interactions by all users is important; it helps to create curiosity and may be used to address specific physical goals like reaching, bending, standing, and lifting. Plants with seasonal interest and that attract birds, insects, and other wildlife into the garden should be included. As previously discussed, plants chosen for HT activities have functional, sensory, and responsiveness qualities; hence these factors should be considered when choosing plants for the garden.

Greenhouse and Grow Lights

An activity conducted in the greenhouse offers the indoor experience of protection from the elements, provides a safe and nurturing environment, and at the same time offers the visual experience of being outdoors. Greenhouses can be a component of any HT program but more often are found in vocational horticulture programs where production continues throughout the year. Greenhouses require adaptations to be accessible to both the structure itself and the interior design. Like the therapy garden, it is designed to meet the needs of the participants. Doorways can be a minimum of 32″ to meet the ADA requirements though most clients carry items or a bag and 36″ can accommodate all without difficulty. It is recommended that greenhouse aisles are 60″ wide to allow for moving in and out and turning around but are only required to be a minimum of 36″ wide to meet the ADA. Benches must have a minimum of 27″ clearance underneath. Spigots should be placed without obstruction a minimum of 15″ from the floor, though 24″ is a more comfortable height. Hanging baskets can be adapted by a pulley system to allow a client to move them up and down thus keeping them above eye level. A work area is also necessary for potting, mixing soil, transplanting, sowing seeds, and other related activities. This space should allow for both clients who use wheelchairs and clients who do not (Figure 7.8).

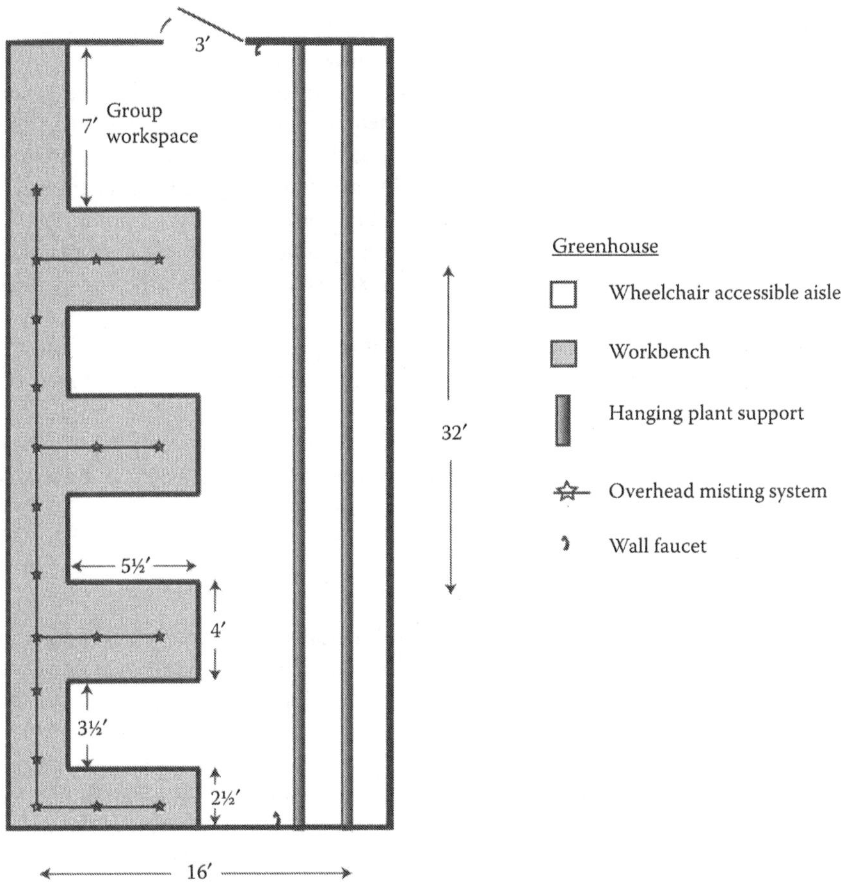

FIGURE 7.8 Greenhouse interior designed for accessibility and horticultural therapy programming, Florida State Hospital, Chattahoochee, Florida, 1990. (Designed by Leigh Anne Starling.)

When designing an accessible environment inside the greenhouse, it is important to note that in reference to wheelchairs, the ADA Standards are based on the dimensions of a standardized wheelchair. Personal wheelchairs have varying dimensions and can be manual, motorized, or electric with the associated controls placed on arms or sides of the chair. Adjustable head rests, foot rests, and wheels all affect the dimensions of the chair. Doorways and aisles should generally be wider than the minimum ADA requirements to accommodate all types of wheelchairs. Adjustable benches can also be in the greenhouse interior to accommodate personal wheelchairs as personal wheelchairs can be higher, or lower, than the "standard" wheelchair height. A survey of the types of wheelchairs used by clients is advisable in designing the greenhouse interior.

Grow light stands are ideal for programs with little to no access to outdoor space. Grow light stands are a combination of lights and shelves available in different sizes with adjustable shelves to meet plant growing requirements. Table top grow lights are a smaller version of the light stand, usually with only one adjustable light, and are designed to sit on counter tops or table tops. Full spectrum fluorescent lights are the most practical and least expensive for indoor programs; however, they do produce heat. Newer compact fluorescent bulbs (CFL) are available for the table top stands making them more efficient with slightly less heat output. More expensive LED light tubes are available for both types of light stands; these use less energy and produce less heat. Small countertop hydroponic growing systems using CFL lights are recent additions to indoor growing and, as self-contained soilless units, are appropriate for use in programs where soils and other growing medium are not appropriate. With a fish tank in the activity room, aquaponics systems are available too for growing indoor low light plants.

Grow lights can be used to grow and maintain plant materials, start seedlings, for propagation activities, and can be set to a timer for light schedules. Since HT activities using grow lights are primarily based indoors, projects like propagating houseplants can lead to extended activities like having plant sales or giving plants as gifts. Clients may benefit from the natural light given off by full spectrum lights and the plants remain in the activity space beyond the group session.

Special Considerations

The primary responsibility of the horticultural therapist is to the client. Along with incorporating all the components discussed so far, these two additional considerations apply to gardens, greenhouses, and indoor equipment.

Safety

In all programs, using nontoxic plant materials and knowing if the participant has allergies to plants or bees must be considered. Safety includes avoiding the use of toxic products and eliminating hazardous materials. Considerations specific to the mental health setting include avoiding objects such as tools or other instruments and plant materials to harm one-self or others (Cooper Marcus and Sachs 2014a, 181–182). Gardens should be designed to prevent participants with cognitive impairments from wandering away and at the same time be designed to prevent falls by using raised planters and gently sloped pathways with railings rather than stairs. In vocational programs, greenhouses can be made safe through nonslip concrete flooring, rounded bench edges, and lockable storage. Hot grow lights should have light covers.

Medications

Medications are known to affect a patient's mobility, cognitive functioning, vision, sensitivity to the sun, and ability to recognize and regulate body temperature. Gardens should incorporate shade trees or shade structures and patients should have easy access to the indoors. Greenhouses should have an effective cooling and heating system and use shade cloth in the summer. Drinking water should always be available. If working outdoors, conduct activities during the cool part of the day, ensure participants have sunscreen, hats, and sunglasses, limit the use of sharp tools or power equipment, and make sure walkways are kept clear (Haller 2006, 11).

FUNDING OPPORTUNITIES

Program funding is dependent on the goals and mission of the organization, size and staffing of the program, the status of the organization (private or public), and the population served. External funding generally refers to monetary or physical resources obtained from outside of the program. Internal funding generally refers to revenue generated from providing services or products and is distributed to the program within the organization.

Sources of external funding include federal, state, and local government agencies, foundations, corporate giving programs, community nonprofits, and individuals. Monetary funds are in the form of grants, donations, scholarships, sponsorships, and loans. While physical resources might not be considered direct funding, this can include donated equipment, products, facilities, or even land, which may offset the need for monetary funds. Applying for external funds requires an application process and programs must meet specific eligibility criteria. Mission statements from the program and the funder usually need to match. In most cases, measureable outcomes dependent on populations served and the number of people served are required.

Programs under the umbrella of a larger service, for example, in the occupational therapy department of a hospital or a vocational program in a school, are generally funded within the organization's budget. The hospital or organization receives monetary compensation for providing client services which then contributes to the HT program budget. An organization may receive financial reimbursement from private insurance companies or government agencies like the Department of Rehabilitation, or government health insurance such as Medicare. These are examples of internal funding. However, these programs benefit from generating revenue through activities like plant sales or landscaping businesses to supplement program expenses. In some programs, fee-based services like landscaping can be the primary funding source.

Recognizing community concerns like food sustainability or national issues like unemployment and aligning the mission of the HT program to address public concerns open up opportunities for funding in ways that bring benefits to all involved. For instance, cities have neglected and abandoned lots. A nonprofit agency serving adults with mental illness works with the city to turn the lot into an urban farm. The horticultural therapist works with clients in the garden. The clients are productive, the garden is productive, and the program receives continued funding meeting the broader needs of the public. This simplified example demonstrates how HT programs can be part of urban horticulture.

HT IN URBAN HORTICULTURE

Research related to the benefits of gardens in our communities and gardening to our health has been reviewed in HT literature (Relf 1998, 22–34; Relf and Lohr 2003, 86–88). All studies reviewed confirm the important role of plants in our lives. Research demonstrates that viewing plants and passive interactions with plants result in lower blood pressure, decreasing anxiety and stress, and can lead to shorter recovery time after surgery. Studies show that having plants in community landscapes helps to decrease crime rates. Therapeutic gardens are now considered by many to be a vital component in healthcare facility design; however, as discussed in this chapter, therapeutic gardens can be found in all different types of facilities.

As exemplified in the case studies, HT gardens are often available for public use, though in some settings the gardens are not open to the public but still contribute to the urban environment by their presence and the benefits they provide to patient's family members, the staff, and the volunteers. For example, some VA healthcare facilities like the Miami, Florida, VA Medical Center and the Wilkes-Barre, Pennsylvania, Veterans Affairs Medical Center have therapeutic gardens that are used as therapy gardens and are not public gardens (Kirk et al. 2010, 69–73).

Vocational programs that function as urban farms provide not only local food but also jobs for clients; these gardens may or may not be open to the public. One such urban farm is the Homeless Garden Project (HGP) located in Santa Cruz, California. HGP is a nonprofit organization offering job training, transitional employment, and support services to people who are homeless. Located on a three-acre organic farm, HGP is an example of both a therapeutic and vocational horticulture program serving the broader community through volunteers, interns, and job trainees all working together. HGP operates a community-supported agriculture (CSA) program, a business selling products made from the garden, and a community education program. HGP hosts events and workshops regularly bringing the community together with trainees.

With people living longer lives there are more long-term care and assisted living facilities in our communities. Therapeutic gardens located in long-term care and assisted living programs are generally not public gardens; gardens in these facilities are designed for safety to keep residents from wandering off-site and generally are accessed from indoors. An example of this type of HT program is the Margaret T. Morris Center located in Prescott, Arizona. The center is dedicated solely to caring for those with dementia. Within the one-acre therapeutic garden there are theme-based gardens, raised beds, and a children's playground area. The garden is used by residents, staff, and family.

While some are more visible than others, a vast majority of HT programs and therapeutic gardens are situated in the urban setting. Not only do these programs add green space to the environment, but the additional benefit to society and to the individuals who participate in HT programs is the potential decrease in the stigmatization of persons with disability or illness as people become more familiar with therapeutic gardens in the urban environment.

SUMMARY

Patients worked on the fields and gardens of early institutions. Physicians recognized improvements in patients working with plants and by the 1900s, gardening was used for therapeutic purposes. Horticultural activities were used with recovering WWII veterans in the 1950s, propelling HT into professional recognition. During the 1960s many universities offered degree programs. The National Council for Therapy and Rehabilitation through Horticulture formed in 1973 and is now the American Horticultural Therapy Association.

Horticultural therapy is defined as a professionally conducted client-centered treatment modality that utilizes horticulture activities to meet specific therapeutic or rehabilitative goals of its participants. The focus is to maximize social, cognitive, physical, and psychological functioning and to enhance general health and wellness (Haller 2006, 5).* This definition includes the HT components of client, activities, goals, and therapist.

A horticultural therapist facilitates horticultural activities to meet the needs of a client or group. The therapist is a member of a team of professionals involved in the client's care. A horticultural therapist has an educational foundation in plant science, human science, and horticultural therapy and additional training through an internship. After completing the required coursework and internship, a student may apply for professional registration with the AHTA. AHTA is the professional organization representing the field of HT.

There are three types of HT programs: therapy, therapeutic, and vocational. A therapy program is defined in clinical terms, a therapeutic program is defined in wellness terms, and a vocational program is defined in terms of job skills training and employment. Programs take place in settings like healthcare facilities, nonprofit and profit organizations, schools, rehabilitation facilities, correctional programs, botanic gardens and arboreta, and long-term care facilities, and serve people of all ages and abilities who might have a mental disorder, cognitive disorder, physical disability, developmental disability, substance use disorder, or a life circumstance requiring treatment or rehabilitation services.

Goals and objectives are defined for the participant and the group. Goals are used to measure progress and to determine if the client is benefiting from HT services; objectives are the steps used to achieve the goal(s). Activities are used to meet the identified goal(s) of the participant or the group. Activities are primarily plant based; plants used in HT activities have sensory, functional, and responsiveness qualities. Passive viewing of plants by those unable to participate is considered a therapeutic

---

activity. The benefits of HT cross four human dimensions: social, cognitive, psychological, and physical. These benefits are multidimensional and can occur simultaneously through plant-based activities.

Horticultural therapy gardens are accessible and designed to provide ongoing horticultural activities. Enabling gardens are a type of HT garden. Greenhouses complement an HT program and are frequently found in vocational programs. Grow light stands and hydroponic systems are used in indoor programs. All three horticultural resources are designed, created, and/or arranged in accordance with the AHTA Therapeutic Garden Characteristics, the 2010 ADA Standards, and the Principles of Universal Design.

Program funding is available through external resources such as grants, donations, and sponsorships. Physical resources are also a type of funding as physical donations may offset the need for monetary funds. Programs can be funded internally through insurance reimbursement for client services or from producing products for sale or providing services like landscape maintenance.

Horticultural therapy programs are a part of urban horticulture. Not only do HT programs provide services to the community at large, but HT gardens also provide benefits for all users.

REVIEW QUESTIONS

1. Define horticultural therapy and the four components of the definition. Describe the three program types and give an example of each type.
2. Describe the characteristics of a therapeutic garden. What is the significance of the ADA and the Principles of Universal Design?
3. Discuss the four types of benefits participants might experience from participating in HT and give several examples of each. How do these benefits overlap? Who might experience these benefits and how?
4. What are goals and objectives and why are they important? Do all HT programs have goals? What are some examples of client goals versus program goals?
5. What characteristics are important when selecting plant materials for use in HT programs? Are there other types of characteristics that you think are important?

ACTIVITY IDEAS

1. Interview a horticultural therapist. Find out what population is served, the program type and setting, and what type of garden, if any, is supporting the program. Ask how the program is funded and how clients are referred to the program.
2. Do a HT activity with a group. Complete a floral design using floral foam in a container with a cut flowers and greens. Imagine you are the client. Tape your left hand fingers together and smear Vaseline on the lens of a pair of sunglasses to see what it might be like to have a physical disability and

limited vision. Try to complete the activity. Observe your own experience and write about it.

3. Volunteer for a day at a long-term care facility for residents with Alzheimer disease. Write about your experience and what you learned, observed, what was unexpected, and how you think horticultural activities might benefit the residents.
4. Visit a healthcare facility with a therapeutic garden. Was the garden accessible? How? Where there adapted structures? What type? What plant materials were used? Make a list of plants and describe the qualities with a supporting example.

REFERENCES

American Horticultural Therapy Association (AHTA). 1995. *Therapeutic Garden Characteristics*. King of Prussia, Pennsylvania: American Horticultural Therapy Association.

American Horticultural Therapy Association (AHTA). 2007. *AHTA Definitions and Positions*.King of Prussia, Pennsylvania: American Horticultural Therapy Association.

Cooper Marcus, C. and N. Sachs. 2014a. Gardens for mental and behavioral health facilities. In C. Cooper Marcus and N.A. Sachs (eds.), *Therapeutic Landscapes: An Evidence-Based Approach to Designing Healing Gardens and Restorative Outdoor Spaces*. New Jersey: John Wiley & Sons, Inc., pp. 179–205.

Cooper Marcus, C. and N. Sachs. 2014b. Introduction. In C. Cooper Marcus and N.A. Sachs (eds.), *Therapeutic Landscapes: An Evidence-Based Approach to Designing Healing Gardens and Restorative Outdoor Spaces*. New Jersey: John Wiley & Sons, Inc., pp. 1–5.

D'Andrea, S.J., M. Batavia, and N. Sasson. 2007–2008. Effect of horticultural therapy on preventing the decline of mental abilities of patients with Alzheimer's type dementia. *Journal of Therapeutic Horticulture*, 18, 8–17.

Davis, S. 1998. Development of the profession of horticultural therapy. In S.P. Simson and M.C. Straus (eds.), *Horticulture as Therapy: Principles and Practice*. Binghamton, New York: The Haworth Press, Inc., pp. 3–18.

Davis, S.H. and M.S. De Riso. 1992. Horticulture hiring people with disabilities. *HortTechnology*, 2(2), 183–189.

Department of Justice. 2010. *ADA 2010 Standards for Accessible Design*. ADA.

Detweiler, M.B., T. Sharma, J.G. Detweiler, P.F. Murphy, S. Lane, J. Carman, A.S. Chudhary, M.H. Halling, and K.Y. Kim. 2012. What is the evidence to support the use of therapeutic gardens for the elderly? *Psychiatry Investigation*, 9, 100–110. http://dx.doi.org/10.4306/pi.2012.9.2.100.

Haas, K.L. and R. McCartney. 1996. The therapeutic qualities of plants. *Journal of Therapeutic Horticulture*, 8, 61–68.

Hale, B., G. Marlowe, R.H. Mattson, J.D. Nicholson, and C.A. Dempsey. 2005. Ahorticultural therapy probation program: Community supervised offenders. *Journal of Therapeutic Horticulture*, 16, 38–49.

Haller, R.L. 2006. The framework. In R.L. Haller and C.L. Kramer (eds.), *Horticultural Therapy Methods: Making Connections in Health Care, Human Service, and Community Programs*. Binghamton, New York: The Haworth Press, Inc., pp. 1–22.

Hazen, T. 2014. Horticultural therapy and healthcare design. In C. Cooper Marcus and N.A. Sachs (eds.), *Therapeutic Landscapes: An Evidence-Based Approach to Designing Healing Gardens and Restorative Outdoor Spaces*. New Jersey: John Wiley & Sons, Inc., pp. 250–260.

Kirk, A.P., A. Karpf, and J. Carmen. 2010. Therapeutic garden design and veteran affairs: Preparing for future needs. *Journal of Therapeutic Horticulture*, 20, 66–76.

Lewis, C.A. 1996. *Green Nature/Human Nature: The Meaning of Plants in Our Lives.* Urbana & Chicago, Illinois: University of Illinois Press.

National Council for Therapy and Rehabilitation through Horticulture (NCTRH). 1976. *A Brief History of Horticulture Therapy.* Gaithersburg: NCTRH.

Page, M. 2008. Gardening as a therapeutic intervention in mental health. *Nursing Times*, 104(45), 28–30.

Relf, D. and S. Dorn. 1995. Horticulture: Meeting the needs of special populations. *HortTechnology*, 5(2), 94–103.

Relf, P.D. 1998. People–plant relationship. In S.P. Simson and M.C. Straus (eds.), *Horticulture as Therapy: Principles and Practice.* Binghamton, New York: The Haworth Press, Inc., pp. 21–42.

Relf, P.D. and V.I. Lohr. 2003. Human issues in horticulture. *HortScience*, 38(5), 984–993.

Rice, J.S., L.L. Remy, and L.A. Whittlesey. 1998. Substance abuse, offender rehabilitation, and horticultural therapy practice. In S.P. Simson and M.C. Straus (eds.), *Horticulture as Therapy: Principles and Practice.* Binghamton, New York: The Haworth Press, Inc., pp. 257–284.

Sellers, K.D. 2001. Effectiveness of horticultural therapy activities in a psychiatric hospital. *Journal of Therapeutic Horticulture*, 12, 9–20.

The Center for Universal Design. 1997. *The Principles of Universal Design,* Version 2.0. Raleigh, NC: North Carolina State University.

The Center for Universal Design. 2008. Universal Design History. http://www.ncsu.edu/ncsu/design/cud.

Verra, M.L., F. Angst, T. Beck, S. Lehmann, R. Brioscchi, R. Schneiter, and A. Aeschlimann. 2012. Horticultural therapy for patients with chronic musculoskeletal pain: Results of a pilot study. *Alternative Therapies in Health and Medicine*, 18(2), 44–50.

Wichrowski, M., J. Whiteson, F. Haas, A. Mola, and M.J. Rey. 2005. Effects of horticultural therapy on mood and heart rate in patients participating in an inpatient cardiopulmonary rehabilitation program. *Journal of Cardiopulmonary Rehabilitation*, 25, 270–274.

8 Urban Greening

Amy McFarland

CONTENTS

OBJECTIVES

Upon completion of this chapter, the reader should be able to

- Describe the development of urban green spaces throughout history.
- Discuss different types of urban greening (rooftop gardens, million tree initiatives, green belts, inner-city parks).
- Discuss examples of urban forestry and urban greening through case studies.
- List and describe benefits of urban greening to individuals, the community and the region.
- Recognize funding sources for urban greening programs.

KEY TERMS

- Aquifer
- Carbon sequestration

- Community supported agriculture
- Extensive green roof
- Evapotranspiration
- Green roof
- Green belt
- Green nets
- Green wedges
- Greenways
- Greywater systems
- Ground water
- Fauna
- Flora
- Habitat fragmentation
- Intensive green roof
- Logic models
- Migratory corridors
- Place attachment
- Sense of place
- Stakeholders
- Sustainability
- Triple bottom line
- Rooftop farm
- Rooftop garden
- Urban areas
- Urban ecotourism
- Urban forests
- Urban green tourism
- Urban growth corridors
- Urban heat island effect
- Urban shrinkage
- Urban sprawl
- Wildlife corridors

HISTORY OF GREEN SPACE

The vast majority of the ice-free land surface on earth has been impacted by land-use transformations, much of which has been related to deforestation for urbanization and agricultural purposes in the past 100–200 years (Turner et al. 2007). In 2015, more than 50% of people live in urban environments. However, less than 0.5% of the Earth's total land area is urban (Schneider et al. 2009). In the United States, agriculture and urbanization are responsible for having the greatest impact on the American landscape. Urban green spaces have the opportunity to bring nature into the city and have strong impacts on large populations of people.

While many environmentalists are concerned with the status of protecting large natural spaces and wildlife in nature, fewer people and less political action is focused on urban green spaces, which are the types of nature most people have contact with on a daily basis. That nature is the space created within our urban and built environments. Just as environmentalists are concerned with air quality in urban areas, green quality in urban areas also needs attention. *Urban areas* are "places dominated by the built environment" (Schneider et al. 2009). These spaces are important to focus on since they are located in proximity to large populations of people with the potential to have increasingly critical and mutual impact.

While the focus of this chapter is urban green space in North America, the history of green space in densely populated areas undoubtedly predates the discovery of North America. Evidence dating back to Babylonian times emphasizes the importance gardens played in civilized society (Shimmin 2012). For example, the Hanging Gardens of Babylon were considered to be one of the seven wonders of the ancient world, built by King Nebuchadnezzar for his wife Amytis. The garden stood 75 feet above the ground with a 400 foot base and was watered by an interior irrigation system which siphoned water from the Euphrates River. A second example of ancient gardens in densely populated areas were the ziggurats of ancient Mesopotamia, built between 4000 and 600 BC which also possessed rooftop gardens. Garden trees were used to provide shade in these outdoor living spaces and provided a cool resting place from the harsh Middle Eastern sun. There is also archeological evidence that the Roman city of Pompeii had extensive rooftop gardens prior to its destruction. In Pienza, Italy in 1463, Pope Pius II commissioned a palazzo to be built at his personal residence with a stunning rooftop garden.

Not only were gardens utilized by the wealthy and royal, but common citizens also used green space in their residences, though to a lesser extent. For example, the Norwegian culture engaged green roofs in the construction of their residential structures (Shimmin 2012). *Green roofs* or *rooftop gardens* are a method of urban greening involving the use of structures on rooftops. In modern times, they often are used in particularly dense and built-up areas with high value real estate. The Norwegian roofs were used for more practical purposes when compared to the wealthy and royal documented above; they used sod roofs as a means of insulation from snow and to reduce the rate of decay caused by snow melt and rain water.

By and large, green space techniques found their way to the United States due to the migration out of Europe with the goal of replicating the aristocratic grandeur for common people (Jenkins 1994). Green space for immigrants to North America was a symbol of freedom and democracy from the feudal systems of Europe.

The location of green space in urban areas in the United States dates back as far as the founding of the country. Green space was considered in the designs of many of the earliest cities in the United States. Both George Washington's and Thomas Jefferson's visions of Washington D.C. included trees, spacious grounds, and parks dispersed throughout capital buildings (Choukas-Bradley 2008). In fact, in the years following the turn of the nineteenth century, Jefferson himself designed and supervised the installation of trees along Pennsylvania Avenue between the Capital and the White House. The direct oversight by a person of Jefferson's status exemplifies the importance and impact of green space.

The White House itself is well known for its landscaped grounds with many media events occurring within, for example, the Rose Garden. The case of the White House is interesting—on the one hand, it is a replication of the gardens around royal European estates, but on the other hand, the White House is an icon of democracy and freedom. Numerous presidents throughout US history have been involved with improving the green space surrounding the property including John Adams, James Madison, John Quincy Adams, Martin Van Buren, and many others (Choukas-Bradley 2008). The changing nature of the space around the White House is iconic of America—the grounds were used for a Victory Garden during World War II, planted by Eleanor Roosevelt to aid in the promotion of home food production during the war, and has been used in a similar fashion by Michelle Obama to assist in the fight against obesity and Type II Diabetes (Whitehouse 2009; Obama 2012).

The National Mall—the large area of green space between Capitol Hill and the Lincoln Memorial—was also a part of Washington and Jefferson's plan for the city and was included in the city blueprint designed by Pierre Charles L'Enfant. A large avenue connecting the capitol with a planned memorial for George Washington was a centerpiece of L'Enfant's vision for the city (Sherald 2009). Andrew Jackson Downing was hired in the early 1850s to design a landscape for this area of the capitol city which included Victorian and naturalistic styled gardens throughout the National Mall area (Halnon n.d.).

The National Mall (Figure 8.1) is another symbol of freedom. Memorials throughout the expanse of lawn space celebrate icons of freedom such as George Washington, Abraham Lincoln, and Martin Luther King Jr. Civil rights protests were held at the National Mall and Lincoln Memorial, and Martin Luther King Jr. delivered his famous "I have a dream" speech at the site.

In the mid-1850s, wealthy New Yorkers sought to replicate the vast expanse of public grounds in London and Paris. A landscape design contest was held to select a plan for the new park. The senior Frederick Law Olmsted's plan was selected with a design featuring pastoral, romantic landscapes (Rosenzweig and Blackmar

FIGURE 8.1 Looking across the national mall at the White House. (Photograph by Dr. Amy McFarland.)

1992). Olmsted's philosophy promoting parks and green space in urban areas as a means to improve public health and promote democracy coupled with Central Park's initial success following its 1859 opening encouraged leaders in other cities to replicate the model of adding green space in urban environments (Muller 2010).

The turn of the twentieth century saw a reform movement—the City Beautiful Movement—to improve the availability of green spaces through parks and planned urban development embracing beauty (Wood 2010). Frederick Law Olmsted Sr. was hired in the 1870s and 1880s to plan grounds surrounding Capitol Hill. Olmsted's plan included planting groves of trees in places to offer shade while not obstructing the view of the Capitol building. Today, more than 3000 trees are planted in the Capitol Hill area, many labeled, making it a quite extensive arboretum (Choukas-Bradley 2008). In 1902, the Senate Park Commission devised a plan, known as the McMillan Plan, which inspired the current incarnation of the area the National Mall, which includes an expansive grassy lawn with rows of American elm trees (Halnon n.d). In 1935, Frederick Law Olmsted Jr. was hired to develop a formal landscaping plan for the White House grounds which included additional tree plantings around the perimeter of the property (Choukas-Bradley 2008).

TYPES OF URBAN GREEN SPACE

Urban green space can manifest itself in many different forms depending upon the size and density of the urban area. Of course, one of the most obvious forms is urban parks. Urban parks can take many forms—they can be large and well-known, such as Central Park in New York City, Millennium Park in Chicago, or Stanley Park in Vancouver, British Columbia. These large parks can serve as tourist attractions and destinations, potentially serving as an income source for cities. However, large urban parks are just one side of the coin. Large cities can also have numerous smaller parks which also serve the same ecological and social functions as large parks. The Chicago Park District, for example, is steward to 570 parks as well as other natural spaces (Chicago Park District 2014). New York City Department of Parks and Recreation is steward to 14% of the landmass of New York City with numerous parks located throughout the city such as Battery Park on the southern tip of Manhattan and Washington Square directly on 5th Avenue.

Urban green space can also be found through *urban forests*. Several large cities around the world have implemented "million tree initiatives" within the past decade. The Los Angeles initiative, for example, promotes the stated mission of increasing the urban canopy by planting one million new trees to enhance energy savings, reduce the urban heat island effect, improve air quality, supply fresh fruit, create green jobs, and make Los Angeles greener and more beautiful (Million Trees LA 2006). These trees include street trees, trees in residential yards, trees in parks, and trees around businesses. The City of Los Angeles supports community efforts by providing city residents up to seven free trees to plant on their property or the public space between the sidewalk and street. Similar initiatives exist in New York City, New York in the United States, London, Ontario in Canada, and Shanghai in China.

Green belts are another method urban areas use to increase green space in urban environments. Originally, *green belts* were defined as a policy protected nonsettlement area on the perimeter of an urban area for the purpose of conservation and prevention of urban sprawl (Amati 2008). Urban sprawl is the physical growth in cities and includes the development of suburbs. These spaces date back to biblical periods, but popularity in the modern era peaked in the first 20 years post-World War II, particularly in the United Kingdom. Green belts fell out of favor with policy makers as urban populations increased and the need for expansion grew. Urban sprawl began to simply jump over the protected areas. The desire to protect these spaces for conservation remained high, however, and the definition of a green belt has changed to include other urban green spaces as well such as *green nets, greenways, green wedges, urban growth corridors* which do not necessarily surround the city to prevent sprawl (Amati 2008).

Large expanses of protected tracts of land are not feasible in all urban environments, however. Particularly dense and built-up areas with high value real estate have often had difficulties finding space for such areas. However, the desire and need for nature contact in areas without legally protected lands is extensive. One method for increasing green space without taking up additional land is through green roofs. Green roofs are simply green spaces, gardens, or even farms on the roof of a building. Green roofs can be quite expensive with regards to up-front costs as soil, plants, and water all increase the weight load on the roof structure. However, many environmentalists and architects argue that the costs of a green roof are paid for with the increased life span of the roofing materials and mitigating cooling and heating costs (Weiler and Scholz-Barth 2009).

Green roofs are classified in one of two ways depending on their architecture (Weiler and Scholz-Barth 2009). "*Extensive*" *green roofs* have shallow soil profiles and are primarily used for ecosystem benefits of managing stormwater runoff by allowing rainwater to penetrate into soil. The rainwater is then either used rather quickly by plants with shallow root systems or collected into a cistern for a variety of purposes such as irrigation or for use similar to *greywater systems*. Greywater systems reuse water that is not contaminated with human waste for various appropriate uses including irrigation. These systems often use succulent plants with shallow root systems that can survive the higher heats on roofs and irregular irrigation. Extensive green roofs are relatively inexpensive to develop, require little building modification since they have lesser weight loads compared to the alternative system. "*Intensive*" *green roofs* have much deeper soil profiles that can support larger trees in addition to shallow rooted plants. These systems usually require supplemental irrigation due to the increased diversity in plant material supported. These roofs are often used for purposes beyond the ecosystem services they provide and may include recreational and esthetic uses (Weiler and Scholz-Barth 2009).

The desire for fresher and healthier food choices has sparked the age of the urban farmer. In the past, efforts to address these needs were often initiated by nonprofit urban renewal organizations and community garden projects. However, the current local foods movement is seeing an increase in for-profit urban farmers. In larger cities, this movement has produced urban rooftop farms. Rooftop farms are similar to rooftop gardens, but usually have the intent to produce edible goods for consumption or sale. Rooftop farms cannot support the use of heavy equipment such as tractors (which saves money in exchange for increased labor) or large extensive irrigation systems to water. Much of the work is done with hand tools and the irrigation duties fall upon workers hand-watering with the everyday garden hose. Large size farms like the Brooklyn Grange and Eagle Street rooftop farms are turning Brooklyn, NY into an agricultural hot bed (Foderaro 2012). These large scale rooftop farms are promoting a healthier life style, educating the general public through local volunteer initiatives, and creating paid jobs in the inner-city while providing all of the traditional ecological and social benefits of a green roof.

Urban farms have also been popping up around the United States utilizing smaller tracts of land than traditional farms. Some utilize a number of vacant lots. Urban Roots Farm in Grand Rapids, Michigan runs a *community supported agriculture* (CSA) farm providing food grown on numerous small lots they have obtained from various sources including neighborhood associations and homeowners with unused space. CSA farms bring produce directly to their customer. Customers purchase subscriptions to the CSA farm early in the season prior to planting and share in the risk and reward of the upcoming season. The proximity of these farms to their markets has added value. The ability to supply fresh produce without an extensive supply chain and the cost that accompanies it can help mitigate increased prices from labor and help contribute to urban greening. Refer to Local Foods chapter for more reading about urban food production.

BENEFITS OF URBAN GREEN SPACE

Many of the benefits of urban greening projects align with the benefits of people interacting with nature in various ways and are documented elsewhere in this text. Urban populations are particularly in-need of reconnecting with nature, and benefits of urban greening projects are amplified in consideration of needs of urban populations.

Crime

One important determinant in the quality of life of urban populations is crime. Evidence suggest vegetation and large trees in urban areas are associated with lower crime rates (Kuo and Sullivan 2001; Donovan and Prestemon 2012). Unfortunately, public perception conveys a sense of fear of crime in areas with vegetation—relating vegetation with criminal hiding places. Research does not support these fears and community outreach work is needed to overcome these fears and garner support for urban greening projects. Instead, research has found that areas with vegetation and trees tend to be better cared for and attended to and less likely to be subjects of either property or violent crimes (Snelgrove et al. 2004). Additionally, after the urban greening projects, residents may become more watchful over their neighborhood and may begin to witness criminal activity in the area and report it to the police.

Mental Health

Urban environments are stressful in many ways and simply viewing the urban environment can invoke stressful feelings (Ulrich 1979, 1981; Hartig et al. 1991). Natural environments within urban areas can help alleviate such stresses (Tennessen and Cimprich 1995). Visiting green areas in cities can counteract stress, renew vital energy, and speed healing processes (McPherson 2000). People with access to nearby natural settings or parks are healthier overall when compared to other individuals, and long term, indirect impacts of "nearby nature" included increased levels of satisfaction with one's home, job, and life in general (Kaplan and Kaplan 1989).

Green space is particularly beneficial for children. Green space immediately outside the home in inner cities helped female children score more positively on measures of concentration, inhibition of impulses, and delay of gratification (Taylor et al. 2002). These measures are linked with later career success in life (Mischel et al. 1988). Children who spent more time in natural environments also felt more positive about their self-worth and less stressed in stressful situations (Kohlleppel et al. 2002).

Physical Health

Urban green space has the opportunity to impact physical health in a variety of ways. One way is by improving air quality. Trees remove airborne dust and chemical matter, or particulate matter from the air, where it is stored on leaves, twigs, and trunks (Beckett et al. 1998). Higher values of aerosol concentrations, related to vehicular pollutants, are present on the tree leaves on the roadside compared to values on the tree leaves on the side of the tree further from the roadside (Matzka and Maher 1999). Tree cover or "green belts" around factories and other industrial locations reduce air pollution by absorbing or otherwise filtering pollutants and dust (Rao et al. 2004). Planting around industrial areas reduces air pollution by as much as 63%, including a reduction of sulfur dioxide by 39%, nitrogen oxides by 40%, 37% of particulate matter, and a 93% reduction in carbon monoxide levels. In 1984, it was estimated in urban areas that planting one million new trees would remove 200 tons of particulate pollution each day after the trees reached 10 years of age (Petit et al. 1998) providing a research basis supporting million tree initiatives. Utilizing grasses and shrubs on rooftops reduces air pollution in urban areas by increasing the amount of vegetation in areas with limited space (Currie and Bass 2008). An increase in tree density is related to a decrease in the prevalence of health problems such as asthma in urban areas (Lovasi et al. 2008).

Moreover, urban greening projects can benefit the physical development of children. Outdoor play is beneficial to the physical development of children; however, forest or natural areas used for play provides increased benefits for childhood physical development (Fjortoft 2001; McFarland 2011). Children using forested and natural areas for play performed better on motor skill tests when compared to children playing on traditional playgrounds (Fjortoft 2001). Furthermore, schools reporting more green elements also reported more physical activity among students when compared to schools reporting fewer green elements such as trees, rocks, flowers, and water features (Dyment and Bell 2007).

Sociological Benefits

Urban parks and forests also provide residents of these areas with the ability to gather and have social interactions with other people (Figure 8.2). One explanation for a reduction in *place attachment* (also called *sense of place*) among residents of urban areas is the lack of feelings of community within the framework of the inner-city. Place attachment is the sense of connectedness to a physical location and has consistently been reported to be lower in residents of urban areas (Williams et al. 1992). Kuo et al. (1998) noted that community bonding was inhibited where crowding, crime, and noise proliferate in large cities. Natural spaces within the urban environment can help mediate these issues as urban green space reduces crowding and crime (Maller et al. 2009). This can also be related to the improvement in quality of life with access to and use of green spaces (McFarland et al. 2008).

In urban parks, people spend quality time with family members such as conversing with their spouse without as many technological interruptions and playing with their children. Parents have the opportunity to meet other parents as they jointly watch their children play. These activities promote the development of a "sense of place" or attachment to the community which is often lacking in urban populations (Maller et al. 2009). Sense of place and community is vital to ongoing community development and participation (Chavis and Wandersman 1990). This sense of

FIGURE 8.2 The Banyan tree at the courthouse square in Lahaina, HI was planted in 1873 and is the center of the community both physically and in the hearts of the citizens. In its shade, there are frequent art, music, and craft shows. (Photo by Dr. Tina M. Waliczek.)

connection is further developed through the use of festivals and events which can be hosted in the urban park or forest. Participation in these events creates a sense of shared identity and belonging (Duffy and Mair 2014).

Urban Revitalization

While the overall urban population is growing, many cities are facing a different problem known as urban shrinkage. *Urban shrinkage* refers to the sustained population loss in older industrial cities as the American economy becomes more service and technologically oriented. An accompanying problem with urban shrinkage is an increase in the number of vacant lots and abandoned properties (Schilling and Logan 2008). Vacant lots and abandoned properties can cause a decrease in feelings of safety and attract vandalism (Wachter et al. 2007). Properties located adjacent to abandoned and vacant lots have a 20% reduced property value compared to lots further from the vacant or abandoned lots.

Green space is a community asset. Urban revitalization plans can integrate green space located in proximity to neighborhoods rather than relegated to the periphery. Urban greening programs increase property values and rebuild an engaged resident population (Schilling and Logan 2008). Urban greening in these vacant lots can recover nearly 30% of the lost value through the reclamation and revitalization of adjacent vacant lots (Wachter et al. 2007).

ENVIRONMENTAL AND ECOSYSTEM SERVICES

The *urban heat island* effect is a phenomenon where cities experience higher air and surface temperatures when compared to surrounding suburban and rural areas due to the removal of vegetation which is replaced with heat absorbing nonpermeable materials such as concrete and asphalt (Environmental Protection Agency 2013). Urban green space can help mitigate these effects, providing cooler microclimates for wildlife and improving human comfort (Larson et al. 2009). Rooftop gardens or green roofs, in particular, can reduce buildings' energy consumption by providing winter insulation and summer cooling through increased evapotranspiration and shading from solar radiation (Oberndorfer et al. 2007).

Urban green space, including green roofs, provides environmental and ecosystem services beyond mitigating temperature extremes. Urban green space can help manage storm water runoff by slowing the speed of moving water, allowing water to percolate into the ground. This water then may be returned to the atmosphere through *evapotranspiration* from vegetation, become *ground water* and recharging *aquifers*, or drain slowly after the peak rainfall event (Oberndorfer et al. 2007).

Urban greening programs can improve the diversity of both *flora* (vegetation) and *fauna* (animals) in cities. The ways in which urban corridors transect and divide *wildlife corridors* is well documented, especially with regards to *migratory corridors* for birds and mammals and winter nesting for insects. For example, the Louisiana Black Bear is listed federally as a threatened species due to *habitat fragmentation* (US Fish and Wildlife Services 2013). Wildlife corridors are the paths along which animals travel to feed during a season, while migratory corridors are paths animals travel as they move across seasons. Habitat fragmentation is the division and separation of spaces capable of supporting the species. "Artificial" ecosystems created by urban greening and urban forestry projects can help address these issues and restore these habitats to improve biodiversity in urban areas. These projects can provide habitat and migratory paths when intersected that have previously been compromised by habitat destruction from increased urbanization (Figure 8.3). Urban forests provide cover, nesting places, and a diverse food source for wildlife to improve the biodiversity of species in cities (Savard et al. 2000).

Another benefit green space provides is *carbon sequestration* (Groffman et al. 1995). Carbon sequestration is the storage of carbon in various ways. The rise in atmospheric carbon in the form of carbon dioxide from the combustion of stored carbon (fossil fuels, lumber) is thought to play a key role in climate change. The potential for carbon sequestration is a motivating factor for many urban areas with

FIGURE 8.3 A man walks his dog in winter in Corporate Woods office park in Overland Park, KS. The award-winning office park has office buildings nestled into 294 acres of wooded parkland including walking and jogging trails and frequent wildlife sitings. (Photo by Dr. Tina M. Waliczek.)

excessive pollution. While many of the benefits documented previously pertain to any urban greening, carbon sequestration is particular to urban forests because large trees have the ability to sequester more carbon when compared to small trees and other vegetation (Nowak and Crane 2002).

Trees (and other vegetation) have the ability to turn atmospheric carbon dioxide into carbohydrates through photosynthetic processes. Due to larger biomass, on an individual basis, over a lifetime, trees produce and store more carbohydrates compared to smaller plants like shrubs (using more carbon dioxide—providing more carbon sequestration). In 2009, the United States emitted 5447 million metric tons of carbon dioxide. In 2002, annual carbon sequestration by urban trees was estimated at almost 23 million metric tons (Nowak and Crane 2002; US Energy Information Administration 2011). This is less than 1% of the total emissions sequestered, leaving room for improvement with additional trees. Carbon sequestration has been a driving force behind million tree initiatives.

INITIATING AND RUNNING A SUCCESSFUL URBAN GREENING PROJECT

Initiating an urban greening project can be as simple as a couple of people planting a tree or a series of trees. But, growing and running a program with long-term viability and success requires thoughtful planning, action, and funding. A long-term successful urban greening project requires the cooperation of the private sector, the public sector, and other *stakeholders* (people who have an interest in the project).

Johnston et al. (2013) highlight the importance of leadership in their top 10 tips for green cities. Writing from the perspective of city governance, they understand the role that leadership plays on inspiring the private sector to follow suit. While examples of San Marcos, Texas, and Philadelphia, Pennsylvania exemplify cities where the private sector has forged the way for urban greening; these were volunteer efforts from concerned citizens who then garnered government support and collaboration. Ultimately, greening projects require support from all stakeholders to be successful, though Johnston et al. (2013) argue that if the city government is not doing it, the private sector will not prioritize it. This argument applies to many facets of city management, but perhaps especially with regards to urban greening.

In order to get city governing committees on-board with urban greening, Johnston et al. (2013) provide the advice to utilize competitive instincts by highlighting examples of urban greening in nearby cities, while understanding that compromise will occur and perfection is not the goal. Small steps to improve urban greening can multiply and have big impacts. Moreover, sharing stories of success to media and social outlets can assist the broader goal of national greening by engaging those competitive instincts in cities nearby.

Accompanying the message of sharing success stories, maintaining urban green space over time is of quintessential importance to the long-term success of these projects. New York City experienced a city-wide decline in the 1980s characterized by high levels of violent and property crimes. During this period, the state of Central Park also deteriorated and the area was widely associated with high crime. While the entire city was impacted by this broad decline in the 1980s, the crimes perpetrated in

Central Park led to a stigma that still persists relating green space and vegetation to crime (Rosenzweig and Blackmar 1992). Just as media attention surrounding urban greening success stories can lead to new urban greening projects, nonmaintenance of green space and urban greening projects can have negative impacts on the development of other urban greening projects.

Ongoing maintenance of these spaces are labor intensive and expensive, so it is important that these spaces align with multiple goals, goals of other stakeholders, and achieve the "*triple bottom line*." The triple bottom line is the concept where *sustainability* strategies and programs address multiple problems—economic, environmental, and social—in one effort. Addressing the ways in which urban greening programs can lead to job creation, address needs of low-income populations, address crime rates, impact carbon emissions, and generate income are important in gaining support for the ongoing needs of an urban greening project (Johnston et al. 2013).

In order to share success stories and properly identify and address challenges, a system of measurement and reporting is critical. Understanding the key metrics and issues that stakeholders in the urban greening project are interested in addressing with the project are required to develop metrics. Metrics should focus on short-term goals (trash removal, tree planting), long-term goals (reducing energy use in adjacent buildings, decreasing neighborhood crime rates), how progress toward these goals are to be measured (obtaining energy usage figures from the power company, obtaining crime rates per capita from the police department), where the bar for success lies and at what point in time is the program expected to meet that level (Is 1% reduction in energy successful? Is 10% reduction in energy successful? What if energy use is not reduced but new businesses opened in adjacent buildings?). If the bar for success is not met, then action plans should be in place for identifying and addressing the problem.

Logic models can be useful in constructing a framework to identify goals, plan for achieving goals, and plan for measuring progress (Innovation Network n.d.). Logic models offer a visual map-like representation of short-term, mid-term, and long-term goals, outcomes, stakeholders, and plans for achieving and measuring. Logic models can assist planners in developing a strategy as well as a tool to present to potential investors (Figure 8.4).

FUNDING IDEAS

Logic models addressed above serve as a blueprint to transition between planning and funding. The development of the logic model can show potential investors such as private donors, foundations, and federal granting programs that the program has a plan for addressing a multitude of stakeholder needs and is a worthwhile investment.

While having a solid plan in place can assist in funding, organizations are often stressed to find sufficient philanthropic funds to fully finance comprehensive programs. Turning to successful programs and methods used to establish unique financial streams is important in regards to urban greening. Toronto's Green Tourism Association (GTA), the Brooklyn Botanic Garden, and the Pennsylvania Horticulture Society are programs that have taken advantage of the increase in consumer demands toward "green" living to finance urban greening projects.

Program										
Organization										
Date										
		Inputs	Outputs		Outcomes and Impacts			Measurement Plan		
Situation	Priorities	Investments	Activities	Participation	Proximal: short term	Distal: Medium Term	Distal: Long Term	Success Indicator	Outcome Measurement Tools	Evaluation Design
		What we invest	*What we do*	*Who we reach*	*What the shorterm results are*	*What the medium term results are*	*What the ultimate results are*	*Evidence of success*	*Surveys; standard tests; measures*	*Data Mgmt. Design & Schedule*
Literature teachers lack effective tools to engage students in hands-on learning experiences.	Provide teachers a space and activities in which to engage students in literature.	Developed and maintained garden area with seating and "storybook" features	Storybook readings in the garden, and activities related to the storybook	Teachers. Children grades Pre-K through 8	Improved grades in reading and spelling for primary education.	Improved high school graduation rates. More students are college bound.	More productive adults and better able to communicate effectively.		Survey teachers; TAKS scores	
Families lack opportunities to spend time with each other in natural settings.	Provide families a space and activities in which to spend time together.	Volunteer storytellers	Storytelling in the garden and activities related to the story	Families in the community	Increased family time				Survey parents	
College students in fields such as education, psychology, horticulture, and agriculture education lack opportunities to work with youth.	Provide college students a space, activities, and children with whom they can work to develop future job skills.	Activity development	Acting-out of stories in the garden and activities related to the story	College students	College students are provided with more opportunities to work with children.	College students acquire better jobs due to their experiences.	Students who participated are more effective at working with children and make bigger impacts.		Survey college students	

FIGURE 8.4 An example logic model for potential garden-related projects.

Tourists in urban areas often feel lost and overwhelmed. Offering guided tours highlighting urban greening projects can provide a funding source for additional urban greening projects. Toronto's GTA has pioneered an industry in urban ecotourism (Gibson et al. 2003). *Urban ecotourism* or *urban green tourism* is travel and exploration within urban environments highlighting the city's ecological resources and natural areas (Gibson et al. 2003). Toronto's GTA argues that by increasing exposure of the ecological and natural areas of the city, demand for those types of environments will increase. In other words, the conservation of Toronto's natural areas can lead to economic growth—with economic impact being a contested area for any environmental decision. Toronto's GTA works by educating tourists about the natural areas within the city through a green map of Toronto, a guide to green tourism in Toronto, and a website dedicated to green tourism (Gibson et al. 2003).

Annual events with admission fees can also help promote and fund continued greening projects. Brooklyn Botanic Garden's annual cherry blossom festival—the Sakura Matsuri—is a prime example of an annual event in a built, natural environment which attracts one of the largest, diverse, and unique audiences of any cultural event in New York City. Planners interested in developing annual events, such as the Sakura Matsuri, can use the marketing strategies of promoting events in a variety of ways to a diversity of backgrounds attracting large audiences who pay a premium admission fee for the event (Brooklyn Botanic Garden 2014). Pennsylvania Horticulture Society's funding for their urban greening projects in Philadelphia come primarily from their annual Philly Flower Show. The Philly Flower Show is the largest, indoor flower show with a specific purpose and constituency. The show attracts repeat annual visitors from across the northeast, as well as visitors from across the nation.

There are numerous small grants available for gardening and tree planting programs, especially for community and nonprofit organizations. For example, there is a page dedicated to small grants and funding opportunities on the National Gardening Association's website (National Gardening Association 2014). These grants are targeted toward children's gardens which could be designed in conjunction with urban greening and tree planting programs. Community-based organizations also often have tree planting grants or giveaways, sometimes around Arbor Day, that organizations and individuals could use for smaller projects.

CASE STUDY 1 Chicago, Illinois

Early in Chicago's history, the phrase "urbs in horto" (meaning city in a garden) was adopted as a motto for the burgeoning city and government. Mayor Richard Daley recognized the importance green space can have toward improving the overall urban experience and co-opted the ideology into his efforts to transform the city into one that leads the nation in rooftop gardens and green streets. Today, there are over six million square feet of rooftop gardens in Chicago. In 2003, Mayor Daly hired urban greening expert Sadhu Johnston to lead Chicago's green initiatives (Johnston et al. 2013). Mayor Daly believed that urban greening could improve the global environment, quality of life of people, and provide

Chicago with important economic advantages to aid in recessions. Moreover, Mayor Daly recognized the importance of governmental examples in encouraging citizens to take similar greening actions at their homes. In 2010, National Geographic published an article labeling Chicago as "America's Green City" (Conaway 2010) for the efforts in greening the streets of Chicago by replacing lanes of traffic with planted medians and installing rooftop gardens. Chicago's Park District currently manages over 8000 acres of open spaces (Chicago Park District 2014).

One of Chicago's most notable green spaces is the 24.5 acre Millennium Park, winning numerous awards including the 2005 Travel and Leisure Design Award for "best public space" (Ryan 2005). Millennium Park includes an outdoor music pavilion, theatre, works of art, and Lurie Garden, a 5 acre homage to the city's motto "urbs in horto" which highlights sustainable design and garden practices such as refraining from chemical use, using mass plantings of bee pollinated flowers, and planting drought resistant native plants (Kinzer 2004; Lurie Garden n.d.). Though over budget and late in completing construction, Millennium Park has been heralded as a success by Chicago residents, to urban greening enthusiasts, and as a model for other cities.

CASE STUDY 2 New York City

New York City is home to 29,000 acres of green space, accounting for over 14% of the city's overall physical size (NYC Parks n.d.). Most notably is New York's Central Park, which opened in 1859 following a desire from wealthy New Yorkers to replicate the vast expanse of public grounds in Europe (Figure 8.5).

FIGURE 8.5 Central Park offers city dwellers and tourists a reprieve from the stresses of New York City. (Photo by Dr. Amy McFarland.)

FIGURE 8.6 Trinity Church, New York City. (Photo by Dr. Amy McFarland.)

The story of Central Park demonstrates the contested nature of public spaces. The stated purpose of the park was to provide a place for everyone to enjoy—an alternative for the working class to the saloon. However, after opening, the upper echelons of New York had different ideas about to whom Central Park belonged. The building of the park displaced African American neighborhoods and at times, when the park was used most frequently by people of color, the association between the park and crime was exaggerated. The necessity of continued upkeep in urban greening projects became apparent when, in the 1970s, Central Park fell into disrepair and actual crime rates were higher in the park. In the 1980s and 1990s, private money was used to restore the park, again bringing the question of a park with democratic access into question.

Walking through Central Park today, it becomes clear that the founding principle of a park for everyone has, possibly, been realized. The park has attractions for children and adults alike including locals and tourists, with playgrounds, museums, jogging paths, and interpretive signage. The reprieve offered by this park is more extensive than the 2 mile by 1/2 mile area would suggest. Immediately upon entering the park, the tourist who thought they would walk straight through Central Park, realizes the enormity of the task. Designed with meandering paths, forested areas, and water features that serve as obstacles to a direct walk through the park, Central Park forces city life to slow-down within its boundaries.

While Central Park is certainly the most notable area of green space within New York City, it is by no means the only. Small pockets of green space are located throughout Manhattan—parks, street trees, cemeteries (Figure 8.6). For example, the 9/11 Memorial (Figure 8.7) was designed with green space at the forefront of the concept with over 400 trees planted around the memorial to convey a sense of peacefulness at the site (9/11 Memorial 2014). Washington Square Park (Figure 8.8) can serve as another example of a pocket of widely accessible green space in Manhattan covering over 8 acres. Although this particular park

FIGURE 8.7 Looking through the trees at the 9/11 Memorial to view the new One World Trade Center building. (Photo by Dr. Amy McFarland.)

FIGURE 8.8 Washington Square Park, New York City. (Photo by Dr. Amy McFarland.)

is often criticized for too much paving, recent redesign efforts have included removing pavement in favor of additional green space (Williams 2005).

CASE STUDY 3 Vancouver, British Columbia

Vancouver, British Columbia adopted an ambitious goal in 2011—to become the greenest city in the world. The Greenest City 2020 Action Plan was adopted in an effort to prepare the city's population for expansion and growth while achieving this green goal (Johnston et al. 2013). Throughout the development and implementation of this action plan, the City of Vancouver has sought green ideas from residents, making this request accessible to all residents with signage on city buses (Figure 8.9). As a part of the initiative, the City of Vancouver embarked on a social media campaign to encourage green behavior such as urban agriculture projects including the development of community gardens (Figure 8.10) and urban orchards on city owned lands (City of Vancouver 2014a).

Vancouver's Stanley Park is one of the largest and oldest examples of urban greening in North America (Kheraj 2013). Since the park's creation, a struggle between restoring nature void of human activity and encouraging use of the park for the urban population has been problematic. The park consists of 1000 acres of forest that is often described as "impenetrable." However, Stanley Park is an integral part of the city of Vancouver as they were designed in tandem. A three-lane highway occupies the center of the park in such a way that fuses the city and the park into a single entity. Jogging and walking paths, swimming pools, and playgrounds are frequently used on the exterior of the park along the waterfront—and interior paths offer over 16 miles of a

FIGURE 8.9 Vancouver public transit advertising requesting green ideas input for the Greenest City 2020 Action Plan. (Photo by Dr. Amy McFarland.)

FIGURE 8.10 Davis Village Community Garden, Vancouver, British Columbia, Canada. (Photo by Dr. Amy McFarland.)

quieter, cooler, and closer to nature experience for locals and tourists (City of Vancouver 2014b).

CASE STUDY 4 Philadelphia, Pennsylvania

Philadelphia is a case of urban deterioration with many abandoned lots from old warehouses and empty vacant lots blighted by dumping. What happens to these places? In Philadelphia, the Pennsylvania Horticultural Society developed a number of programs to aid in the restoration of these blighted, vacant properties.

The Pennsylvania Horticultural Society (PHS) was founded on November 24, 1827 by a group of various farmers, botanists, and plant enthusiasts. At that time, 80 members convened "to establish a Horticultural Society in the City of

Philadelphia for the promotion of this interesting and highly influential branch of Science" (PHS 2014). Over the past nearly two centuries, the organization has flourished, now burgeoning over 23,000 members and pursuing numerous greening projects throughout the greater Philadelphia area (PHS 2014). For example, the LandCare program focuses on neighborhood redevelopment by transforming vacant lots into simple landscapes with a park-like feeling. LandCare currently provides management for 10 million square feet in Philadelphia. In 2013, the Pennsylvania Horticultural Society revealed a new program, Civic Landscapes, designed to transform seven key areas in Philadelphia to create spaces that redefine the expectations of residents and tourists of urban landscapes.

Philadelphia is an interesting case because it is driven by a volunteer organization rather than the city government. The primary goal of PHS is to improve quality of life of the residents of Philadelphia. PHS and associated programs are funded primarily through donations and their annual Philly Flower Show.

CASE STUDY 5 San Marcos, Texas

Thus far, case studies have focused on urban greening and forestry in very large cities with strong influence from city governance. Smaller cities and those with less influence from city governance can also benefit from the urban greening movement. A case in point is San Marcos, Texas. San Marcos is a less developed city, but local organizations have recognized the importance in conserving natural areas as the city grows. In San Marcos, the development of the San Marcos

FIGURE 8.11 Schulle Canyon is one of several greenbelt parks maintained by San Marcos Greenbelt Alliance in San Marcos, TX. (Photo by Dr. Alice Leduc.)

FIGURE 8.12 Ringtail Ridge is a greenbelt park in San Marcos, TX maintained by San Marcos Greenbelt Alliance. The park was the former site of an animal slaughterhouse and processing plant. (Photo by Dr. Alice Leduc.)

Greenbelt Alliance has had an integral role in the creation and conservation of parks and natural areas (Figures 8.11 and 8.12). This organization works closely with developers, the city and county governance, and residents to secure natural areas and turn these into parks with access via trails. The organization has a volunteer base to help in maintaining areas in such a way to invite use of these natural areas while minimizing impacts from use (San Marcos Greenbelt Alliance 2014). This consortium of interested organizations and individuals working in collaboration with local and larger governments had direct positive attention and action toward developing and conserving green space.

SUMMARY

Most people in today's world have very little contact with the pristine, virgin nature of the wilderness ethic. Most people connect with nature on a daily basis through small plots of protected trees and constructed parks in urban environments. It is important to provide attention and environmental focus for these places as they provide numerous services for the growing urban population.

Urban gardens and green spaces have been developed in areas of higher population densities since ancient times. However, modern governing boards are recently beginning to understand the multiplicity of benefits urban greening programs can provide. Some of these benefits are environmental, but the "triple bottom line" of sustainability, the economic, environmental, and social benefits have all been identified through peer reviewed academic research. The benefits of urban horticulture documented elsewhere in this text are included in the benefits of urban greening projects. Unique to urban green spaces with an abundance of trees is the ability to sequester substantial amounts of carbon dioxide from the atmosphere and reduce carbon emissions. Urban greening programs also have the potential to help mitigate the urban heat island effect by absorbing and using solar energy.

Urban greening projects have the possibility to help reverse the conditions many cities that were built on manufacturing are experiencing with regards to urban shrinkage. Urban shrinkage often leaves vacant or abandoned lots near residential properties, causing declines in property values and increase in crime rates. Urban greening projects help clean up and revitalize these areas resulting in increasing property values and decreasing crime rates, both of which are important urban metrics.

Developing a logic model can help stakeholders of urban greening projects understand the ways in which purposes and outcomes can be interconnected, helping to gain government and financial support. Ongoing maintenance of urban greening projects is critical, not only for the success of the individual program, but also for the long-term success of national greening. Monitoring, measuring, reporting, and responding to successes and challenges is also critical in the long-term success of urban greening.

REVIEW QUESTIONS

1. What are some of the earliest examples of urban green space?
2. What earliest US cities were designed with green spaces in the original plans?
3. What are rooftop gardens and what is the difference between an intensive system and an extensive system?
4. What is a million tree initiative and how do they benefit urban populations?
5. What were two primary intents for green belts?
6. Give three reasons why the maintenance of urban green space is critically important to the long-term success of these areas.
7. What services do urban green spaces provide animals?
8. What is the "triple bottom line" of sustainability?
9. How does urban green space address the triple bottom line?

10. How might green space in smaller towns be managed differently from green space in large cities?

ENRICHMENT ACTIVITIES

1. Locate a park in your city that lacks maintenance and develop a master plan to revitalize it.
2. Locate a vacant lot in your city and develop and implement a plan to green it.
3. Visit an urban farm in your city. Have a conversation with the farmer about the history of the plot, how they receive permission to farm, and problems they have had in producing off this plot.
4. Identify wild spaces in your city that could be preserved by a volunteer organization.
5. Develop by-laws for a volunteer organization devoted to the development of urban green space or forests.
6. Identify volunteer organizations in the community devoted to preserving and/or developing green spaces.

REFERENCES

9/11 Memorial. 2014. Design Overview. Retrieved September 1, 2014 http://www.911memorial.org/design-overview.

Amati, M. 2008. *Urban Green Belts in the Twenty-First Century*. Burlington, VT: Ashgate Publishing Limited.

Beckett, K.P., P.H. Freer-Smith, and G. Taylor. 1998. Urban woodlands: Their role in reducing the effects of particulate pollution. *Environmental Pollution*, 99, 347–360.

Brooklyn Botanic Garden. 2014. Sakura Matsuri: A celebration of Japanese Culture at BBG. Retrieved September 7, 2014. http://www.bbg.org/press/sakura_matsuri_a_celebration_of_japanese_culture_at_bbg.

Chavis, D.M. and A. Wandersman. 1990. Sense of community in the urban environment: A catalyst for participation and community development. *American Journal of Community Psychology*, 18, 55–81.

Chicago Park District. 2014. History. Retrieved August 13, 2014. http://www.chicagoparkdistrict.com/about-us/history/.

Choukas-Bradley, M. 2008. *City of Trees: The Complete Field Guide to the Trees of Washington, D.C.*, 3rd edition. University of Virginia Press, Charlottesville.

City of Vancouver. 2014a. Start a community garden or orchard. Retrieved September 6, 2014. http://vancouver.ca/people-programs/start-a-new-community-garden.aspx.

City of Vancouver. 2014b. Stanley Park trails. Retrieved September 6, 2014. http://vancouver.ca/parks-recreation-culture/stanley-park-trails.aspx.

Conaway, J. 2010. Chicago: America's Green City. National Geographic. September.

Currie, B.A. and B. Bass. 2008. Estimates of air pollution mitigation with green plants and green roofs using the UFORE model. *Urban Ecosystems*, 11, 409–422.

Donovan, G.H. and J.P. Prestemon. 2012. The effect of trees on crime in Portland, Oregon. *Environment and Behavior*, 44, 3–30.

Duffy, M. and J. Mair. 2014. Festivals and sense of community in places of transition. In: A. Jepson and A. Clarke (eds.), *Exploring Community Festivals and Events*. New York: Routledge.

Dyment, J.E. and A.C. Bell. 2007. Grounds for movement: Green school grounds as sites for promoting physical activity. *Health Education Research*, 23, 952–962.

Environmental Protection Agency. 2013. Basic Information. Retrieved July 9, 2014. http://www.epa.gov/heatisland/about/index.htm.

Fjortoft, I. 2001. The natural environment as a playground for children: The impact of outdoor play activities in pre-primary school children. *Early Childhood Education Journal*, 29, 111–117.

Foderaro, L.W. 2012. *Huge Rooftop Farm Set for Brooklyn*. New York Times. Retrieved August 5, 2015. http://www.nytimes.com/2012/04/06/nyregion/rooftop-greenhouse-will-boost-city-farming.html.

Gibson, A., R. Dodd, M. Joppe, and B. Jamieson. 2003. Ecotourism in the city? Toronto's green tourism association. *International Journal of Contemporary Hospitality Management*, 15, 324–327.

Groffman, P.M., R.V. Pouyat, M.J. McDonnell, S.T.A. Pickett, and W.C. Zipperer. 1995. Carbon pool and trace gas influxes in urban forest soils. In R. Lal, J.M. Kimble, E. Levine, and B.A. Stewart (eds.), *Soil Management and Greenhouse Effect*. Boca Raton, FL: CRC Press, pp. 147–158.

Halnon, M. n.d. The Mall: The Grand Avenue, the Government, and the People. University of Virginia. Retrieved July 7, 2014. http://xroads.virginia.edu/~CAP/MALL/chron.html.

Hartig, T., M. Mang, and G.W. Evans. 1991. Restorative effects of natural environment experiences. *Environment and Behavior*, 23(1), 3–26.

Innovation Network. n.d. Logic model workbook. Retrieved September 8, 2014. http://www.innonet.org/client_docs/File/logic_model_workbook.pdf.

Jenkins, V. 1994. *The Lawn: A History of an American Obsession*. Washington, DC: Smithsonian Institution Press.

Johnston, S.A., S.S. Nicholas, and J. Parzen. 2013. *The Guide to Greening Cities*. Washington DC: Island.

Kaplan, R. and S. Kaplan. 1989. *The Experience of Nature: A Psychological Perspective*. Cambridge, New York: Cambridge University Press.

Kheraj, S. 2013. *Inventing Stanley Park: An Environmental History*. Vancouver, British Columbia: UBC Press.

Kinzer, S. 2004. *Letter from Chicago: A Prized Project, a Mayor, and Persistent Criticism*. The New York Times. July 13.

Kohlleppel, T., J.C. Bradley, and S. Jacob. 2002. A walk through the garden: Can a visit to a botanic garden reduce stress? *HortTechnology*, 12(3), 489–492.

Kuo, F.E. and W.C. Sullivan. 2001. Environment and crime in the inner city: Does vegetation reduce crime? *Environment and Behavior*, 33, 343–367.

Kuo, F.E., W.C. Sullivan, R.L. Coley, and L. Brunson. 1998. Fertile ground for community: Inner-city neighborhood common spaces. *American Journal of Community Psychology*, 26, 823–851.

Larson, K.L., D. Casagrande, S.L. Harlan, and S.T. Yabiku. 2009. Residents' yard choices and rationales in a desert city: Social priorities, ecological impacts, and decision tradeoffs. *Environmental Management*, 44, 921–937.

Lovasi, G.S., J.W. Quinn, K.M. Neckerman, M.S. Perzanowski, and A. Rundle. 2008. Children living in areas with more street trees have lower prevalence of asthma. *Journal of Epidemiology and Community Health*, 62, 647–649.

Lurie Garden. n.d. Lurie Garden. Retrieved August 20, 2014. http://www.luriegarden.org.

Maller, C.J., C. Henderson-Wilson, and M. Townsend. 2009. Rediscovering nature in everyday settings: Or how to create healthy environments and healthy people. *EcoHealth*, 6, 553–556.

Matzka, J. and B.A. Maher. 1999. Magnetic biomonitoring of roadside tree leaves: Identification of spatial and temporal variations in vehicular-derived particulates. *Atmospheric Environment*, 33, 4565–4569.

McFarland, A. 2011. Growing minds: The relationship between parental attitude about nature and the development of fine and gross motor skills in children. (Doctoral dissertation.) Retrieved from Texas Digital Library. Available electronically from http://hdl.handle.net/1969.1/ETD-TAMU-2011-05-9067.

McFarland, A.L., T.M. Waliczek, and J.M. Zajicek. 2008. The relationship between student use of campus green space and quality of life. *HortTechnology*, 18, 232–238.

McPherson, E.G. 2000. Urban forestry: The final frontier? *Journal of Forestry*, 101(3), 20–25.

Million Trees LA. 2006. About Million Trees LA. Retrieved August 13, 2014. http://www.milliontreesla.org/mtabout.htm.

Mischel, W., Y. Shoda, and P. Peake. 1988. The nature of adolescent competencies predicted by preschool delay of gratification. *Journal of Personality and Social Psychology*, 54, 687–696.

Muller, E.K. 2010. Building American cityscapes. In M.P. Conzen (ed.), *The Making of the American Landscape*. New York: Routlege.

National Gardening Association. 2014. Grants and Awards. Retrieved December 4, 2014. http://grants.kidsgardening.org/.

Nowak, D.J. and D.E. Crane. 2002. Carbon storage and sequestration by urban trees in the USA. *Environmental Pollution*, 116, 381–389.

NYC Parks. n.d. About the New York City Department of Parks and Recreation. Retrieved August 13, 2014. http://www.nycgovparks.org/about.

Obama, M. 2012. *American Grown: The Story of the White House Kitchen Garden and Gardens Across America*. New York: Crown Publishers.

Oberndorfer, E., J. Lundholm, B. Bass, R. Coffman, H. Doshi, N. Dunnett, S. Gaffin, M. Kohler, K. Liu, and B. Rowe. 2007. Green roofs as urban ecosystems: Ecological structures, functions, and services. *Bioscience*, 57, 823–833.

Petit, J., D.L. Bassert, and C. Kollin. 1998. *Building Greener Neighborhoods: Trees as Part of the Plan*, 2nd edition American Forests National Association. Washington, DC: Home Builders.

PHS. 2014. Retrieved September 8, 2014. http://phsonline.org/.

Rao, P.S., A.G. Gavane, S.S. Ankam, M.F. Ansari, V.I. Pandit, and P. Nema. 2004. Performance evaluation of a green belt in a petroleum refinery: A case study. *Ecological Engineering*, 23, 77–84.

Rosenzweig, R. and E. Blackmar. 1992. *The Park and the People: A History of Central Park*. Ithaca, New York: Cornell University.

Ryan, K. 2005. Chicago's new Millenium Park wins Travel and Leisure Design Award for "Best Public Space" and the American Public Works Association "Project of the Year" award. City of Chicago. Retrieved September 1, 2014. http://web.archive.org/web/20080910132600/.http://www.millenniumpark.org/newsandmedia/pdfs/4.1%20Millennium%20Park%20Awards.pdf.

San Marcos Greenbelt Alliance. 2014. About. Retrieved September 6, 2014. http://www.smgreenbelt.org/About.htm.

Savard, J.L., P. Clergeau, and G. Mennechez. 2000. Biodiversity concepts and urban ecosystems. *Landscape and Urban Planning*, 48, 131–142.

Schilling, J. and J. Logan. 2008. Greening the rust belt: A green infrastructure model for right sizing America's shrinking cities. *Journal of the American Planning Association*, 74, 451–466.

Schneider, A., M.A. Friedl, and D. Potere. 2009. A new map of global urban extent from MODIS satellite data. *Environmental Research Letters*, 4(4), 44003.

Sherald, J. 2009. Elms for the Monumental Core: History and Management Plan. Washington, DC: Center for Urban Ecology, National Capital Region, National Park Service. pp. 2–5. Natural Resource Report NPS/NCR/NRR—2009/001. Retrieved July 7, 2014.

http://www.nps.gov/nationalmallplan/Documents/Studies/ElmsoftheMonuCore_HistandMgmtPlan_122009.pdf.
Shimmin, H. 2012. A brief history of roof gardens. Heather Shimmin Photography. Retrieved July 9, 2014. http://www.heathershimmin.com/a-brief-history-of-roof-gardens.
Snelgrove, A.G., J.H. Michael, T.M. Waliczek, and J.M. Zajicek. 2004. Urban greening and criminal behavior: A geographic information systems perspective. *HortTechnology*, 14, 48–51.
Taylor, A.F., F.E. Kuo, and W.C. Sullivan. 2002. Views of nature and self-discipline: Evidence from inner city children. *Journal of Environmental Psychology*, 22, 49–63.
Tennessen, C.M. and B. Cimprich. 1995. Views to nature: Effects on attention. *Journal of Environmental Psychology*, 15, 77–85.
Turner, B.L., E.F. Lambin, and A. Reenberg. 2007. The emergence of land change science for global environmental change and sustainability. *Proceedings of the National Academy of Sciences of the United States of America*, 104, 20666–20671.
Ulrich, R.S. 1979. Visual landscapes and psychological well-being. *Landscape Research*, 4, 17–23.
Ulrich, R.S. 1981. Natural versus urban scenes: Some psychophysiological effects. *Environment and Behavior*, 13(5), 523–556.
US Energy Information Administration. 2011. Emissions of greenhouse gases in the US Retrieved September 7, 2014. http://www.eia.gov/environment/emissions/ghg_report/ghg_carbon.cfm.
US Fish and Wildlife Services. 2013. Federally threatened, endangered, and candidate species in Mississippi. Retrieved January 9, 2015. http://www.fws.gov/mississippiES/pdf/MS_Species_Habitat_Descriptions_2013.pdf.
Wachter, S.M., K.C. Gillen, and C. Brown. 2007. Green investment strategies: How they matter for urban neighborhoods. Retrieved September 7, 2014. http://gislab.wharton.upenn.edu/Papers/Green%20Investment%20Strategies%20How%20They%20Matter%20for%20Urban%20Neighborhoods.pdf.
Weiler, S.K. and K. Scholz-Barth. 2009. *Green Roof Systems: A Guide to the Planning, Design, and Construction of Landscapes Over Structure*. Hoboken, NJ: Wiley.
Whitehouse. 2009. Inside the White House: The kitchen garden. Retrieved September 7, 2014. http://www.whitehouse.gov/video/Inside-the-White-House-The-Garden.
Williams, T. 2005. *Washington Square Park, Haven for Eccentricity, Is Set to Fall Into Line*. The New York Times, May 10.
Williams, D.R., M.E. Patterson, J.W. Roggenbuck, and A.E. Watson. 1992. Beyond the community metaphor: Examining emotional and symbolic attachment to place. *Leisure Sciences*, 14, 29–46.
Wood, J.S. 2010. Creating landscapes of civil society. In M.P. Conzen (ed.), *The Making of the American Landscape*. New York: Routlege.

FURTHER READING

Conzen, M.P. 2010. Introduction. In M.P. Conzen (ed.), *The Making of the American Landscape*. New York: Routlege.

9 Local Food

James Buratti and Ronald Hagelman III

CONTENTS

OBJECTIVES

Upon completion of this chapter, the reader should be able to

- Discuss and compare definitions of local food.
- Identify components of local food systems.
- Describe the origins of the local food movement.
- Understand consumer motivation for local food.
- Describe the benefits of local food.

KEY TERMS

- Alternative food networks
- Biodynamic agriculture
- Certified organic
- Community/allotment gardens
- Community supported agriculture
- Direct to consumer
- Edible landscapes
- Farmers market
- Farm to school
- Farm to institution
- Food desert
- Foodie
- Food insecure
- Food miles
- Foodshed
- Foragers
- Frankenfoods
- Gleaning
- Global Industrial Food System
- Hyper-local sourcing
- Locavore
- Local food network
- Local/regional food hub
- Producers
- Short food supply chain
- Sustainable agriculture
- Square foot gardening
- Urban farming

INTRODUCTION

Prior to the nineteenth and twentieth centuries, nearly all food consumed in the United States was sourced locally and limited by seasonality in all regions. With little processed food and slow or expensive transportation, perishable foods traveled less than a day's journey to market (Giovannucci et al. 2010). Consumers were often in direct contact with the farmer, rancher, or dairyman and had knowledge of seasonal availability of various foods. With improved food preservation techniques and faster and cheaper transportation options, retailers were able to offer agricultural products from more distant locations and producers were able to sell their products to a larger market area. A *producer* in the agricultural world is farmers, ranchers, dairyman, or anyone who grows or produces food for sale.

Like many consumer items during this period, American agricultural food products became industrialized throughout WWI and WWII. Building on manufacturing and distribution techniques established before the wars, innovative processing, packaging, preserving, and preparation approaches further expanded the reach of food production and consumption following WWII. Packaged foods were marketed as easing the burden on the domestic homemaker. Cake mixes and canned soups did not take the place of the homemade meal, but created an affordable, quick substitute. Fast food also entered the American diet and eating out became more commonplace.

Industrialization, globalization, and inexpensive international shipping created a *global industrial food system*, a highly organized network of agricultural production, distribution, marketing, and sales that produces most food consumed in the world. It is highly dependent on inexpensive inputs of petroleum, water, land, and labor. With this system came a bounty of food choice. Consumers no longer had to wait for strawberries in the spring and squash in the summer. They could be shipped from South America and sold cheaper than a local farmer could produce them. Food miles, the distance it takes for a product to get from its point of origin to a retail outlet, skyrocketed. Pirog et al. (2001) showed produce delivered by truck traveled an average of 1245 miles in 1981 to reach Chicago and increased to 1518 miles in 1998. In just a few decades, Americans' agricultural products went from locally sourced to globally sourced.

Not everyone welcomed the industrialization and globalization of food. With it came food designed for convenience and profit, not nutrition and health. Food was built around packaging and shelf life, not quality and freshness. The past two decades have seen increasing demand for locally produced food in the United States; so much so a new term was coined for those trying to eat locally—the *locavore*.

Consumer research shows a wide variety of motivations to eat locally produced foods. These include distrust of corporations responsible for globally sourced food products, food safety concerns, rising rates of obesity and diabetes, a desire to avoid genetically modified organisms (GMOs) and crops (the so-called *Frankenfoods*), concerns for the environment and sustainability, the loss of local farms in communities around the country, increasing demand for organic foods, boycotting of exploitive labor practices, and the view that industrial food production can lead to various types of animal cruelty (Bougherara et al. 2009; Bean and Sharp 2010).

Farmers, ranchers, and retailers have reacted to the demand, offering more choices for the local food consumer. The United States Department of Agriculture

FIGURE 9.1 Austin farmers market, Austin, Texas. (Photo by James Buratti.)

(USDA) counted 14,540 organic farms in the United States in 2007 (USDA 2010), 12,549 community supported agriculture (CSA) operations (USDA 2007), and 3600 farmers markets in 2010 (USDA 2010). Most major grocery store chains now offer a selection of organic and local products from Whole Foods Market to Walmart. The trend toward large national and international food distribution retailers offering food products labeled local has spurred debate on what exactly constitutes locally produced food with definitions being issued from entities ranging from the USDA to farmers markets around the nation (Figure 9.1).

DEFINING "LOCAL FOOD"

The definition of what qualifies as local food is both relative and subjective. Producers, consumers, retailers, nongovernmental organizations (NGOs), farmers markets, state and federal governments may all have different and conflicting definitions. To further complicate matters, in the minds of most consumers local and organic food are often interchangeable, although the term *certified organic* has very strict requirements for its usage by producers as defined by the USDA (Berlin et al. 2009; Blake et al. 2010). Certified organic prohibits the use of synthetic pesticides, herbicides, and GMOs. Often, local is also certified organic, but just as often certified organic is far from local. Many government and private entities have chimed in on the definition of locally produced food. These are discussed according to type of organization below.

US Federal and State Definitions

The U.S. federal government first defined local food in the Food, Conservation, and Energy Act of 2008 (HB 2419 2008). It stated that, "The term locally or regionally produced agricultural food product means any agricultural food product that is raised, produced, and distributed in—(I) the locality or region in which the final product is marketed, so that the total distance that the product is transported is less than 400 miles from the origin of the product; or (II) the State in which the product is produced." By including mileage in the definition, the U.S. federal government attempted to make the concept of local food production and consumption much less subjective. However, this definition ignores the varying size of U.S. states. A local food product in California can travel 1287 km (800 miles) to the consumer. If transported along Interstate 10 in Texas, local produce could travel 1415 km (879 miles). Alternately, food produced in Delaware could be transported to 15 different states within 400 miles and still be considered local by the federal government.

Since its initial definition, the federal government has refrained from further defining local, although it has recognized its importance to the nation's farmers, economy, and health. Multiple federal programs now offer assistance to farmers, public schools, and agricultural extensions in support of local food production. In order to popularize gardening and local food consumption, First Lady Michelle Obama has vigorously promoted eating healthy and local and has hosted numerous events in the White House garden.

The idea that food grown within a state should be considered local is a common definition. Surveys of both consumers and producers in Maine showed the state was the largest region that they would consider local (Hunt 2006). Only one-third of consumers in a national survey defined local as, "grown in their state or region" (Pirog and Rasmussen 2008). Many states reinforce this notion with state-run campaigns and advertising encouraging the purchase of products made within a state. Patterson (2006) counted 43 states with state-branding agricultural campaigns.

These campaigns, however, often target both in-state and out-of-state consumers and are designed primarily to build brand awareness. A good example is the Texas Department of Agriculture's program called GO TEXAN that, "celebrates, promotes and supports the business savvy and plainspoken grit Texas agriculture is known for throughout the world" (GO TEXAN 2014). It has an annual budget over $1 million (General Appropriations Act 2013) and can be considered an international branding campaign, designed as much to sell Texas produce internationally as to incentivize local food systems. The advertising is ubiquitous in Texas and presumably does generate attention toward locally sourced products among Texas shoppers.

More true to the local food movement, Wisconsin promotes the concept of state as local though the Buy Local, Buy Wisconsin promotion. Buy Local, Buy Wisconsin is a competitive grant program for producers designed to, "reduce the marketing, distribution, and processing hurdles that impede the expansion of sales of Wisconsin's food products to local purchasers." The state estimates its investment of $625,400 between 2008 and 2010 generated $5.9 million in local food sales, $1.49 million in investments and the creation of 63 new jobs benefitting 2062 producers (Wisconsin Department of Agriculture, Trade and Consumer Protection 2014).

Consumer Definitions

The term "local" connotes a limited geography, but pair the word "local" with "food" and geography is only one of many factors that consumers consider when they talk about local food. Research shows consumers consider social, economic, environmental, political, supply chain, and even intangible factors such as freshness and taste factors in their personal definition of what constitutes local food. It is not simply a matter of miles traveled from farm to plate. Geography is probably the easiest relationship to define when speaking of local food systems, but that does not mean there is any wide-scale agreement. Common consumer definitions of local vary by region. Some common definitions include within 100 miles of the place a product was grown; within a political boundary—a county (i.e., Napa County, California) or state (i.e., State of Florida); or from within a region (i.e., New England), or something more arbitrary like within a days' drive (Hunt 2006; Peters et al. 2008; Berlin et al. 2009; Blake et al. 2010).

Surveys have shown that the majority of consumers feel the state is too big to be considered local (Brown 2003; Hunt 2006) and define a much smaller geographic area. In another national survey Pirog and Rasmussen (2008) found 67% of consumers define local as traveling 100 miles or less from farm to point of purchase, while 38% chose a more limited 25 miles or less. The idea of local being within 100 miles of a consumer was recently popularized by the book *The 100-Mile Diet: A Year of Local Eating* (Smith and MacKinnon 2007). In the book, Canadian writers Alisa Smith and J.B. MacKinnon recount their experience of eating a diet procured from within only 100 miles of their home. The book was well timed with the growing local food movement and became a best seller, prompting many others to try their hand at a purely local diet.

As mentioned, a number of other factors are at play when consumers define local food. Many of these relate to the relationship the consumer has with the producer. Meeting a farmer, visiting the farm, buying at a farmers market, or even following them on social media channels builds a relationship that is missing from the global industrial food complex. In consumer interviews, Berlin et al. (2009) found a sense of pride toward locally produced products among those who buy local. Knowing where the product came from and the person producing it allows consumers to see and question the process first-hand and tends to instill a hyper-local view. In this case local becomes, "the story behind the food," (Martinez et al. 2010, p. 4) more than location for which it originates.

Retailer Definitions

While local food advocates tend to talk about one-to-one relationships with farmers, the reality is most consumers buy their food in regional or national grocery store chains. These stores have seen the increasing demand for local products and have often moved to meet the demand. This is now the case from international health food icon Whole Foods Market to the largest retailer in the United States, Walmart. However, research shows that retailers' definition of local is very different than that of the consumers and producers they serve.

For retailers, the definition of local also varies widely ranging from miles traveled, political boundaries, relationships with growers, geographic regions, where a product is distributed or processed, and even quality (Hunt 2006; Blake et al. 2010;

Dunne et al. 2010). For example, Whole Foods Market defaults to states' borders but leaves it up the individual store if they want a narrower boundary (Whole Foods Market 2015). Whole Foods Market also advertises a dedicated team of "foragers," whose job it is to find and source local products.

Dunne et al. (2010) found political boundaries were most often used by retailers studied in Oregon's Willamette Valley bioregion. Beyond that commonality retailers' parameters for local varied including food safety considerations, quality, methods of production, farm size, local ownership, and local operation. Likely contrasting with consumer and farmer sentiment, food grown elsewhere, but processed in the region, was also considered local by 70% of retailers. They reasoned the processing centers contributed to the local economy and their products were, therefore, local. Yet only 33% considered food grown in the region, but processed outside the local boundaries, to be local. This is despite the fact that farmers and ranchers are often required to make these arrangements in order to use a certified organic slaughter house to maintain organic designation after processing. Eighty-six percent of the retailers Dunne et al. (2010) interviewed agreed that they should display how they define local food to customers. However, less than half felt they should be required to have a strict definition.

Nonprofits and Nongovernmental Organization Definitions

Many nonprofits and NGOs are responsible for defining local food. There are thousands of organizations with very diverse missions including promoting healthy lifestyles, maintaining community gardens, improving urban residents' diets, managing farmers markets, preserving farmland, decreasing global warming, supporting small farmers, protecting farm labor, preventing animal cruelty, and protecting biological diversity to name just a few. All comment on and influence the local food movement in some way.

For example, the Pennsylvania Association for Sustainable Agriculture manages the Buy Fresh Buy Local® trademark through the FoodRoutes Network, LLC. It describes the program as the "premier trademark of the local foods movement in the United States" (Pennsylvania Association for Sustainable Agriculture 2014). FoodRoutes Network states there are Buy Fresh Buy Local® campaigns currently underway in 20 states with over 75 local chapters. Chapters pay dues to be part of the campaign and the FoodRoutes Network "provides communications tools, technical support, networking and information resources to organizations nationwide that are working to rebuild local, community-based food systems" (FoodRoutes Network 2015). The website is incomplete but it appears to be the largest local food promotion of its kind.

Producer Definitions

Producers—farmers, ranchers, and dairyman among others—are where local starts. Although producers need consumers and consumers need producers, there is no local without local producers. According to the 2007 Census of Agriculture and the 2007 and 2008 Agricultural Resource Management Survey, an estimated 107,200 farms were engaged in local food farm sales, or about 5% of all U.S. farms (Low

FIGURE 9.2 Boggy Creek Farm fields, Austin, TX. (Photo by James and Jennifer Buratti.)

and Vogel 2011). Producers are considered to be selling locally if they sell through direct-to-consumer channels (e.g., farm stands, farmers markets, community supported agriculture (CSA) or pick-your-own) or intermediate channels (sales to chefs/restaurants, retailers, and local and regional distributers).

Producers are often less rigid when it comes to food miles, seeing the one-to-one interaction with, and education of, consumers being the ultimate definition of local. While urban farms, like Boggy Creek Farm in Austin, Texas and Growing Power's Farms in Milwaukee, Wisconsin and Chicago, Illinois, are literally inside the city limits, this is not the case for most producers. By their very nature many producers live in rural settings not abutting an urban area and may drive many miles to get their product to customers. Farmers markets, many of which have strict distance guidelines on what is local, acknowledge this reality and allow "hard to get" products from further afield. This is especially true for proteins such as beef, pork, and lamb or specialty fruits such as citrus and olives (Figure 9.2).

CHARACTERISTICS OF LOCAL FOOD SYSTEMS

Local food networks include the multiple components of local agricultural production, distribution and sales of agricultural products. Approaches to local food production, distribution, and consumption are as varied as its definitions. The variations result from different regional opportunities for food production, unique business models employed by food producers and retailers specializing in local foods, and the unique dietary preferences or requirements of the consumers who support local food consumption. Even though national and multinational corporations are beginning to capitalize on the local food movement, local food production and consumption remain a largely grassroots effort. The most prevalent characteristics of these approaches are discussed below.

Short Food Supply Chains

Short food supply chains (often referred to as local food chains or local food webs) occur when the producers (farmers, ranchers, dairymen, beekeepers, etc.) sell directly, or within a single transaction, of the consumer taking on the roles of marketing, processing, storage, advertising, transportation, distribution, and sales that would be handled by commercial food distributors in the global food system. The phrase, "know your food, know your farmer," embodies short food supply chains.

The use of intermediate marketing channels, such as local grocers, chefs/restaurants, craft food makers, local institutions (schools, colleges, hospitals, and senior centers) or local/regional distributors and farm delivery services also play an important role in the short food supply chain. From a social standpoint, local food chains aim to change the dynamics between the consumer and the producer, adding value and meaning to the transaction, rather than simply being about the product, or the lowest cost (Marsden et al. 2000). In 2008 alone, local food system sales in the United States accounted for $4.8 billion (Low and Vogel 2011).

Local/Regional Food Hubs

Local/Regional food hubs are facilitators in the local food movement and serve as local/regional middleman between the producer and consumer, often taking on the roles of storage, processing, advertising, transportation, distribution, and sales. They may be commercial or not-for-profit organizations that provide permanent facilities for coordination of local food sales; as opposed to national distribution chains with wide-ranging, capital-intensive networks. In the local food web they facilitate consumer access, freeing up resources for both producers and consumers, including retailers and institutional buyers (Barham et al. 2012).

Direct to Consumer

Direct to consumer is a term used to describe the one-to-one producer-to-consumer transaction that is practiced extensively in the *local food network*. Sales methods include CSA distribution, farm stands, pick-your-own, farm-to-school/institution, farmers markets, special events, and direct order such as websites and catalogs (Johnson et al. 2012; Sage and Goldberger 2012). Eighty-one percent of those engaged in direct to consumer sales are small farms (less than $50,000 in income) (Johnson et al. 2012). Direct to consumer sales accounted for $1.2 billion in 2007 (Low and Vogel 2011).

Community Supported Agriculture

With *community supported agriculture*, referred to as CSAs, consumers buy advance shares or subscriptions in the potential harvest of a single farm or a group of local producers. CSAs usually require an upfront payment that allows the farmer to have the resources and time to buy what is needed to manage the farm's crops and sustain the farm and family. Participants pay the same amount regardless of the success or failure of the crops, sharing the risk and reward with the farmer. *Biodynamic agriculture*, a holistic approach to agriculture combining spiritual, ethical, and ecological considerations to create a self-sustaining farming system, as practiced by the followers of Rudolf Steiner and economist, E. F. Schumacher, influenced the formation of the first CSAs in the United States in the 1980s (McFadden 2014). The first CSAs were inspired to combine environmental and economic sustainability with farm production. In 2010, there were over 1400 CSAs in operation in the United States, up from 400 in 2001 and 2 in 1986 (Martinez et al. 2010). Federal government data puts the number much higher at 12,549 CSA operations in 2007 (USDA 2007).

Most CSAs provide the personal connection with the farmer and farm that local food supporters desire. Visits to the farms, education events, field days, and on-farm dinner events increase the connection between farm and plate. Many CSAs also offer the option of working shares, where a member works on the farm for a portion of their share. This can not only relieve some of the financial burden of a share but also provide the connection with the land and food that many CSA members desire (Tegtmeier and Duffy 2005). CSA shares often include more than vegetables offering fruit, meat, dairy, flowers, herbs, nuts, and value-added products such as jams, jellies, cheeses, and sauces. In recent surveys, both farmers and consumers felt they were being environmentally and socially responsible by participating in a CSA and the majority of CSAs practiced some form of organic or biodynamic agriculture (Lass et al. 2001; Strochlic and Chelley 2004).

Farmers Markets

Farmers markets are traditionally once-a-week gatherings of local producers who sell their product directly to the consumer. The popularity of farmers markets in the United States has increased tremendously in the last decade increasing from 1755 in 1994 to 2756 in 1998 and to 5274 in 2009, a 200% increase (USDA 2007; Martinez et al. 2010).

Hunt (2006) studied eight farmers markets in Maine and reported that, for shoppers at farmers markets, freshness was most important, with quality, availability of specialty products, supporting local farmers, and farmer contact being cited as important as well. Market shoppers were interested in "connecting their purchases with helping farmers, improving the rural economy, supporting their communities, and supporting agricultural open space" (Hunt 2006, p. 64). Based on these and other research results, there are two factors at play among farmer market shoppers—the quality of the product (fresh, local, organic, seasonal) and the social component (fun, friendly, interaction with and support of local farmers, community organization) (Kezis et al. 1998; Hunt 2006; Brown and Miller 2008).

Farm to Restaurants

Restaurants purchasing from local sources is a popular trend in food retailing. The National Restaurant Association surveyed professional chefs, members of the American Culinary Federation, in 2012, 2013, and 2014 and the top two trends 3 years running were locally sourced meats and seafood and locally grown produce

(National Restaurant Association 2013, 2014a,b). *Hyper-local sourcing* (a trend among restaurants with their own farms or gardens on-site) and farm/estate branded items also made the top 10 (National Restaurant Association 2013, 2014a,b). Chefs cite freshness and quality as their top reasons to buy local (Martinez et al. 2010). This approach to local food distribution can increase profitability for local food producers who are able to share in the value-added income resulting from the restaurant's retail pricing. Whereas a tomato purchased directly from a local grower may earn the farmer a fair price, the same tomato served by a chef in a restaurant setting and prepared in the form of a specialty dish will garner a much higher profit that can be shared by the farmer and the chef.

Farm Delivery Services

Farm delivery services are a rather recent business model and often considered part of the local or regional food hub. These services act as a middleman between the farmer and consumer, often taking on the roles of marketing, storage, advertising, transportation, distribution, and sales. Farmers may be paid more by a farm delivery service than they would be by a traditional distributor who earns only a small margin being one link in the global food distribution chain. Farm delivery services like Greenling and Farmhouse Delivery in Austin, Texas, rely and advertise heavily on their direct connection to farmers. They offer the connection of a farmers market with the convenience of a grocery store.

Farm to School

Farm to school, the practice of K-12 school buying locally produced food products, is part of the larger farm to institution segment of the local food movement. Farm to school usually refers to the K-12 portion of school with farm to university being considered a different segment. Farm to school programs encompass "efforts that bring local or regionally produced foods into school cafeterias; hands-on learning activities such as school gardening, farm visits, and culinary classes; and the integration of food-related education into the regular, standards-based classroom curriculum" (USDA Farm to School Factsheet 2014). The National Farm to School Network states that it "empowers children and their families to make informed food choices while strengthening the local economy and contributing to vibrant communities" (National Farm to School Network 2015). The UDSA estimates there are 4322 school districts, 40,328 schools, and over 23 million school children participating in farm to school programs in all 50 states and that $385 million in local food was served in the 2011–2012 school year (USDA Farm to School Census Website 2015). For producers, a farm to school program can supply a dependable, steady source of income (Martinez et al. 2010) and introduce potential new shoppers to their farms.

Farm to Institutions

Farm to institution, the practice of institutional food providers (hospitals, senior centers, colleges, etc.) buying locally produced food products, while not being organized under one national umbrella organization like farm to school, is gaining in popularity. The planning and organization for institutional purchasing of local food products is currently mostly at the state and regional level. Local food networks; state extension agencies; agencies fighting hunger, poverty, obesity, and *food deserts* (urban or rural areas with lack of access to fresh, health food); healthcare organizations; and hospitals are taking the lead promoting farm to institution local food networks.

Edible Landscapes/Home Gardening

Edible landscapes revolve around the concept of replacing an urban lawn and all the inputs that maintain it, with edible vegetables, herbs, and fruit trees. Square foot gardening and container or roof top gardening are the most common implementations. *Square foot gardening* is an intensive method of gardening that promotes the interplanting of many plant varieties in a small space popularized by Mel Bartholomew (2013) and his book *All New Square Foot Gardening.*

The National Gardening Association estimates 42 million people participated in growing their own food in home and community gardens in 2013, or one in every three households (National Gardening Association 2014). *Community or allotment gardens* are plots of land set aside for growing vegetables that is managed by the gardeners themselves. Many urban areas have eased their restrictions on small livestock to accommodate the edible landscaping and grow-your-own movement, allowing chickens and honeybees. However, city zoning regulations and homeowners' association restrictions often have those who try to convert their lawn to edibles running afoul of community standards (see Austin, Texas case study below).

Gleaning Networks

Gleaning is harvesting a crop that otherwise would have gone to waste. Gleaning networks usually donate their food to the food insecure and those lacking access to fresh, healthy food through food banks and community pantries. City Fruit in Seattle, Washington, "promotes the cultivation of urban fruit in order to nourish people, build community, and protect the climate" (City Fruit 2015), while The Gleaning Network of Texas works with growers to harvest excess crops in fields and orchards postharvest (Gleaning Network of Texas 2015).

Jail/Prison Farming

Prison farm programs are not new and were not uncommon in the recent past. Taking advantage of free labor, many prison systems were nearly self-sufficient when it came to feeding inmates. However, with the increased mechanization and move toward massive agribusiness farming operations, the time and expense of training prison labor made farming unprofitable (Winters 2013). The increased reliance on highly processed foods supplied by large institutional vendors is a concern to correctional staffers, who know that poor-quality food can create discipline and health issues. A few prisons are again experimenting with intensive agriculture not only to feed inmates, but also to teach farming skills. Two experimental programs—one in

California's San Quentin State Prison and one at New York City's Rikers Island—have had experienced prison reentry rates of only 5%–10% among their inmate farmers. This is compared to over 60% of the general prison population in California (O'Connor 2014).

Community Gardens

Simply put, a community garden is a plot of land set aside for growing vegetables that is managed by the gardeners themselves. They come in many shapes and sizes, are located in rural, suburban, or inner city settings, and may serve many different functions including sustenance gardens, food pantries, job training, therapeutic gardens, school gardens, or market gardens (McKelvey 2009) (Figure 9.3).

The community garden or allotment as it is known in the United Kingdom, had its origins in the United Kingdom but was quickly established in the U.S. urban environment as early as 1893. Community gardens have flourished at times of hardship in the United States, including the Great Depression, WWI and WWII, and in the 1970s as one solution toward addressing blighted urban areas (McKelvey 2009). Recent interest in local food has created a renewed interest in community gardens. The American Community Garden Association estimates there are 18,000 community gardens in the United States and Canada as of 2015 (American Community Garden Association 2015).

Foragers

Forager is a new term meant to encompass the role of connecting local producers with local businesses. Whole Foods Market employs regional foragers to find unique, local, high-quality food products, as do some high-end restaurants. Other foragers are

FIGURE 9.3 Festival beach community garden, Austin, Texas. (Photo by James Buratti.)

freelancers, working for a number of small, local restaurants, and grocers. The forager helps ease the burden of working with a number of small producers that can take more time than simply buying from a regional wholesaler or distributor. Producers also save time selling to one forager instead of a number of individual establishments.

Urban Farming

Urban farms can be found in many urban areas throughout the United States. These are often small-scale intensive agriculture operations, usually less than 5 acres in size. Urban farm locations vary and include traditional urban neighborhoods, rural areas that have been encroached upon by the expanding city, former industrial areas that were abandoned as industries moved out, urban areas no longer considered desirable due to urban blight and neglect, or patchworks of modern urban yards. Most urban farms sell direct to the consumer on-site or via a CSA. Income is often supplemented by hosting events such as weddings, parties, or educational events. Production is often limited due to the small scale of the operations. Because contemporary Americans are rarely familiar with or sensitized to farm activities, especially in their own neighborhoods, urban farms can come into conflict with their neighbors due to farm noises, animals, smells, increased visitor traffic, or conflicts over changing patterns of occupancy in gentrifying neighborhoods (see Austin, Texas case study below).

BENEFITS OF LOCAL FOOD

As previously discussed, local food systems are not a new idea, but they are being reconceived to fit the goals of contemporary consumers and modern urban landscapes. Features of local systems are often proposed as components of urban sustainability initiatives, community development goals, or local economic development planning. The most prevalent benefits to local food systems identified in scholarly literature and urban planning proposals are discussed below.

Social Benefits

Community gardens and farmers markets have a long history of being employed as tools for reclaiming blighted, abandoned urban lots and providing a platform for improved social interaction. In addition to providing local, fresh, nutritious food, community gardens have been shown to decrease criminal activity, build a sense of community, engender community pride, and to provide inviting green space and help stabilize a community from further decline (Schukoske 2000). Community gardens provide a place for neighbors to come together, make connections and have positive social interactions (Wakefield et al. 2007). In short, community gardens augment community and increase social interaction.

For participants, the mental and physical health benefits are also significant. Research has shown working in a garden decreases stress, increases exercise, provides motivation and a peaceful retreat, and was found to be relaxing and calming (Wakefield et al. 2007; Kortright and Wakefield 2011). Garden therapy is a recognized practice and while not focusing on vegetables directly, many programs do incorporate food plots. A significant portion of the recommended daily intake of calories, vitamins, and minerals can be provided by a relatively small garden plot. This can ease the financial burden of buying fresh food (Thaman 1995). These mental and physical benefits occur regardless of the location of the garden, be it an inner-city community garden, a rural church garden, or a private square-foot garden.

Consumer Benefits

Local food consumers are concerned about benefits beyond quality and price—environmental and social credence attributes, such as supporting local farmers and farm workers, investing in the local economy, supporting rural communities, and environmentally friendly production methods, are important as well (Cooley and Lass 1998; Stephenson and Lev 2004; Bougherara et al. 2009; Bean and Sharp 2010) (Figure 9.4).

Consumer benefits often benefit others or the environment as a whole—reducing their carbon footprint, helping keep a local farm in business, or lowering pesticide use and, therefore, worker exposure. The benefits of connecting a consumer with their food, the grower who produced it, and the land on which it was produced are hard to measure. What can be measured is a significant number of consumers who are willing to pay more for the benefits they perceive from participating in the local food network (Jekanowski et al. 2000; Stephenson and Lev 2004; Bean and Sharp 2010). More tangible benefits have been documented in regards to CSAs and farmers markets. These include access to certified organic produce that may not be available via normal retailers and education on farming, cooking, and preserving. Research has also shown cost savings associated with CSA shares; computing the retail cost

FIGURE 9.4 Green Gate Farms, a certified organic farm, Austin, Texas. (Photo by James Buratti.)

of the share's produce was more than double the share cost (Cooley and Lass 1998). Quality and freshness are often at the top of the local consumers' reasons to buy local, and consumers perceive the products at farmers markets to be both higher quality and lower in price (Brown 2003).

Economic Benefits

The economic impact of local foods is substantial and well documented. Sales of locally produced food accounted for 1.6% of U.S. agricultural products, or $4.8 billion in 2008 (Low and Vogel 2011). This is a small but growing part of the U.S. food network. The USDA tracks food sales in the local food network in two ways: direct-to-consumer and intermediate marketing channels. Small and medium-sized farms accounted for 95% of all farms reported as engaged in local foods or 101,928 farms (Low and Vogel 2011). Small farms were much more likely to be engaged in direct-to-consumer sales than medium or large farms and their numbers have increased (58%) between 1992 and 2007 to 136,000. Since supporting small farms is a key reason cited by many local food consumers, this sustained increase in participating small farms demonstrates a potentially positive impact. On the opposing end of the spectrum, large farms dominated the use of intermediated channels, accounting for 93% of the value of goods sold. While this may go against the popular notion of knowing your local farmer, it could have the effect of making local foods more accessible to more people in grocery stores and restaurants.

Numerous states have promoted the increased purchase of locally grown food for economic reasons. These programs aim to shift existing food dollars away from the global food system back to the local food system. Illinois created the Illinois Local and Organic Food and Farm Task Force through the Illinois Food, Farms, and Jobs Act of 2007. The purpose of the task force was to retain some of the $48 billion Illinois consumers spend on food annually. The goal was for "20% of Illinois food expenditures to be grown, processed, and distributed in-state by 2020" (The Illinois Local and Organic Food and Farm Task Force 2009, p. 3). It was estimated that this would generate $20 to $30 billion of new economic activity annually within Illinois and create thousands of new jobs. North Carolina has the 10% campaign in which participants pledge to spend 10% of their existing food dollars locally. Since July 2010, 7458 people and 976 businesses have spent over $64 million on North Carolina local food (The 10% Campaign 2015).

Producer Benefits

Producers of local food cite numerous benefits. They too enjoy the direct relationship with customers and many decide what to produce and how to produce based on customer input. Farmers have greater control over what they grow and the prices they can command when selling direct to consumer, leading to greater independence and control of their business (Hunt 2006). Farming is a challenging profession, but farmers that are part of the local food network are satisfied with their profession, enjoying their quality of life contributing to the well-being of their customers, community, employees, and the environment (Lass et al. 2001; Ross 2005; Conner et al.

2007). Additionally, many farmers want to practice methods of *sustainable agricultural* that employ ecological systems to sustain a healthy, self-sustaining system with minimal external inputs.

Retailer Benefits

Retailers have embraced local food for a variety of reasons and see many benefits to their participation in the network. When surveyed, supporting the local economy was the most common response for being part of the local food network. Whether this is because it makes good financial sense and supports the store's bottom line, provides quality, cheaper food for customers, or they are simply being a good member of the local financial community is not known. Other retailers see local foods as essential to their identity, such as Whole Foods Market, numerous local grocery stores, and co-ops throughout the country. These retailers cite similar reasons as local food consumers including supporting local farmers, environmental protection, and offering the highest quality and safest food possible (Dunne et al. 2010; Martinez et al. 2010). Others may be simply reacting to demand and offering what consumers want.

Nongovernmental Organization Benefits

NGOs that work with the *food insecure* (those lacking consistent access to an adequate supply of food) see many benefits to local food networks. As previously mentioned, community gardens, school gardens, gleaning networks, and individual gardeners often provide food to local food banks and pantries. Many provide educational classes on cooking and eating local, fresh, healthy, and seasonally abundant produce. The Women, Infants and Children (WIC) Farmers' Market Nutrition Program (WIC-FMNP) and the Senior Farmers' Market Nutrition Program (SFMNP) provide federal grant funds to states and allow farmers' markets to accept Supplemental Nutrition Assistance Program funds increasing local food access to the food insecure (Johnson et al. 2012).

Environmental Benefits

The environmental benefits of participating and expanding the local food web are many. As the majority of local food producers practice some form of organic or biodynamic agriculture (Lass et al. 2001; Strochlic and Chelley 2004), synthetic chemical input for pesticides, herbicides, and fertilizers are lower. This decreases exposure for farmers, farm workers, consumers, and the environment.

Short supply food chains are just that—short. Products travel shorter distances from farm to plate decreasing carbon emissions. The act of gardening can help impact global climate change by sequestering carbon dioxide. One study estimated that the 10,000 community gardens in the United States have sequestered 190,000 tons of carbon in 10 years (Okvat and Zautra 2011). Farmland and gardens also provide a variety of other environmental benefits associated with green space including providing habitat, increasing biodiversity, decreasing groundwater runoff, increasing groundwater recharge, removing airborne pollution, and lowering local temperatures. Local food networks may also be contributing to preserving the genetic diversity of food crops. Many farmers market and CSA owners cite the use of heirloom varieties to differentiate themselves from the supermarket and other growers. A study of apple growers in Ohio confirmed growers with pick-your-own or local sale operations selected for heirloom varieties that would not be acceptable in the commercial conventional market system (Goland and Bauer 2004). Growers could select for taste or versatility of use instead of ability to be shipped long distances, withstand bruising, or provide a long shelf life.

IMPEDIMENTS TO LOCAL FOOD SYSTEMS

There are obstacles to participation in the local food systems. Retailers cite problems sourcing from so many different producers and small farmers. It is time and resource intensive, potentially impacting a store's profits. However, businesses are springing up to meet this challenge to act as distributors specifically for small, local producers.

One key problem for farmers and ranchers is a lack of processing options for meats. Animals often have to be shipped a long way to be processed in a USDA-certified slaughter house. Certified organic slaughter houses are even harder to find. Mobile processing trucks are one innovative solution to try and address this problem. Additionally, grains and oil-producing plants are not grown and processed in every region so most are shipped in from elsewhere in the country or internationally.

Not everyone lives where food grows or is grown abundantly. Locavores in Michigan pointed out during the long harsh winter that many locally produced foods are simply not available (Bingen et al. 2011). When shopping out of season, they will then try to substitute the next best alternative that supports their beliefs like certified organic or fair trade. However, it is more common that local food is simply not available to many consumers where they live and shop.

Survey research has consistently shown purchasers in the *alternative food networks* are generally white, college educated, higher income, older, and female (Hunt 2006; Berlin et al. 2009; Bean and Sharp 2010). Alternative food networks are those whose agricultural production, distribution, and sales do not rely on the current global food supply chain. These networks often utilize small farms, organic production methods, and direct to consumer sales. A survey of Maine farmers market shoppers had a mean household income of $70,000 with 60% holding a degree compared to the state mean of $45,000 and 25% with a degree for overall Maine residents (Hunt 2006). That being said, all ethnicities, education levels, incomes, and ages are represented in the alternative food network. Some survey work that has shown support for local food has no association with education or income (Stephenson and Lev 2004) although support does not necessarily translate into participation.

This local food consumer is in stark contrast to one who lives in a food desert. The term food desert was first coined by a Scottish resident of public housing in the 1990s during an interview regarding the conditions he faced with access to quality food (Cummins and Macintyre 2002). The phrase was originally meant to convey a situation where urban residents did not have access to healthy and affordable food. Since its initial use, a wide variety of definitions have been proposed by those trying to study, document, define, and solve the phenomenon of food deserts. Russel and Heidkamp (2011, p. 1197) in their detailed examination of the loss of the only supermarket in New Haven, Connecticut, defined a food desert as, "an urban or rural area with significantly limited access to retail sources of healthy and affordable food, due to a combination of socioeconomic disadvantages and physical distance." Additionally, the USDA estimates that "23.5 million people live in low-income areas…that are more than 1 mile from a supermarket or large grocery store" (Ver Ploeg et al. 2009, p. iii). Many potential consumers who are concerned about their health simply do not have access to fresh, local food.

Within underserved communities, the use of abandoned property for urban community gardens has the potential to provide fresh produce at a low cost (Corrigan 2011). The rise in farmers markets paired with the increased acceptance of electronic debit cards (EBT) SNAP and WIC cards at farmers markets has created new sources of fresh fruit and produce in traditional urban deserts. The 2008 Farm Bill set aside approximately $500,000 in competitive grant funding for FY 2009 for new EBT projects at farmers markets (Ver Ploeg et al. 2009). The City of New York created the Health Bucks program which as part of the SNAP program added $2 in Health Bucks for every $5 spent using SNAP EBT at farmers' markets for the purchase of fresh fruits and vegetables (Ver Ploeg et al. 2009).

Retail chains often cannot source a desired product, in a large enough quantity, or at a high enough quantity to satisfy consumer demand. Dunne et al. (2010, p. 51) found that despite a stated desire to buy local, "consumers often exhibit a poor understanding of their local food systems." They fail to understand why they cannot get strawberries in December or Brussels sprouts in August. Even with access, farmers and retailers have found some consumers are not willing to pay the higher prices often charged for local items. Institutional and large vendors also need consistency to their supply not only in quantity but even size and volume (Cascade Harvest Coalition 2015).

Lass et al. (2001) reported almost one out of four CSA farmers in their national survey did not own the land they were operating on, making their tenure tentative. Similar research in the Midwest found a higher percentage of land ownership by CSA farmers, 85% (Tegtmeier and Duffy 2005). As producers in the local food movement tend to be younger than traditional farmers, they likely have less personal financial resources to draw on while starting their farms or to maintain them through rough financial patches.

CASE STUDY 1 Local Food Systems in East Austin, Texas

The City of Austin, Texas, prides itself on its commitment to the environment and the well-being and advancement of its citizens. The city's firm belief in being green and sustainable—in jobs, the economy, transportation, housing, water conservation, air quality, energy, waste management, and the environment—makes Austin a successful, attractive, and globally competitive city.

One portion of Austin's sustainability efforts is the local food sector. An analysis commissioned by the City of Austin showed Austin's food sector annually accounted for, "$4.1 billion in total economic activity," an economic equivalent of the coveted Austin creative sector (City of Austin, Texas 2013) which includes entertainment events like South by Southwest and Austin City Limits Music Festival. Only a portion of that $4.1 billion stays in Austin, much of it heading out in to the global industrial food network. However, Austin's local food network could serve as a model for other metropolitan areas that are trying to increase their local food supply.

Austin's unique high-tech, high-education culture has a thriving foodie scene, supporting a dynamic array of organic and local farmers, ranchers, distributors, consumers, advocates, nonprofits, bloggers, educators, entrepreneurs, foragers, chefs, restaurateurs, and retailers. Austin's foodies are constantly seeking out new food experiences. Annual local food events include a robust Eat Drink Local Week, Funky Chicken Coop Tour, East Austin Urban Farm Tour, Beekeeping Seminar, and Austin Community Gardens Tour. The local chapter of Slow Food International, Slow Food Austin, actively promotes local farms through farm tours, educational slow food sessions, local happy hours, and a grub trivia contest with live butchering at halftime.

Access to local food is often cited as a barrier by consumers in purchasing local. In Austin, consumers have a wide variety of access options for local foods including supermarkets, farmers markets, CSAs, farm stands, cooperatives, community gardens, school gardens, backyard gardening, home delivery, and locally sourced restaurants and food trucks. Supermarkets and grocery stores catering to local food markets include four Whole Foods Markets, two HEB Central Markets, three Sprouts Farmers Markets (a grocery chain, not an actual farmers market), not to mention a plethora of locally owned options like Wheatsville Coop, Mr. Natural, Fresh Plus, Farm to Market Grocery, Peoples Rx, and many more. For those who prefer to order in, Austin has three farm to home delivery services. Thanks to its mild winters, Austin also enjoys at least 7 year-round farmers markets, more than a half-dozen urban farm stands and CSAs, and almost 50 community gardens.

One organization addressing access for Austin's food insecure is the nonprofit Sustainable Food Center. It operates five area farmers markets and believes that strengthening the local food system is fundamental to improving access to nutritious, affordable food. It provides year-round education on growing, processing, and cooking healthy local food through its culturally aware La Cocina Aleger/The Happy Kitchen. It also serves as a local food hub for delivery of farm to school, work, and home.

One of the most visible and well-known examples of Austin's local food network is Whole Foods Market, the largest retailer of organic and natural foods worldwide. Starting with one Austin store in 1978, Whole Foods Market has grown to 367 stores in the United States. and United Kingdom with $12.9 billion in annual sales in 2013 (Whole Foods Market 2013 Annual Report 2013). Whole Foods is a strong supporter of local farmers in all the markets it operates reporting purchasing produce from more than 2000 different farms amounting to ~25% of the produce sold in its stores originating from local farms. Since 2007, they have also supported local production through their Local Producer Loan Program, loaning $10 million in low interest loans to local producers.

Another significant part of the local food network is Austin's urban farms. More than a half-dozen commercial farms operate within Austin's city limits. Most of Austin's urban farms are located east of Interstate 35. This is due to a combination of geography, politics, and economics. Geographically, east Austin contains the rich blackland soils best suited for agriculture, while west Austin transitions to the limestone hill country with shallow soils. East Austin has also traditionally been home to Austin's minority populations and, until recently, not considered a desirable place to live. It was this area, with its less expensive real estate, that some early innovators of urban agriculture saw an opportunity. Boggy Creek Farm purchased 5 acres in 1992 (Boggy Creek Farm 2014). They were followed by a number of other farms including Rain Lily Farm, Springdale Farm, and Hausbar Farm.

Urban farming has been officially governed and encouraged by the City of Austin since at least 2009 under the direction of the Austin/Travis County Sustainable Food Policy Board (SFPB) and the Sustainable Urban Agriculture and Community Garden Program (SUACG). Their purpose is to "improve the availability of safe, nutritious, locally, and sustainably grown food at reasonable prices for all residents, particularly those in need, by coordinating the relevant activities of city government, as well as nonprofit organizations, and food and farming businesses" (City of Austin, Texas 2008). However, as the city discovered in 2013, attempting to integrate agriculture into a rapidly changing social, economic, and geographic urban landscape has its challenges.

As the Austin economy heated up, east Austin real estate was discovered by many entrepreneurs and developers. Land prices increased dramatically and many new business and developments moved into what were traditionally minority residential neighborhoods. According to City of Austin demographer Ryan Robinson, "The average price of a single family home in the 78702 (central east Austin) zip code has tripled since 2007" (MyFoxAustin.com 2013). Residents began to feel the effects of gentrification—increased taxes, changing neighborhood demographics, and the inability to control the changes around them.

As the farms' popularity increased and more Austinites became interested in eating locally, so did on-farm events such as the annual East Austin Farm Tour, on-farm sales, special events such as weddings, and other events meant to diversify the income streams keeping the farms afloat financially. Farm diversification included sales to local chefs and restaurants as well as the raising of animals for slaughter. While many urban farms sold eggs, few were offering the sale of locally raised chicken, rabbit, or fish. In an effort to increase sales and minimize their environmental impact, Hausbar Farm began raising, slaughtering, processing, composting, and selling chickens onsite.

In November 2012, a neighbor of Hausbar Farm reported the presence of a bad smell emanating from the farm. The city investigated and discovered the farm's animal composting system had gotten out of balance, creating the smell, but it had since been rectified. By mid-December, the farm had been visited by three city departments and amid disagreement between the departments whether it was in compliance, shut down its chicken operation. Over the course of the next year, the SFPB held four public meetings to update the city's code related to urban farms. A number of east Austin community activist groups and neighborhood associations including People Organized in Defense of Earth and Her Resources (PODER), a group organized around environmental, economic, and social justice issues, recommended urban farms be removed from all areas zoned single family, slaughter not be allowed, and that parking must be provided on-site. The SFPB, being in favor of urban farming, recommended an expansion of urban farming throughout the city with no restrictions. After a mediated arbitration failed to build any consensus, the Austin City Council held a vote on the proposed code changes. Citizens signed up to give over 7 hours of testimony, of which 6 hours and 12 minutes was in support of urban farms.

In the end, both sides gained and lost. The city voted to support and expand urban agriculture throughout the city, allowing even smaller urban farms and increasing their ability to hire laborers. Addressing neighborhood concerns, the city banned the slaughter and composting of animals in single family zoning, and special events were limited to six per year requiring each to secure a temporary use permit. The fact the City of Austin has an Urban Farm Code demonstrates its commitment to, and acknowledgement of, the importance and legitimacy of the local food movement. However, the recent tensions that arose between neighborhood associations and the urban farms and their supporters show that urban farms face both the good and ills of being urban. Regardless of the future of Austin's urban farms, the local food system in Austin continues to thrive and innovate.

SUMMARY

Until the late period of the Industrial Revolution, nearly all food was locally sourced out of necessity. Building on techniques and technologies developed during the nineteenth and early twentieth centuries, American agriculture and food production underwent rapid industrialization and globalization. Innovations in packaging, preserving, and shipping agricultural products made it possible to source products at distant locations and use them where and when consumers wished. Modern manufacturing food products at an industrial scale and utilizing chemical preservatives to enable them to be shipped around the world rendered many foods less perishable, but also less nutritious and appealing to discerning consumers. In response, many communities are seeking methods to reintroduce locally grown foods and to support farmers and retailers who engage in local food production methods. These include, but are not limited to, organic farming, urban farming, community gardening, CSA, roof gardens, and a host of emerging techniques aimed at bringing agricultural production back into our urban environments and providing all consumers with locally sourced food options.

As local food production has reemerged in the United States, different definitions of what locally produced food is have emerged as well. Government officials, community leaders, farmers, and consumers have all weighed in on the question of "What is local?" Some definitions, such as the one issued by the U.S. government, are based exclusively on distance to market or food miles. Others promote a community-based definition that is less geographic and more social in its parameters. Many retailers, who promote locally sourced food options, are more flexible in their definitions of local food because of the necessity to source a variety of products, not all of which can be grown in the immediate area. The definition of local food production becomes important when government entities issue subsidies in support of local food, when urban managers design zoning and land-use regulations to support or dissuade the activity, and when consumers make choices based on a preference for locally sourced foods.

Producers and consumers have driven a variety of innovative approaches to local food production in recent years. These range from volunteer-based community gardens, to public investment in farm-to-school programs, to entrepreneurial farming in urban and peri-urban settings. Each of these approaches has its challenges and opportunities, but collectively they are intended to provide healthier and more varied food options, community-based consumerism, and to augment efforts to reduce the environmental impacts of farming and broader sustainability efforts. The degree to which locally produced foods achieve these goals is actively debated and will continue to be studied and discussed by government officials, scholars, producers, retailers, and community activists. One example of this interaction between the various stakeholders in the local food movement can be found in Austin, Texas, where local food production has gained widespread popularity with consumers and community development organizations. In particular, urban farming has recently run afoul of urban zoning regulations in Austin, despite the fact that the city has one of the more progressive zoning ordinances among U.S. cities in support of urban farming. The support for, and conflicts over, urban farming in Austin illustrate the challenges that are faced in incentivizing and supporting local food production, while the rapid economic and social expansion of local food in the city illustrates the opportunities locally sourced foods can provide.

REVIEW QUESTIONS

1. When and how did agriculture and food production in the United States become globalized?
2. Why has local food production reemerged in many U.S. communities?
3. How, and by whom, is local food defined?
4. Who are the participants in the local food movement in the United States?
5. Discuss the challenges and benefits of local food from the perspective of the following.
   a. Governments (federal, state, urban, etc.)
   b. Community developers and activists
   c. Economic developers and retail businesses
   d. Environmental managers and sustainability proponents
6. How does Austin, Texas' local food movement serve as both an example of success and a cautionary tale for local food proponents?

ENRICHMENT ACTIVITIES

1. A *food shed* is the geographic region that produces food for a specific population, such as an individual city (i.e., Chicago, Illinois). Calculate and compare local foodsheds for a local farmers market by interviewing consumers and producers to assess how far each has traveled to participate

in the market. Calculate the average for both groups and draw concentric circles on a map of your community around the farmers market to illustrate each market's consumer/producer foodshed. What are the differences between consumers and producers? What is "local" based on this assessment?

2. Interview community gardeners, farmers market participants, CSA managers, institutional food buyers, and local chefs to explore why they value local foods. Compare and contrast these options and propose improvement to your community's local food system based on the comparisons.
3. Propose a community garden and/or CSA for your school. Engage school officials in planning and propose its implementation.
4. Using online search tools generate a list of retail establishments in your community that advertises local food system participation. Create a map of the retailers using an online computer mapping tool or hard copy maps of your community. Compare your map to census maps of your community and explore their relationship between race, ethnicity, average income, average home age, etc.

REFERENCES

American Community Garden Association. 2015. https://communitygarden.org/resources/faq/ (accessed January 13, 2015).

Barham, J., D. Tropp, K. Enterline, J. Farbman, J. Fisk, and S. Kiraly. 2012. *Regional Food Hub Resource Guide*. Washington, DC: U.S. Department of Agriculture, Agricultural Marketing Service.

Bartholomew, M. 2013. *All New Square Foot Gardening*. Minneapolis, Minnesota: Cool Springs Press.

Bean, M. and J. Sharp. 2010. Profiling alternative food system supporters: The personal and social basis of local and organic food support. *Renewable Agriculture and Food Systems*, 26(3), 243–254.

Berlin, L., W. Lockeretz, and R. Bell. 2009. Purchasing foods produced on organic, small and local farms: A mixed method analysis of New England consumers. *Renewable Agriculture and Food Systems*, 24(4), 267–275.

Bingen, J., J. Sage and L. Sirieix. 2011. Consumer coping strategies: A study of consumers committed to eating local. *International Journal of Consumer Studies*, 35, 410–419.

Blake, M., J. Mellor, and L. Crane. 2010. Buying local food: Shopping practices, place, and consumption networks in defining food as "local." *Annals of the Association of American Geographers*, 100(2), 409–426.

Boggy Creek Farm. 2014. About Us. Boggy Creek Farm Website. http://www.boggycreekfarm.com/main/about-us/ (accessed December 2, 2014).

Bougherara, D., G. Grolleau, and N. Mzoughi. 2009. Buy local, pollute less: What drives households to join a community supported farm? *Ecological Economics*, 68, 1488–1495.

Brown, C. 2003. Consumers' preferences for locally produced food: A study in southeast Missouri. *American Journal of Alternative Agriculture*, 18(4), 213–224.

Brown, C. and S and Miller. 2008. The impacts of local markets: A review of research on farmers markets and community supported agriculture. *American Journal of Agricultural Economics*, 90(5), 1296–1302.

Cascade Harvest Coalition. 2015. Farm-to-Institution Strategies. http://cascadeharvest.org/ (accessed January 9, 2015).

City Fruit. 2015. http://www.cityfruit.org/ (accessed January 13, 2015).

City of Austin, Texas. 2008. *Ordinance No. 20081120-058—An Ordinance Amending Chapter 2-1 of the City Code Relating to City Boards and Commissions to Add Section 2-1-170 Establishing the Sustainable Food Policy Board.* City of Austin: Austin, Texas.

City of Austin, Texas. 2013. *The Economic Impact of Austin's Food Sector.* City of Austin: Austin, Texas.

Conner, D., V. Campbell-Arvai, and M. Hamm. 2007. Value in the values: Pasture-raised livestock products offer opportunities for reconnecting producers and consumers. *Renewable Agriculture and Food Systems*, 23(1), 62–69.

Cooley, J. and D. Lass. 1998. Consumer benefits from community supported agriculture membership. *Review of Agricultural Economics*, 20(1), 227–237.

Corrigan, M. P. 2011. Growing what you eat: Developing community gardens in Baltimore, Maryland. *Applied Geography*, 31(4), 1232–1241.

Cummins, S. and S. Macintyre. 2002. "Food deserts"—Evidence and assumption in health policymaking. *BMJ*, 325, 436–438.

Dunne, J. B., K. J. Chambers, K. J. Giombolini, and S. A. Schlegel. 2010. What does "local" mean in the grocery store? Multiplicity in food retailers' perspectives on sourcing and marketing local foods. *Renewable Agriculture and Food Systems*, 26(1), 46–59.

FoodRoutes Network. 2015. http://foodroutes.org/home/our-mission/ (accessed January 9, 2015).

General Appropriations Act for the 2014–15 Biennium, Eighty-third Texas Legislature, Regular Session, 2013, Text of Conference Committee Report on Senate Bill No. 1. 2013.

Giovannucci, D., E. Barham, and R. Pirog. 2010. Defining and marketing "Local" foods: Geographical indications for U.S. products. *Journal of World Intellectual Property, Special Issue: The Law and Economics of Geographical Indications*, 13(2), 94–120.

Goland, C. and S. Bauer. 2004. When the apple falls close to the tree: Local food systems and the preservation of consumer interactions and influences on farmers' market vendors diversity. *Renewable Agriculture and Food Systems*, 19(4), 228–236.

The Gleaning Network of Texas. 2015. http://www.gleantexas.org/ (accessed January 13, 2015).

GO TEXAN. 2014. http://www.gotexan.org/ (accessed November 17, 2014).

HB 2419. 2008. Food, Conservation and Energy Act of 2008. https://www.govtrack.us/congress/bills/110/hr2419 (accessed December 4, 2014).

Hunt, A. 2006. Consumer interactions and influences on farmers' market vendors. *Renewable Agriculture and Food Systems*, 22(1), 54–66.

The Illinois Local and Organic Food and Farm Task Force. 2009. Local Food, Farms & Jobs: Growing the Illinois Economy, A Report to the Illinois General Assembly.

Jekanowski, M., D. Williams II, and W. Schick. 2000. Consumers' willingness to purchase locally produced agricultural products: An analysis of an Indiana survey. *Agricultural and Resource Economics Review*, 29(8), 43–53.

Johnson, R., T. Cowan, and R. A. Aussenberg. 2012. The Role of Local Food Systems in U.S. Farm Policy. Congressional Research Service. http://www.ams.usda.gov/AMSv1.0/getfile?dDocName=STELPRDC5097249 (accessed November 17, 2014).

Kezis, A., T. Gwebu, S. Peavey, and T. H. Cheng. 1998. A study of consumers at a small farmers' market in Maine: results from a 1995 survey. *Journal of Food Distribution Research*, 29(1), 91–99.

Kortright, R. and S. Wakefield. 2011. Edible backyards: A qualitative study of household food growing and its contributions to food security. *Agriculture and Human Values*, 28(1), 39–53.

Lass, D., A. Bevis, G. Stevenson, J. Hendrickson, and K. Ruhf. 2001. *Community Supported Agriculture Entering the 21st Century: Results from the 2001 National Survey.* Madison, WI: Center for Integrated Agricultural Systems (CIAS), College of Agricultural and Life Sciences, University of Wisconsin. http://www.cias.wisc.edu/csa-across-the-nation-findings-from-the-1999-and-2001-csa-surveys/ (accessed November 26, 2014).

Low, S. and S. Vogel. 2011. Direct and Intermediated Marketing of Local Foods in the United States, ERR-128. U.S. Department of Agriculture Economic Research Service.

Marsden, T., J. Banks, and G. Bristow. 2000. Food supply chain approaches: Exploring their role in rural development. *Sociologia Ruralis*, 40(4), 424–438.

Martinez, S., R. Ralston, and L. Luanne. 2010. *Local Food Systems: Concepts, Impacts, and Issues, ERR-97.* U.S. Department of Agriculture Economic Research Service.

McFadden, S. 2014. *The History of Community Supported Agriculture, Part I.* Rodale Institute. http://www.newfarm.org/features/0104/csa-history/part1.shtml (accessed December 2, 2014).

McKelvey, B. 2009. *Community Garden Toolkit.* Columbia, Missouri: University of Missouri Extension.

MyFoxAustin.com. 2013. Cost of East Austin Transformation into a Hip Neighborhood. MyFoxAustin.com. May 09, 2013. http://www.myfoxaustin.com/story/22212259/cost-of-east-austin-transformation-into-a-hip-neighborhood#ixzz2mF48ETIZ (accessed December 2, 2014).

National Farm to School Network. 2014. http://www.farmtoschool.org/about (accessed January 13, 2015).

National Gardening Association. 2014. Food Gardening in the U.S. at the Highest Levels in More Than a Decade According to New Report by the National Gardening Association. http://assoc.garden.org/press/press.php?q=show&id=3819&pr=pr_nga (accessed November 24, 2014).

National Restaurant Association. 2013. What's Hot 2013 Chef Survey.

National Restaurant Association. 2014a. What's Hot 2014 Chef Survey.

National Restaurant Association. 2014b. What's Hot 2015 Chef Survey.

O'Connor, Lydia. 2014. How A Farm-To-Table Program Could Revitalize Prisons. Huffington Post. http://www.huffingtonpost.com/2014/05/27/california-inmate-farm-program_n_5400670.html (access January 9, 2015).

Okvat, H. and A. Zautra. 2011. Community gardening: A parsimonious path to individual, community, and environmental resilience. *American Journal of Community Psychology*, 47, 374–387.

Patterson, P. 2006. State-grown promotion programs: Fresher, better? *Choices and the American Agricultural Economics Association*, 21(1). Available at: http://www.choicesmagazine.org/2006-1/grabbag/2006-1-08.htm (accessed November 05, 2014).

Pennsylvania Association for Sustainable Agriculture. 2014. Buy Fresh Buy Local. https://www.pasafarming.org/about/pennsylvania-buy-fresh-buy-local/pennsylvania-buy-fresh-buy-local-r-splash-page (accessed November 05, 2014).

Peters, C. J., N. L. Bills, J. L. Wilkins and G. W. Fick. 2008. Foodshed analysis and its relevance to sustainability. *Renewable Agriculture and Food Systems*, 24(1), 1–7.

Pirog, R. and B. Rasmussen. 2008. *Food, Fuel, and the Future. Consumer Perceptions of Local Food, Food Safety and Climate Change in the Context of Rising Prices.* Ames IA: Leopold Center for Sustainable Agriculture.

Pirog, R., T. Van Pelt, K. Enshayan, and E. Cook. 2001. *Food, Fuel, and Freeways: An Iowa Perspective on How Far Food Travels, Fuel Usage, and Greenhouse Gas Emissions.* Ames, IA: Leopold Center for Sustainable Agriculture.

Ross, N. 2005. How civic is it? Success stories in locally focused agriculture in Maine. *Renewable Agriculture and Food Systems*, 21(2), 114–123.

Russell, S. and C. Heidkamp. 2011. "Food desertification": The loss of a major supermarket in New Haven, Connecticut. *Applied Geography* 31(4), 1197–1209.

Sage, J. and J. Goldberger. 2012. Decisions to direct market: Geographic influences on conventions in organic production. *Applied Geography*, 34, 57–65.

Schukoske, J. E. 2000. Community development through gardening: State and local policies transforming urban open space. *Legislation and Public Policy*, 351, 351–392.

Smith, A. and J. B. Mackinnon. 2007. *The 100-Mile Diet: A Year of Local Eating.* Toronto: Random House.

Stephenson, G. and L. Lev. 2004. Common support for local agriculture in two contrasting Oregon communities. *Renewable Agriculture and Food Systems*, 19(4), 210–217.

Strochlic, R. and C. Chelley. 2004. *Community Supported Agriculture in California, Oregon and Washington.* Davis, California: California Institute for Rural Studies.

Tegtmeier, E and M. Duffy. 2005. *Community Supported Agriculture (CSA) in the Midwest United States: A Regional Characterization.* Ames, Iowa: Leopold Center for Sustainable Agriculture.

Thaman, R. R. 1995. Urban food gardening in the Pacific Islands: A basis for food security in rapidly urbanising Small-Island States. *HABITAT International*, 19(2), 209–224.

The 10% Campaign. 2015. http://www.ncsu.edu/project/nc10percent/ (accessed January 20, 2015).

USDA. 2007. 2007 Census of Agriculture Data Release Powerpoint. USDA National Agricultural Statistics Center. http://www.agcensus.usda.gov/Newsroom/2009/2007_Census.ppt (accessed October 25, 2014).

USDA. 2010. Organic Production Survey (2008). 2007 Census of Agriculture. Volume 3, Special Studies, Part 2.

USDA Farm to School Census. 2015. http://www.fns.usda.gov/farmtoschool/census/ (accessed January 13, 2015).

USDA Farm to School Factsheet. 2014. http://www.fns.usda.gov/sites/default/files/F2S_FarmtoSchool_March2014.pdf (accessed November 26, 2014).

Wakefield, S., F. Yeudall, C. Taron, J. Reynolds, and A. Skinner. 2007. Growing urban health: Community gardening in South-East Toronto. *Health Promotion International*, 22(2), 92–101.

Whole Foods Market. 2013. Annual Report 2013.

Whole Foods Market. 2015. Locally grown, raised and produced. http://www.wholefoodsmarket.com/local (accessed January 21, 2015).

Winters, R. 2013. Evaluating the Effectiveness of Prison Farm Programs. Corrections.com. http://www.corrections.com/news/article/33907-evaluating-the-effectiveness-of-prison-farm-programs (accessed January 9, 2015).

Wisconsin Department of Agriculture, Trade and Consumer Protection. 2014. Buy Local, Buy Wisconsin 2014 Impact Report. http://datcp.wi.gov/Business/Buy_Local_Buy_Wisconsin/ (accessed November 15, 2014).

Ver Ploeg, M., V. Breneman, T. Farrigan, K. Hamrick, D. Hopkins, P. Kaufman, S. Kim et al. 2009. Access to affordable and nutritious food—Measuring and understanding food deserts and their consequences: Report to Congress. USDA Economic Research Service, Ap-036.

10 Volunteerism

Ann Marie VanDerZanden

CONTENTS

OBJECTIVES

Upon completion of the chapter, the reader should be able to

- Compare and contrast the three common categories of volunteer motivation.
- Create the framework for developing a new volunteer program.
- Describe the benefits of a volunteer program to an organization.
- Describe four components of volunteer management.

KEY TERMS

- Needs assessment
- Position description
- Program evaluation
- Recruitment strategy
- S.M.A.R.T. goals
- Volunteer
- Volunteer manager
- Volunteer motivation

VOLUNTEERISM

Volunteers are defined as staff who give time and expertise without receiving or expecting monetary pay. Millions of people in America and around the world participate in volunteer activities. They are motivated to share their time and talents with organizations for a variety of reasons, and the end result is that both the organization and volunteer benefit from this service. Some individuals volunteer hundreds of hours a year to the same organization while others volunteer for specific projects across a few different organizations.

A report by the Corporation for National and Community Service (2013) provides a profile of volunteerism in America. The report states that in 2012:

- One in four adults (26.5%) volunteered through an organization, demonstrating that volunteering remains an important activity for millions of Americans.
- Altogether, 64.5 million Americans volunteered nearly 7.9 billion hours in 2012. The estimated value of this volunteer service is nearly $175 billion, based on the Independent Sector's estimate of the average value of a volunteer hour.
- The top volunteer activities included fundraising or selling items to raise money (25.7%); collecting, preparing, distributing, or serving food (23.8%); engaging in general labor or transportation (19.8%); or tutoring or teaching (17.9%).
- Volunteering has trended upward among Generation Xers (born between 1965 and 1981) over the past 11 years, increasing nearly 5.5 percentage points during that period.
- The volunteer rate of parents with children under age 18 (33.5%) remained higher than the population as a whole (26.5%) and for persons without children (23.8%).

BENEFITS OF VOLUNTEERS TO AN ORGANIZATION

Harnessing the power and enthusiasm of volunteers is an effective way to extend the overall impact and reach of an organization. A well-designed and managed volunteer program provides benefits to both the organization and to the volunteer. The volunteers enable the organization to leverage work of the paid staff in order to reach a larger audience, or expand to new audiences. And, volunteers are able to gain personal and/or professional satisfaction as a result of their service.

Urban horticulture volunteer programs may serve many different purposes. In public gardens, for example, the volunteer program often supports activities in public education, horticulture, research, visitor services, special events, and other aspects of garden operations (APGA 2014). Volunteers provide support for similar projects at zoos and amusement parks. With specialized training, volunteers can be an asset to horticulture therapy programs and to horticulture programs in prisons.

PLANNING A VOLUNTEER PROGRAM

Successful volunteer programs require a significant amount of planning and organization before the first volunteers are recruited and trained. The size of the volunteer program being established as well as the duties the volunteers will complete may influence the planning process. An overarching goal of a volunteer program is to help focus the talents of volunteers where they can be of most assistance and to leverage the work of paid staff.

Figure 10.1 is a checklist that can be used to determine what components of a volunteer program are already in place, and it can also be used as a guide to develop a new volunteer program.

Consistent training for new volunteers regarding specific duties and responsibilities				
Designated supervisors for all volunteer roles				
Periodic assessments of volunteer performance				
Periodic assessments of staff support for volunteers				
Consistent activities for recognizing volunteer contributions				
Consistent activities for recognizing staff support for volunteers				
Regular collection of information (numerical and anecdotal) regarding volunteer involvement				
Information related to volunteer involvement is shared with board members and other stakeholders at least twice annually				
Volunteer resources manager and fund development manager work closely together				
Volunteer resources manager is included in top-level planning				
Volunteer involvement is linked to organizational or program outcomes				

FIGURE 10.1 Volunteer program check list for new and established programs. (From *A Guide to Investing in Volunteer Resources Management: Improve Your Philanthropic Portfolio*, by Paige Tucker, UPS Foundation, 2003. https://www.nationalserviceresources.gov/online-library/items/r4096#.U9bMTUA3cnY.)

A number of resources are available that can aid in developing a volunteer program including the Hands On organization (2010) and the Maine Commission on Community Service (MCCS 2008). These organizations have leveraged expertise from professionals across the volunteer sector and compiled resources that can easily be adapted to urban horticulture programs. See the section "Further Reading" at the end of this chapter for a listing.

DYNAMICS OF PAID STAFF AND VOLUNTEERS

Building investment among staff early during the planning process in the value and benefit of a volunteer program is essential to the short- and long-term success of the program. Staff are usually well-positioned to provide input on which unmet needs the volunteers can assist with, how the volunteers can further the organization's mission, and possibly even the design of the program. Once the planning and development phases start moving forward, it is important to keep the staff updated.

After the program is operational, sharing successes and asking for help in resolving issues is a valuable way to continue to keep staff engaged with the program (MCCS 2008). Studies show if paid staff has buy-in to the volunteer program, they will create a welcoming environment for volunteers. If not, they may give volunteers the impression that they are not valued. It is helpful to reassure paid staff that volunteers will be brought in to support and enhance their work, not to replace them (MCCS 2008).

INITIAL PLANNING AND DEVELOPMENT

A well-executed planning and development phase allows the stakeholders involved with, and impacted by, the volunteer program to have a voice in the program development. These stakeholders usually include the organization's leaders, paid staff, an advisory board, and the clients the organization serves. In some cases, there may be others who should be brought into the initial planning such as community representatives, or members from related or allied volunteer programs in the area. The key issues to address at this stage include both big picture concepts for the volunteer program, as well as details such as volunteer position descriptions and budget.

A thorough planning process will include the following elements: completing a needs assessment, developing a mission statement, setting goals and objectives, determining the program's overall structure, determining a budget, and writing position descriptions (MCCS 2008).

Needs Assessment

The needs assessment phase can be framed in a couple of different ways. One question to ask might be "What organizational needs will the volunteer program address?" This can identify specific unmet needs. A second question could be "If we had more time or expertise we would…" Answering this question might highlight potential areas where a volunteer program could expand to in the future, but that may not be an immediate need for the organization. Answering both of these questions will create a helpful structure for the rest of the planning process.

Two key groups who should have input during the needs assessment are the organization's paid staff, and the audience(s) to be served by the volunteers. Providing ample opportunity for input from the paid staff is critical since some may be working directly with the volunteers. Paid staff may have concerns about how the volunteers will impact their daily work, how they will be expected to coordinate or supervise volunteers, or how the volunteers can meet the unmet needs of the organization. One outcome of their input may be that the paid staff desire training themselves in volunteer management. In some cases, the audience(s) to be served may already be identified, but in other cases identifying the audience(s) is the initial step in the needs assessment. Once the audience is identified, their needs can be determined relative to the volunteer program.

Gathering stakeholder input during the needs assessment is necessary to develop a program that will meet the needs of those involved. Input can be collected in a variety of ways including interviews, surveys (online, mail, or telephone), and focus groups. Focus groups have the advantage of providing opportunity for a fluid conversation around a few select topics. Using a fixed set of questions across all of the stakeholder groups provides data that can be compared across the different groups to determine where there are commonalities and possible differences.

Developing a Mission Statement

A mission statement is a sentence or short paragraph that talks about meeting a need, solving a problem, or defining the purpose of an organization. The urban horticulture organization that the volunteer program will serve likely have a mission (and vision) statement in place. It is important for the mission statement of the volunteer program to align with that of the larger organization. Specifically, the volunteer program's mission statement should impart a sense of purpose among paid and volunteer staff, helping each to understand the importance of the work they do, and how each complements the other (MCCS 2008).

A sample mission statement for a public garden might read:

> Through their efforts, volunteers contribute to the beauty of the gardens, the success of our programs, and the enjoyment and experience of our guests (excerpted from Longwood Gardens).

Setting Goals

The mission statement provides an overarching direction for the volunteer program, and establishing clear and measureable goals is essential to the volunteer program meeting its mission. Using the S.M.A.R.T acronym (*S*pecific, *M*easureable, *A*chievable, *R*elevant, and *T*ime-bound) is an effective way to set goals. The benefits of setting measurable goals that feed into the overall mission are twofold. First, it gives volunteers a clear picture of what needs to be accomplished. Second, it provides the organization with benchmarks it can use to evaluate the volunteer program and its impact. Figure 10.2 illustrates how a set of action-oriented tasks supports a goal within a public garden. See the section "Further Reading" at the end of this chapter for goal-setting resources.

Determining the Volunteer Program's Overall Structure

The next step after the program's goals are determined is to establish the organization and overall structure of the volunteer program. A number of volunteer program resources (many included in the further reading section at the end of the chapter) include questions to be addressed during this planning phase. Examples of these questions include

- How will the volunteer program fit into the organization's structure?
- How will the volunteer program be structured?
- What personnel will be involved in the program?
    - Will there be a volunteer manager?
    - Will there be a program assistant?
    - What will the duties be for paid staff involved with the program?

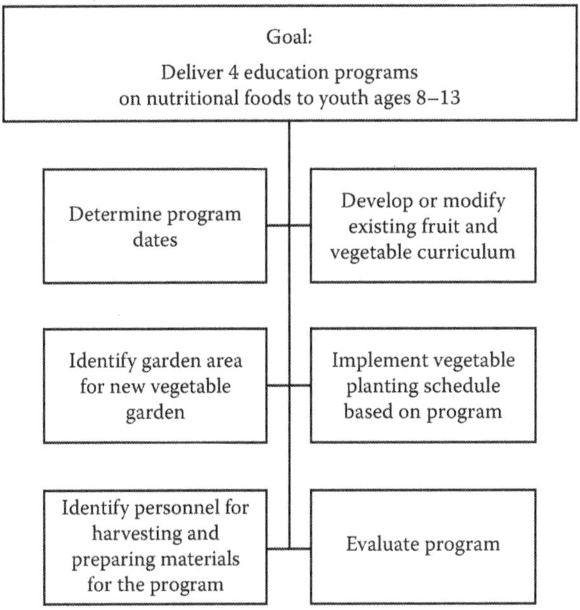

FIGURE 10.2 Example of S.M.A.R.T. goal with related actionable tasks.

- What will the training program entail?
    - Will a fee be charged for the training?
    - What is the initial training requirement?
    - Will there be an exam after training? Is a minimum score required on the exam?
    - Will ongoing/recertification training be required?
- What is the volunteer service commitment?
    - After initial training?
    - Ongoing?
- How often will volunteers be recruited?
- How will the program be evaluated?

An effective way to determine the structure of a new volunteer program is to review program materials for existing successful programs. A good place to start is by contacting professionals in the American Public Gardens Association (APGA 2014) since many gardens and arboreta that belong to the association have long-standing volunteer programs.

Determining a Budget

Successful volunteer programs require an investment of both financial and human capital. Financial sources may come from the organization that supports the volunteer program (public garden, zoo, a city's urban forestry commission, etc.) or the audience being served (church group, youth support organization, etc.). Also, many urban horticulture programs receive in-kind support from other organizations, or companies such as landscape supply companies, food and hospitality groups or companies that provide specialized equipment (e.g., medical) that supports volunteer work.

Determining a short-, mid-, and long-term budget is helpful during the initial planning phase. Initial budget items might include staff training, legal assistance in developing policies and procedures, producing and disseminating recruitment materials, conducting background checks, and developing training supplies. Longer-term budget items might include disseminating recruiting materials, conducting background checks, refining training materials, and hosting recognition events.

Writing Position Descriptions

Position descriptions are critical to the success of a volunteer program because they clearly outline what the volunteers will do. Every volunteer should receive a written position description. In cases where the volunteer may be doing multiple different duties within the organization, they should receive multiple position descriptions. Sample position descriptions are included in Appendix 1. A volunteer position description should include

- Purpose of the position
- A job title
- Who the position reports to; type of supervision
- Qualifications/experience required (e.g., computer knowledge; driver's licence; A/V experience, etc.)
- Key responsibilities and tasks involved
- Time commitment (e.g., timeframe, number of hours per day or for the total event, etc.)
- Where the volunteer service will occur
- Benefits—what is in it for the volunteer

CREATING POLICIES AND PROCEDURES FOR THE VOLUNTEER PROGRAM

Once the decision to establish a volunteer program has been made, and the big picture components of the program have been determined, it is time to operationalize the program. Policies and procedures are important to onboard new volunteers and ensuring they have a successful and rewarding volunteer experience. The inherent flux of volunteer programs, new volunteers joining, other volunteers leaving, makes well-written policies and procedures essential to program management.

One analogy that works to differentiate a policy from a procedure is that a policy is the destination, and a procedure is the road map to get there. A policy is a principle or course of action. A procedure, however, is a series of steps that directs people on what to do in order to complete or comply with the policy.

Policies provide structure and organization to both the volunteer program as a whole, and to the daily operations of the program. At a minimum, policies should connect the volunteer program to the larger organization and its mission, formalize program related decisions that have been made, ensure program continuity, equity, and standardization over time, clarify responsibilities, and define lines of communication and accountability (MCCS 2008). Many volunteer organizations also note that well-written policies contribute to increased volunteer and staff satisfaction, and productiveness, as well as volunteer retention.

There is no standard type or number of policies that should be developed. However, volunteer programs do require a set of policies that address issues unique to the program and that are likely outside of those already established for the overall organization. Graff (1997) suggests starting with the following list of policies for a volunteer program:

- Risk management
  - Risk assessment
  - Risk management culture
  - Risk management committee
- Position development
- Recruitment
- Screening
- Orientation
- Training
- Supervision
  - Format
  - Frequency
  - Documentation
- Corrective action
  - Informal methods
  - Formal methods
  - Progressive discipline
  - Dismissal
  - Immediate dismissal
- Information technology
- Volunteer/Paid staff/ Labor relations
- Volunteer performance and behavior

Once policies and procedures are established, it is important to review them annually to ensure they are still relevant and to make any modifications necessary. This is particularly important for new volunteer programs. Initial policies created during the development phase may need to be modified, or new policies and procedures added, to reflect how the program actually operates.

A sample policy and set of procedures is shown below.

Recording Volunteer Service Hours Policy

An accurate accounting of volunteer hours is important in order to document outreach and impact of the volunteer program. At the completion of each volunteer service event, record the length of time you volunteered as well as the volunteer activity you did.

Procedure:

- At the completion of your volunteer service, complete the volunteer service form located on the clipboard hanging above Volunteer Coordinator's desk.
- Enter the date, your name, tasks or duties completed, and length of time volunteered.
- If appropriate, include notes to update the Volunteer Coordinator about progress made on a long-term project, or if there were any issues during your volunteer service.
- Return clipboard to the Volunteer Coordinator's desk.

RECRUITING VOLUNTEERS

Before the first volunteers are recruited, it is essential to know what tasks volunteers will be doing for the organization. The program may need volunteers with particular skill sets (horticulture expertise such as plant propagation, landscape management, weed and pest identification, etc., or computer skills such as spreadsheet proficiency, or desktop publishing, experience working with youth, or availability to volunteer at a certain time of day or day of the week). Having a clear vision of the organization's needs will determine what the recruitment plan entails. The goal of recruitment is to motivate people to get involved with the organization and to help them see how they can contribute to the organization's goals.

Create a Compelling Recruitment Message

The first place to start in recruiting volunteers is to create a compelling recruitment message. The opening of the message must be interesting enough to entice the potential volunteer to continue reading or listening. The body of the message must be appealing enough to interest the potential volunteer to consider the volunteer opportunity or, at least, to contact the agency to get more information (McCurley 2003). The goal is to help potential volunteers understand how they can contribute to the organization and what benefits they will gain from their involvement.

Potential volunteers are motivated by a variety of factors including a desire to be of service, an interest in learning new skills through a volunteer role, or an interest in meeting and being with people (501 Commons.org, 2014). (This is discussed in more detail later in this chapter.) Addressing these motivations in the message is an effective strategy. Many volunteer organizations also personalize the recruitment message by including testimonials of past volunteers, as well as testimonials from people who have interacted with the program's volunteers and benefited from the services they provided.

Larger urban horticulture programs often have multiple recruitment messages which allow them to tailor their message. Some messages are targeted to specific groups of potential volunteers such as professionals, students, or family members of the audience to be served. Some messages are general and not focused toward a

Volunteers Provide the Energy that Powers Reiman Gardens

It's as simple as this; without volunteers there would be no Reiman Gardens. There are many unique programs that rely on volunteers include gardening-related jobs, as well as administrative, operations and facilities maintenance duties. We could use your help!

Reiman Gardens' volunteer education program provides volunteers with a dynamic learning environment. Training and continuing education classes are the cornerstone of our volunteer program. As a volunteer learns more, they in turn can teach more to others.

Student groups such as fraternities, sororities, clubs, dorm floors, learning communities and more are encouraged to volunteer at Reiman Gardens. Volunteering is a great addition to any student resume. Reiman Gardens even offers an online volunteer scheduling tool called VIC.

FIGURE 10.3 Sample recruitment message for a public garden, Reiman Gardens in Ames, Iowa. (From http://www.reimangardens.com/careers/volunteers/.)

specific volunteer activity (Figure 10.3), while other messages are targeted to fill a specific volunteer need.

Regardless of the volunteers being targeted, each message should be interesting enough to catch someone's attention, identify the need, describe how the volunteer can meet that need, and where to get more information. Below is a generic recruitment message outline (Business Volunteer Unlimited Maryland 2011) followed by a specific urban horticulture example.

The generic outline:

> **[Motivational appeal/goal]** by **[task]** for **[persons or goal]** for **[time required]** in/at **[location]**. **[Reward]**. Training provided. **[Any requirements/qualifications]**. For more information contact **[recruiter's name]** at **[organization/program]** by phone **[phone number]** or email **[email address]**. Web: **[website]**.

Urban Horticulture Example 1.

You can help seniors remain independent in their homes by delivering fresh produce twice a week. You will clean and prepare delicious fruits and vegetables from Wednesday and Saturday Prosser Farmer's Market and deliver them to seniors' homes. Training provided. Must have car. For more information, contact Sally Schroeder at 555-1234 or sal@prossermarket.org. Web: www.prossermarket.org

Establish a Recruitment Strategy

Many organizations have a recruitment strategy that determines both where and how the recruitment message is targeted, and how frequently volunteers are recruited.

The position descriptions for the needed volunteers may help determine the recruitment target audience. And a volunteer who is a talented public relations expert can help in designing and implementing the overall recruitment plan.

The two most common strategies used to recruit volunteers are "nontargeted" recruitment and "targeted" recruitment (MCCS 2008). Nontargeted recruitment means looking for people with general skills, such as volunteers to distribute food baskets to clients at a farmer's market, or volunteers to staff a registration table at a garden event. Targeted recruitment, however, involves looking for people with specific skills, such as horticulturists, landscape designers, or public relations experts. In addition to recruiting for specific skills, volunteer programs should also recruit for diversity. A diverse volunteer cohort benefits both the volunteers and the organization. In addition to race and ethnicity, consider other components of diversity, such as age, gender, education, income levels, religious beliefs, physical abilities, and skills (MCCS 2008).

Some organizations have an annual, quarterly or even monthly recruitment effort, while others support ongoing recruitment through their website. Figure 10.4a is a screen capture of ongoing recruitment through a public garden's website. Potential volunteers can select different volunteer positions from the menu on the left to see a description of what that position entails (Figure 10.4b), and then complete the volunteer application online.

Potential volunteers can be reached in a number of ways and often the particular audience being targeted determines the best approach. Examples of effective recruitment approaches for urban horticulture programs include using

- Print and digital media (newspaper, newsletters, websites, blogs)
- Broadcast media (local radio)
- Social media (Facebook, Twitter, etc.)
- Outreach through the organization's existing membership
- Participating in local face-to-face or online volunteer fairs

Just as a compelling recruitment message is important, so too is who delivers the message. Depending on the volunteers being sought, it might be helpful to have someone from the organization with community connections deliver the message in a face-to-face venue. In other instances, peers such as students recruiting students, can be an effective approach. Current volunteers can also be very successful in recruiting new volunteers particularly if they have had a rewarding volunteer experience. Existing volunteers can provide valuable insight on how to tailor the recruitment message, as well as effective ways to spread the word and whom to target.

REVIEWING APPLICATIONS AND SCREENING VOLUNTEERS

The goal of reviewing applications and screening volunteers is to get the right volunteer into the right position, so both the volunteer and the organization benefit. A well-designed volunteer application can provide valuable information about the volunteer's interests, skills, and motivation. Consider having a standard application as well as an addendum with specific questions related to the type of volunteer being

(a)

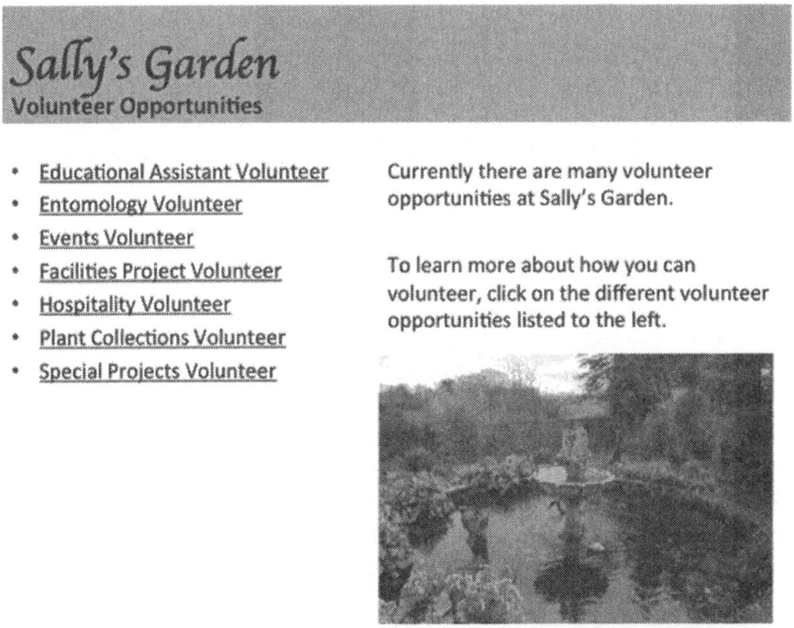

(b)

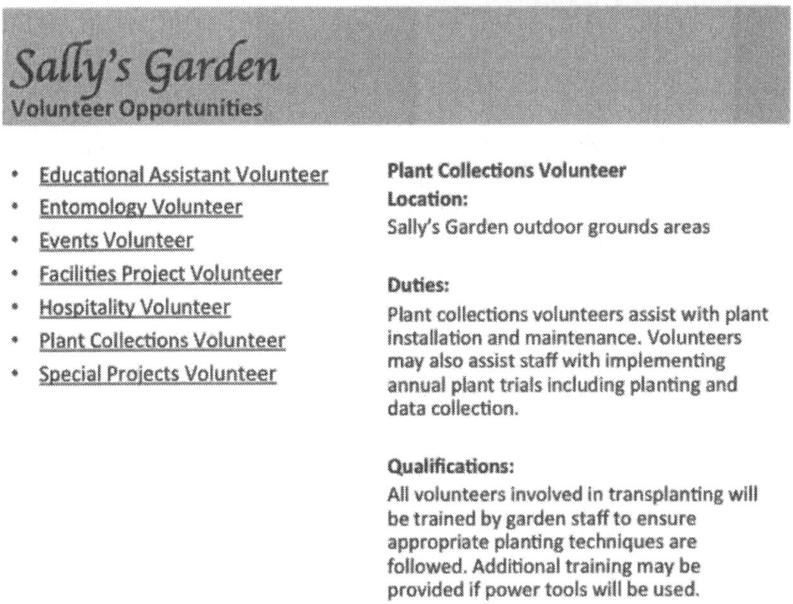

FIGURE 10.4 (a) Sample online listing of volunteer opportunities at a public garden. On a live website, potential volunteers would be able to select volunteer positions from the left menu, and details about the position would be displayed to the right (see Figure 10.4b). (b) Sample online listing of volunteer opportunities featuring a position description for a specific volunteer position.

recruited. For example, there might be additional questions for volunteers who will be working with a youth garden program, compared to volunteers working on a spring garden clean-up project. Volunteer program managers can use many of the same tools human resource managers use when hiring staff including position applications, face-to-face and/or phone interviews, reference checks, and background checks (where appropriate) (MCCS 2008).

Initial Contact

The initial contact is the first step in the process of determining the fit between a potential volunteer and the organization (MCCS 2008). Depending on the recruiting strategy, the contact will occur online through email or an online form, via phone, or face-to-face. The goal during this step is to provide the potential volunteer basic information about the organization and the volunteer opportunities that are available. If possible, this initial contact can also be used as a screening opportunity to determine what the individual is interested in doing for the organization and why he or she is interested in getting involved. If it appears there may be a fit for the individual and organization, they should be directed to complete a volunteer application.

The Volunteer Application

A prospective volunteer application form is necessary for two reasons: (1) to assist in the interview and screening process; and (2) to document basic information about the individual volunteer (Energize, Inc. 2014a). A short list of items to include on the application includes name, mailing address, telephone number(s), e-mail address, preferred method of contact, referral source (how the potential volunteer heard about the program), specific skills the individual has, and volunteer activities or programs in which the individual is interested. Appendix 2 is a sample volunteer application. See the section "Further Reading" at the end of this chapter for additional resources on volunteer applications.

When developing a new application form, (or reviewing existing documents) it is important to consider the tone conveyed by the application. The volunteer management organization Energize, Inc. (Energize, Inc. 2014a) suggests considering the following questions:

- What does the application convey to a prospective volunteer about what's important to your organization?
- What might a potential volunteer learn about you from completing this form?
- Are the questions very formal or more colloquial?
- Does there seem to be interest in who the applicant is as a person beyond the facts of his/her credentials?
- Do you want to know about past volunteering as well as about past paid employment?
- What level of education or literacy is implied by the vocabulary in your questions or the number of essay answers expected?

Volunteer Interviews

A face-to-face interview provides an opportunity for a more detailed discussion of the organization's mission, vision, and goals, as well as the volunteer's interests, motivations, and needs (MCCS 2008). Often it is helpful to have more than one person involved in the interview process because each interviewer will bring a slightly different perspective to the interview. The interview committee might include the volunteer manager, another paid staff member who interacts significantly with the volunteer program, and one or two current volunteers.

The application itself can be used as a starting point for the interview. Some volunteer managers include a role-play or case study question to learn more about the volunteer applicant in a situational context. Other volunteer managers use a scripted list of questions to streamline the process and ensure that they are able to garner enough information to determine a good fit within the organization during the interview. At the completion of the interview, it should be evident whether the prospective volunteer is a match with the organization's needs.

Background Checks

Depending on the nature of the organization, the clients served, and the work to be done by volunteers, additional screening may be required before placing a volunteer (MCCS 2008). The need for additional screening should be discussed early in the volunteer program planning phase and include input from the organization's legal and risk management staff. Because of the expense associated with certain types of screening (background check, driving records, credit check, substance abuse test, etc.) many volunteer programs complete the face-to-face interview process first and then complete the additional screening for those individuals who they have selected to volunteer.

VOLUNTEER ONBOARDING

The initial orientation and training prepares volunteers for their new role within the organization, and is essential to their short- and long-term success. Volunteers will require initial start-up orientation and training, as well as ongoing in-service training. Volunteers who understand what is expected of them do a better job and feel satisfied by performing their duties and serving the organization (MCCS 2008). One way to ensure volunteers have the information they need to be successful is to develop a volunteer handbook (either print or online format) that includes details covered in both the orientation and training processes. A useful approach to developing orientation and training programs is to contact other urban horticulture volunteer program managers to learn about their programs including what training and orientation resources they use, and to learn from their experiences in how best to prepare new volunteers.

Orientation

Orientation provides a broad, general overview of the organization and provides the context of how the volunteer's service will support the organization's mission and goals. Often well-developed policies and procedures can be used to provide a framework for the orientation program. The format of volunteer orientation can vary, but the paid staff member who is the volunteer manager typically facilitates it. A number of volunteer management resources (MCCS 2008; 501 Commons 2014; Community Tool Box 2014; Energize, Inc. 2014b; Hands On 2010) suggest volunteer orientation should include the following:

Overview of the Organization:

- Description and history of the organization, mission, goals, and objectives
- Organizational structure
- Introduction of key staff and explanation of "who's who" and "who does what" (including photographs of these individuals is helpful)
- Description of programs and clients served
- Description of the culture and language of the organization
- Glossary of the organization's terms, codes, abbreviations, and acronyms

Facilities Tour:

- Where to park
- Where, how, and when to access the facility
- Location of rest rooms, supplies, and equipment
- Arrangements for breaks and/or meals
- Where to store personal belongings
- Restricted areas

Volunteer Program Policies and Procedures:

- Types of tasks or other ways in which volunteers contribute
- Check-in/ check-out procedures
- Recordkeeping requirement
- Appropriate use of organization's information technology
- Training expectation (initial and ongoing)
- Service expectation (hours per year; specific program areas)
- Use of the volunteer title/ organization representation
- Volunteer performance and behavior
    - Volunteer supervision
    - Volunteer evaluation procedure
    - Corrective action
        - Informal methods
        - Formal methods
        - Progressive discipline
        - Dismissal
        - Immediate dismissal

People can only absorb and retain so much information during an orientation session. It is helpful to focus on the key points during the face-to-face orientation, and then provide new volunteers with printed or online resources they can access for further information. The printed or online material reinforces the information presented during orientation, helps to address questions that arise during service, and can prove useful as a supervisory tool in dealing with performance issues (MCCS 2008).

An engaging way to make sure volunteers are up to speed on items covered in the orientation is to have a postorientation quiz. Having volunteers work in small groups to answer specific questions, or work through a short case study that requires them to apply various policies or procedures, is a great way to build group rapport and ensure everyone understands volunteer expectations. It may also alert the volunteer manager to areas where additional discussion and clarity is needed.

Training

Training gives volunteers the preparation and skills necessary to carry out assigned tasks related to their position description. The amount and type of training necessary for volunteers will be as variable as each volunteer program and the individual volunteer position descriptions. However, establishing a clear vision of how the volunteers will support the organization, and concrete learning outcomes of the training program, will create a good framework to build the training around. If the nature of the volunteer's work is very basic and routine, then a volunteer might require only a very basic and general introduction to the task they are to complete. In contrast, if the nature of the volunteer's work is rather complex, then the volunteer will likely require a more in-depth training program (AmeriCorps 2014). Creating a training program that is flexible and can meet both of these types of needs is important.

Urban horticulture programs that have a range of volunteer positions often have customized training programs. For example, all volunteers complete a core set of training that is germane to all of the volunteer positions (expectations for client interaction, approved references to use when accessing horticulture information, basic horticulture topics, etc.). Specialized training is then provided to those volunteers who require it based on their position description. Volunteers who need specialized training might include those who will work with youth or at risk populations, or those working with specialized equipment (e.g., landscape equipment, greenhouse systems, etc.). Ultimately, the volunteer's position description will form the basis for the training they require. When volunteers change positions, additional training may be required. Keeping accurate and up-to-date records of training program participants will ensure volunteers are prepared for their service, and in turn that risk is being managed.

Developing the Training Materials

Many urban horticulture volunteer programs use a set of training materials to prepare volunteers for their service. These materials might include a training manual (either electronic or printed), a compilation of outside resources related to the volunteer's duties, and other relevant material. Some organizations have an internal website where training materials are posted, so volunteers can access the materials whenever and from wherever. Short videos, animations, and full color images can easily and inexpensively be incorporated into online training materials as compared to printed materials. Not only are these training materials more dynamic and interactive, it is more cost effective to update online training materials because printing costs can be avoided.

Delivering the Training

Most volunteer training programs have a face-to-face or online interaction component that supports the text-based training manual. The overall structure of the training program may dictate how the interaction occurs. The benefits of including some face-to-face training as part of the volunteer onboarding process is that it helps familiarize the volunteer with the physical location and features of where they will be volunteering, helps him or her establish rapport with the paid staff and others with whom they will be interacting, and builds an "esprit de corps" among the volunteer cohort. Consider the time of day and day of the week face-to-face training is offered. Training done during the workweek might limit who can become a volunteer due to work or other schedule conflicts.

VOLUNTEER MANAGEMENT

Effective volunteer management creates an atmosphere where well-trained volunteers are empowered to complete their work. And as a result of accomplishing their duties, the volunteers have a sense of personal and professional satisfaction. In addition to an appropriate orientation and training program, successful volunteer management requires volunteers receive clear expectations for their service, and that their volunteer performance is evaluated regularly. Some organizations use a volunteer Memorandum of Understanding to delineate expectations/contributions from both parties involved. An additional strategy to ensure volunteers have a successful and rewarding volunteer experience is to provide regular reinforcement and recognition for their work. Retaining good volunteers and keeping them engaged in their service is an important component of volunteer management.

The Volunteer Manager

Most volunteer managers are involved with every aspect of the volunteer program including volunteer recruitment, orientation and training, and ongoing supervision, as well as program evaluation and impact. Successful volunteer managers have excellent communication skills, strong organizational and prioritizing skills, the ability to successfully multitask, and an understanding of human motivation.

Because the role of volunteer manager requires a diverse skill set, many individuals in these positions benefit from participating in ongoing professional development to further their skills and abilities. Volunteer managers, like all professionals, can gain valuable insight and perspective by networking with other volunteer managers. No amount of online research or reading can replace the value to talking with someone who is doing the same work and facing the same challenges (501 Commons 2014). Participating in the Volunteer Engagement Professional Section of American

Public Gardens Association (APGA 2014) is a great way to network with other horticulture program affiliated volunteer managers.

Volunteer Placement

The goal of volunteer placement is to match a volunteer's skills and interests with the needs of the organization. A thorough application and interview process should help a program manager see where the best fit is for the volunteer. Thoughtful placement is the best way to ensure that volunteers are able to perform their duties and that they are satisfied with their volunteer experience. Sometimes the first placement does not work out as planned and the volunteer may need to be moved to a different position, or they may just require some additional training.

Volunteer Supervision

After the initial orientation and training, volunteers need ongoing support and management. Volunteers should have a designated supervisor to whom they can turn for advice, guidance, encouragement, and feedback (MCCS 2008). In some cases, this supervisor may be the overall volunteer program manager; in other cases, the supervisor may be a program assistant who may be either a paid staff member or a volunteer himself or herself. New volunteers may require more supervision to ensure that they are comfortable in their new role. Long-term volunteers whom are familiar with the organization's policies and procedures may require less. However, all volunteers benefit from some level of supervision.

Volunteer Performance Evaluation

Volunteers, and the organization benefit from periodic evaluation of how a volunteer is performing his or her assigned duties. These performance evaluations provide the volunteer systematic feedback and evidence of how their work is contributing to the organization's mission and goals. They also give the volunteer manager an opportunity to determine if the volunteer might benefit from additional training in a specific area and if the overall training program is adequately preparing volunteers for their service duties.

The volunteer's position description should form the basis for the performance evaluation. Other general criteria such as dependability, willingness/ability to make the necessary time commitment, and fit with the overall organization, etc. may also be included. Staff who work with the volunteer, a volunteer self-evaluation, or even input from clients served by the volunteer, can all be used to provide useful background for the evaluation. A number of resources are included in the further reading section at the end of this chapter that provides performance evaluation templates as well as evaluation process descriptions.

What to Do When a Volunteer Is Not Working Out?

Not every volunteer ends up being a good fit for the organization. Regular monitoring of volunteer activity and annual performance evaluations are an effective way to document this potential issue. The orientation and training process should include information to ensure volunteers understand that they may be terminated with or without cause. The program policies should make it clear that infractions of rules and regulations, violations of the law, and other unsafe or inappropriate conduct are all grounds for termination (MCCS 2008).

In nonegregious situations, it may be appropriate to work with the volunteer to correct his or her behavior. It is important to document the issue, clearly outline the changes that need to be made, and communicate the expected and appropriate behavior. Other cases may center on conflict management/resolution and these should also be documented and the desirable outcome determined and communicated. In some cases, it may be necessary to take corrective action immediately and remove the volunteer from the organization in order to maintain the credibility and integrity of the volunteer program and the organization as a whole.

VOLUNTEER MOTIVATION AND RETENTION

A variety of reasons motivate people to volunteer initially, and to continue their involvement as a volunteer over time. Understanding what motivates volunteers in an urban horticulture program, and providing opportunities for volunteers to meet that motivation, is essential to build a strong and sustained volunteer program. This understanding is also key to retaining the most effective volunteers.

Volunteer Motivation

What motivates an individual to volunteer is complex and often relates basic human nature needs. Motivation is further complicated because many times the motivating factors change over time. In 1997, The Independent Sector identified eight reasons why people volunteer:

- To make a difference
- To use a talent or skill
- To gain professional experience or make contacts
- To express religious faith
- To meet people
- To achieve personal growth and enhanced self-esteem
- To seek a more balanced life
- To give something back

There is also an extensive body of research conducted by behavioral psychologists on what motivates people in general. When the eight factors related to volunteerism identified by The Independent Sector (1997) are combined with Maslow's (1970) hierarchy of needs and McClelland's (1988) human motivation descriptors, the motivations for volunteering can be described by three basic categories: achievement, affiliation, and power (Hands On 2010; Schultinik et al. 2009). As a volunteer manager, the key is then to determine what category a volunteer's motivation aligns with and ensure their volunteer experience encompasses it.

Achievement

A person motivated by achievement often seeks to learn new skills through their volunteer service. These individuals tend to set and meet personal goals and this process of goal setting and accomplishment transfers to their volunteer service. Achieving a goal is how they build their self-esteem. These goal-oriented volunteers are an asset when it comes to implementing projects. They tend to work well alone, but can also contribute to a team when everyone is focused on the same goal. Volunteer assignments for achievers could include fund raising, chairing committees/leadership positions, membership campaigns, researching, analyzing, and reporting (McCurley and Vineyard 1988).

Affiliation

Volunteers motivated by affiliation hold a strong belief in the goals or ideals of the organization. What the organization stands for and how the individual can contribute to that mission is important. Affiliation motivated volunteers tend to be social individuals who gain satisfaction from working with other like-minded people toward a common cause. They thrive in environments where they are interacting with others on a regular basis. These volunteers are good in supportive roles and make valuable team members. Volunteer assignments for this affiliation motivated group could include task force membership, public relation activities, hospitality/banquet committees, or other activities done in groups such as program planning and brainstorming (McCurley and Vineyard 1988).

Power

A volunteer motivated by power usually has a need for personal power or a mastery over others. For volunteers who fit in this category, the desire for power itself is neither good nor bad. Some people simply feel the need to have an impact on others. When this feeling of power is used for the purpose of bringing about change that benefits others, this is a positive motivator. However, when a volunteer's actions are overpowering and self-focused, the result can be negative. Volunteers motivated by power may work best independently or where they can have decision-making authority over a particular project, and they generally enjoy projects that bring prestige and recognition to their work. Volunteer assignments for the power motivated group could include public-speaking opportunities, fundraising, program planning and policy committees, and chairing events that bring public recognition to the cause (McCurley and Vineyard 1988).

Leveraging Volunteer Motivation

Developing a basic understanding of what motivates a volunteer to participate in the organization is central to leveraging his or her volunteer service. Some of the motivating factors may come through on the volunteer application or during the interview process. However, building good rapport and ongoing communication with the volunteer helps a volunteer manager continue to leverage the volunteer's motivation. Some volunteers will openly share their suggestions, or concerns about the program, while others may be more reserved. Asking specific questions during the annual performance evaluation about what the volunteer likes or dislikes most about their volunteer assignment(s) can help gauge if the volunteer is suited for the assignment. Moving a volunteer into a position he or she might enjoy more can help retain them and ensure they are having a rewarding volunteer experience.

Volunteer Recognition

Volunteer recognition is a key component of any volunteer program. It is important for the organization to regularly and publicly acknowledge the contributions of volunteers. This recognition can also play a vital role in volunteer retention. Everyone needs an occasional pat on the back to know his or her efforts are valued and appreciated. Determining the most appropriate form of recognition for each volunteer takes some awareness of what motivates the individual. For example, volunteers seeking power may enjoy being thanked by the executive director and board of directors. A simple thank you from the paid staff may be appropriate for volunteers motivated by achievement, while a party is right for the group who volunteers to socialize (MCCS 2008). Regardless of the form the recognition takes, it should be frequent and it should be personal.

The Business Unlimited Volunteers Maryland organization (2011) provides an extensive list of recognition ideas for volunteers. These ideas can be implemented before, during, or after a volunteer's service.

Easy, everyday ways to recognize volunteers:

- Use e-mail to send thank you note/messages.
- Send postcards or thank you cards to volunteers after they attend a project.
- Submit pictures of volunteers in action for publication in the organization's newsletter.
- Post pictures of volunteers on a bulletin board at your organization.
- Provide organizational goodies—hats, shirts, pins, magnets, water bottles, etc.
- Invite them to coffee or lunch.

More involved, intermediate recognition ideas:

- Nominate a volunteer for Star of the Month—award them a certificate, letter, or small gift.
- Sponsor happy hours and social events. Encourage volunteers to meet each other.
- Recognize volunteers on local radio or television programs.
- Invite volunteers to serve as project leaders or committee members.
- Give gift certificates to movies, restaurants, etc. Solicit your community for donations!
- Nominate volunteers for local/national awards such as the Presidential Service Awards.

- Write articles about them in newsletters or newspapers.
- Write a letter to their employer highlighting the volunteer's accomplishments.
- Celebrate major accomplishments.
- Recognize anniversaries with your organization.
- Have them attend a training or seminar at the organization's expense.
- Give them additional responsibilities.
- Create a photo collage or slide show of volunteer activities.

Large-scale means of recognition:

- Hold annual recognition events: dinner, awards ceremony, theme party, etc.
- Recognize long-term volunteers with service awards: a plaque, trophy, certificate, etc.
- Give a volunteer additional responsibilities and a new title.
- Enlist long-serving volunteers to train staff and other volunteers.
- Involve them in the annual planning process.
- Make a donation to the organization of their choice in their name.
- Organize a free outing to an amusement park, sports game, etc., for volunteers.

For some volunteers, recognizing the amount of their service, number of hours annually or number of years volunteering, is important. Sometimes the awards are selected internally based on hours of service or service on a particular project, other times volunteers are nominated for service awards external to the organization. Knowing what motivates a volunteer will help volunteer managers know how best to recognize the volunteer. The right type of recognition is important to keep volunteers motivated and committed to the organization.

VOLUNTEER PROGRAM EVALUATION

Volunteer programs can require a significant investment of time and resources. In order to justify this investment, it is important to document the outcomes and impact of the program. This information can inform the organization's decision makers about the return on their investment into the volunteer program. The data can support the need for continued funding, and it might also be leveraged to make the case for additional funding to further the program's reach. Equally important is that the information gathered through comprehensive program evaluation can be used to secure funding from outside sources.

Program Evaluation

It is helpful to establish a program evaluation plan during the initial program-planning phase to ensure this important element is included. Purposeful program evaluation does not have to be a difficult or cumbersome task, but it does take some planning. Start by answering the question: "What is the goal of doing the evaluation?" Is it to document satisfaction level of the client's served by the program? Or, to document additional organizational reach as a result of volunteers? Or, if clients have changed their behavior (e.g., eating more vegetables) as a result of program efforts?

Some volunteer programs use a predetermined set of questions in the evaluation each year. These questions are then supplemented by additional questions specific to certain projects or other areas within the program. The key is to determine which type of evaluation will provide the most valuable data to meet program goals.

Most volunteer programs use the two basic types of evaluation: formative evaluation and summative evaluation. Formative program evaluation is used to manage ongoing activity. It can provide valuable information that will determine if mid-year (or mid-project) adjustments are needed. Formative evaluation is a great way to ensure that a project is on track and that the associated goals and objectives are being met. It provides a snapshot of the activity/program to date. Summative program evaluation is a year-end (or project-end) evaluation that provides a final overview of the project. Summative evaluation can be used to create a summary of the activity/project's strengths, weaknesses, recommendations, and future plans (MCCS 2008).

Data Collection and Analysis

Once the evaluation goals have been established, the data collection process can be determined. Successful data collection needs to be systematic and the process needs to be user-friendly, particularly if volunteers are being asked to record some of the data. Some volunteer programs use paper-based data collection tools (surveys, log sheets, etc.), while others have either created their own electronic system or adapted an off the shelf system to meet their needs. There are a number of resources that provide information and templates to develop effective data collection tools. (See additional resources at the end of this chapter.)

For new volunteer programs or new projects within established programs, it is valuable to collect baseline data. This provides a comparator of before and after effects of the volunteer program. This before and after effect is a great way to show impact and is particularly useful in year-end reports or when soliciting outside funding. Other quantitative and qualitative data that is often collected from volunteer programs includes: number of volunteers, volunteer hours on an individual and whole program level, number and type of clients served, listing of volunteer projects, hours of community service, and volunteer satisfaction. The goal(s) of the evaluation program will ultimately determine what data is collected.

Data analysis is the critical step in making sense of the data that was collected. Data analysis can illustrate impact of a program or activity, trends in the number of volunteers involved in a program and the hours of service they provide, and even 9 dollar value associated with their service. Ideally, the analysis will show where there is excellence in the volunteer program, and where there are opportunities for improvement. This information can then provide the basis for data-driven decisions about the program moving forward.

Documenting Impact

The data analysis can be combined with a text narrative to generate a program report. It may be desirable to create multiple reports, each with a slightly different packaging to meet the needs of different audiences. Potential audiences include the organization's board of directors, staff and volunteers, clients served by the volunteer program, external funding agencies, and even the local community. Where appropriate, the report should highlight the role volunteers have in the organization achieving its mission. Including testimonials from the clients served personalizes the impact volunteers are having on an individual level. And, including comments from volunteers on the value and personal satisfaction they received from volunteering emphasizes the mutually beneficial impact of the volunteer program.

Evaluating Volunteer Satisfaction

Another area of program evaluation centers on how satisfied volunteers are with their volunteer experience. Although many factors impact a volunteer's satisfaction, there are a few key areas that correlate with overall satisfaction (Business Volunteers Unlimited Maryland 2011). These include

- The volunteer received a clear and concise description of the work they would be doing.
- The volunteer was given an accurate estimate of time required (both in terms of hours per month/week and when the project was expected to be completed).
- The volunteer was told what knowledge and skills they needed to be successful. They received training to meet these needs and were given an opportunity to discuss questions and concerns.
- From initial contact to project completion, the volunteer knew who to call or speak with at any point throughout the experience.

MANAGING AN ESTABLISHED VOLUNTEER PROGRAM

Established volunteer programs still require ongoing recruitment, orientation and training of new volunteers, volunteer performance evaluations, volunteer recognition, and program evaluation. An additional component may be the need to provide recertification or continuing education opportunities for volunteers who have been in the program for multiple years. Facilitating an ongoing dialog with veteran volunteers can ensure that volunteer managers are aware of the needs and contributions of this group. Annual performance evaluations are one way to gather their input. Another way is to ask for their evaluation of the program. Because they have been involved with the program for several years, they can bring a valuable perspective that highlights the program's strengths, weaknesses, and opportunities that may not be evident to new volunteers. As well, the motivations for volunteering change over time for longer serving volunteers. It is important to get their input to determine if changes should be made to their assignments, and if they might benefit from new or additional training.

SUMMARY

Each year, millions of individuals give freely of their time and talents to various organizations. As a result of their service, both the organization and the volunteer benefit. Volunteers help organizations extend their reach and accomplish their mission which otherwise would not be possible with paid staff alone. Developing a new volunteer program requires careful planning and should start with a needs assessment to determine what organizational needs the new program will meet. Other components of the planning process include developing a mission statement, setting goals and objectives, determining the program's overall structure and budget, and writing position descriptions. Successful volunteer programs have a clear set of policies and procedures, and an effective orientation and training program to prepare volunteers for their service. Once volunteers begin their service, they benefit from supervision, evaluation, and recognition to ensure that they are meeting the organization's needs and that they are having a rewarding volunteer experience. Finally, systematic and regular program evaluation informs volunteer managers on the successes and impact of the volunteer program and can be critical in securing continued or expanded funding.

REVIEW QUESTIONS

1. Describe appropriate methods to use when completing a needs assessment for a potential volunteer program.
2. Differentiate between an orientation program and a training program.
3. Differentiate between a policy and a procedure.
4. List benefits a volunteer program can bring to an organization.
5. Describe the type of data usually collected through a volunteer program evaluation. Explain how that data can be used to support the program.

ENRICHMENT ACTIVITIES

1. Create the framework for an urban horticulture volunteer program using the five steps outlined in this chapter: needs assessment, developing a mission statement, setting goals, determining the overall structure of the volunteer program, and creating position descriptions.
2. Develop a compelling recruitment message for each of the following volunteer positions:
   - Public garden docent
   - Horticulture helpline volunteer
   - Community garden volunteer
3. Design a volunteer orientation program for a local food program.
   - What will be covered?
   - What are the expected outcomes?
   - Who will deliver the program?
   - How long will the session(s) be?
4. Describe the multiple components used in managing volunteers and evaluating their effectiveness.
5. Create a recognition program for a prison garden volunteer program.
6. Consider easy-to-implement forms of recognition as well as intermediate and large-scale recognition activities.

APPENDIX 1: SAMPLE JOB DESCRIPTION

Purpose:	The purpose of a Wilkinson Garden Education Assistant is to further the horticulture and youth education outreach of the garden, by supporting garden staff and outside speakers in delivering education programs.
Job Title:	Education Assistant
Example:	After-School Tutor
Location:	The Education Assistant will support programs at the Wilkinson Garden.
Key Responsibilities:	Educational Assistants help at Lunch and Learn lectures and other education workshops by assisting speakers with audiovisual equipment needs, introducing the speakers, and concluding sessions. They also assist with youth education programing in Grandma's Garden by supporting presenters and assisting with craft activities.
Supervisor:	Education Program Manager
Length of Appointment:	The Educational Assistant will serve in this role for two calendar years.
Time Commitment:	The Educational Assistant position requires a minimum commitment of two hours, and no more than four hours per week, each month. In addition, each volunteer must attend a two-hour orientation during the second week in January before the winter educational program series begins.
Qualifications:	Educational Assistant volunteers should feel comfortable talking in front of a group, be positive, helpful, and self-confident. Specific qualifications include Excellent verbal communication skills Excellent interpersonal skills Self-motivated but able to follow directions Ability to work as a team
Support Provided:	Training for this position will be provided at the two-hour orientation session. However, the Education Program Manager is available on an ongoing basis to answer questions and provide other assistance.
Volunteer Benefits:	Free access to many of the garden's educational programs and the opportunity to help children learn more about horticulture and the environment.

Source: Adapted from Nonprofit Risk Management Center, Volunteer Position Description Worksheet and Sample. http://nonprofitrisk.org/tools/volunteer/review/vol_pos_descript.htm.

APPENDIX 2: PROSPECTIVE VOLUNTEER PROFILE

General Information

Name

Mailing address

Phone number

email

Emergency contact: relationship of the contact, address and phone number

Statement and description of prior criminal convictions or offenses

(Continued)

Valid driver's licence number

How did you hear about us?

Personal references with contact information

Professional or work-related reference with contact information

Skills and Experience

Highest education level

Language(s) spoken

Current employer and position

Previous work or volunteer experience

Description of training or experience that may be pertinent to the volunteer position desired

Certifications (First Aid, CPR, etc.) dates of certification and expiration dates

Skills checklist (circle all that apply)
horticulture, computer, tutoring, clerical skills, phone calling, teaching, supervision, event planning, and record keeping
List additional skills:

Volunteer Activity

Preferred volunteer areas

Reason for volunteering

Hours and days available for volunteer work

Physical limitations

Signature

Date

Additional items to include on the application:
Disclaimers from the organization:

- Fair and equal opportunity statement.
- List of requirements for volunteers (reference check, interview, trial period, and required training).
- Statement that all volunteers will be subject a background and/or credit check prior to orientation and training.

Source: Adapted from About.com Nonprofit Charitable Orgs, What is a Volunteer Application Form: Basics of a Volunteer Application. http://nonprofit.about.com/od/volunteers/qt/volunteerapplication.htm.

REFERENCES

American Public Gardens Association. 2014. Volunteer professional section. http://www.publicgardens.org/sections/volunteer-professional-section (accessed July 31, 2014).

AmeriCorps. 2014. Building a high quality AmeriCorps program: From blueprint to implementation: New program start-up guide. http://www.nationalservice.gov/resources/americorps/building-high-quality-americorps-program-blueprint-implementation-new-program (accessed July 31, 2014).

Business Volunteers Unlimited Maryland. 2011. Recruit, train, retain! http://www.volunteercentral.net/downloads/Recruit_Train_Retain_Workbook.pdf (accessed July 31, 2014).

Corporation for National and Community Service. 2013. Volunteering and civic life in America 2013. http://www.volunteeringinamerica.gov/national (accessed July 31, 2014).

Community Tool Box. 2014. Section 3: Developing volunteer orientation programs. http://ctb.ku.edu/en/table-of-contents/structure/volunteers/orientation-programs/main (accessed July 31, 2014).

Energize, Inc.com. 2014a. Volunteer application form. http://www.energizeinc.com/art/apro.html (accessed July 31, 2014).

Energize, Inc.com. 2014b. Volunteer management resource library. http://www.energizeinc.com/art.html (accessed July 31, 2014).

Graff, L.L. 1997. *By Definition: Policies for Volunteer Programs.* Dundas Ontario: Linda Graff and Associates.

Hands on Network. 2010. Starting a volunteer program. http://www.handsonnetwork.org/files/resources/BP_StartingVolunteerProgram_2010_HON.pdf (accessed July 31, 2014).

Independent Sector. 2014. Independent sector's value of volunteer time. http://www.independentsector.org/volunteer_time (accessed July 31, 2014).

Maine Commission for Community Service (MCCS). 2008. Need to know basics of managing volunteers. http://www.volunteermaine.org/shared_media/publications/old/E245B0A4d01.pdf (accessed July 31, 2014).

Maslow, A. H. 1970. *Motivation and Personality.* New York: Harper & Row, Inc.

McClelland, D. 1988. *Human Motivation.* Cambridge: Cambridge University Press.

McCurley, S. 2003. Writing persuasive volunteer recruitment appeals. http://www.energizeinc.com/art/McCurRecr.html (accessed July 31, 2014).

McCurley, S. and S. Vineyard. 1988. *101 Tips for Volunteer Recruitment.* Downers Grove, Illinois: Heritage Arts Publishing.

Schultinik, J., M.J. Riley and G. Schott. 2009. Determining volunteer motivations—A key to success. http://www.msue.msu.edu/objects/content_revision/download.cfm/revision_id.353032/workspace_id.275600/Determining_Volunteer_Motivations.pdf/ (accessed July 31, 2014).

501 Commons. 2014. Volunteer management resources. http://www.501commons.org/resources/tools-and-best-practices/volunteer-management.html (accessed July 31, 2014).

FURTHER READING

City of Bloomington. 2014. 25 elements of volunteer program management. https://bloomington.in.gov/media/media/application/pdf/7718.pdf (accessed July 31, 2014).

Ellis, S.J. 2002. *Volunteer Recruitment (and Membership Development) Book.* 3rd edition. Philadelphia: Energize, Inc.

Ellis, S.J. and K.N. Campbell. 2003. *Proof Positive: Developing Significant Volunteer Recordkeeping Systems.* 21st century edition. Philadelphia: Energize, Inc.

Graff, L.L. 2003. *Better Safe... Risk Management in Volunteer Programs and Community Service.* Dundas, Ontario: Linda Graff and Associates.

Graff, L.L. 2005. *Best of All: The Quick Reference Guide to Effective Volunteer Involvement.* Dundas, Ontario: Linda Graff and Associates.

MacKenzie, M. 1990a. *Curing Terminal Niceness: A Practical Guide to Healthy Volunteer/Staff Relationships.* Downers Grove, Illinois: Heritage Arts Publishing.

MacKenzie, M. 1990b. *Dealing with Difficult Volunteers.* Downers Grove, Illinois: Heritage Arts Publishing.

McCurley, S. 1996. *Teaching Staff to Work with Volunteers.* Downers Grove, Illinois: Heritage Arts Publishing.

McCurley, S. and R. Lynch. 1997. *Volunteer Management: Mobilizing all the Resources of the Community.* 2nd edition. Downers Grove, Illinois: Heritage Arts Publishing.

McCurley, S. and S. Vineyard. 1986. *101 Ideas For Volunteer Programs.* Downers Grove, Illinois: Heritage Arts Publishing.

McCurley, S. and S. Vineyard. 1997. *Managing Volunteer Diversity.* Downers Grove, Illinois: Heritage Arts Publishing.

McKee, J. and T. W. McKee. 2007. *The New Breed: Understanding and Equipping the 21st Century Volunteer.* Carol, IL: Tyndale House Publishers. Points of Light Foundation, 2014. www.pointsoflight.org (accessed July 31, 2014).

Wittich, B. 2003. *Model Volunteer Handbook.* Fullerton, CA: Knowledge transfer publishing.

Index

D

E

F

G

Z

Lightning Source UK Ltd.
Milton Keynes UK
UKHW011558090221
378470UK00005B/80

9 781482 260991